CROP MANAGEMENT AND POSTHARVEST HANDLING OF HORTICULTURAL PRODUCTS

Volume III — Crop Fertilization, Nutrition and Growth

CROP MANAGEMENT AND POSTHARVEST HANDLING OF HORTICULTURAL PRODUCTS

Volume III — Crop Fertilization, Nutrition and Growth

Editors

Dr. Ramdane Dris

World Food Ltd., Meri-Rastilantie 3 C, 00980 Helsinki. Finland

Dr. Raina Niskanen

World Food Ltd., Meri-Rastilantie 3 C, 00980 Helsinki. Finland

Dr. Shri Mohan Jain

Plant Breeding and Genetics Section Join FAO/IAEA, Division of Nuclear Techniques
in Food and Agriculture, Department of Nuclear Sciences and Applications
International Atomic Energy Agency, Vienna, Austria

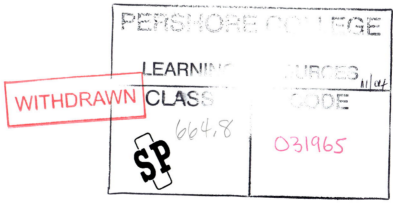
Science Publishers, Inc.

Enfield (NH), USA   Plymouth, UK

SCIENCE PUBLISHERS, INC.
Post Office Box 699
Enfield, New Hampshire 03784
United States of America

Internet site: *http://www.scipub.net*

sales@scipub.net (marketing department)
editor@scipub.net (editorial department)
info@scipub.net (for all other enquiries)

Library of Congress Cataloging-in-Publication Data

Crop management and postharvest handling of horticultural products/editors,
    Ramdane Dris, Raina Niskanen, and Mohan Jain.
        p. cm.
    Includes bibliographical references (p.).
    Contents: v. 3. Crop fertilization, nutrition and growth.
    ISBN 1-57808-278-1
    1. Horticultural crops. 2. Horticultural crops--Postharvest
    technology. I. Dris, Ramdane. II. Niskanen, Raina. III. Jain, Mohan.

    SB318.C76 2003
    635'.046--dc21

                                                          00-063494

Published by Science Publishers Inc., Enfield, NH, USA

Printed in India

General

Food crops require nutrients in order to grow, develop and complete their life cycle. The fertilizer industry faces a permanent challenge to improve the efficiency of their products in relation to sustainable crop production. However, the growth of the world fertilizer industry is increasingly affected by great concerns such as environmental pollution and high energy consumption. Environmental conditions and inadequate or excessive mineral nutrition can have serious implications on the quality of food crops. The future world fertilizer use is expected to grow from 165 million metric tons in 2000 to 227 million metric tons in the year 2020, which would create enviornmental pollution control. Moreover, nearly 700 different types of pesticides and numerous herbicides and fungicides are being used in modern agriculture—many of them regulated in food and animal feeds. Additionally, preservatives and other compounds may be added to the food during the processing and packaging stages. As more compounds are introduced to the market and lower detection limits are regulated, laboratory instrument measurement performance must be improved to meet the evolving needs. The soil fertility and fertilizer use control system encompasses all soil and plant parameters that may contribute to high, stable and profitable yields. Appropriate use of fertilzers maximize the profit to growers, e.g. cost saving on fertilizers. More information can easily be found in www.world-food.net or www.food-farming.info

Additional benefits of the application of correct fertilizer doses improve soil protection and the environment in general and prevent contamination of soil and ground waters with nitrites. The production of good quality food cannot be achieved without the strict control of nutrient supply. Under increased rates of nitrogen fertilizers, the plants are unable to accumulate enough available plant nutrients and become damaged by diseases. Some crops such as leafy vegetables, grown in soils with increased rates of phosphate and potassium fertilizers under moderate nitrogen feeding, are more resistant against the diseases. Selective fertilization may be applied as a corrective measure to remove the deficiency of element(s) as well as improve soil fertility for high crop production. Moreover, water and nutrient stresses have a great impact on a number of metabolic activities in plants. In fact, water stress during the intensive growth of crops causes inhibition of physiological processes, such as nutrient transport, hence decreases the yield and affects the crop quality performance. The application of appropriate fertilizer doses can save soil and ground water quality from pollution with nitrites. Thereby, research and develpoment efforts must continue to prevent contamination of soil and water with heavy metals and microorganisms.

Ramdane Dris
Raina Niskanen
Shri Mohan Jain

Preface

Crop Management and Postharvest Handling of Horticultural Crops is a compilalion of information on management tools. It also reviews the factors affecting the plant mineral nutrition and growth. It highlights the importance of fertilizers and mineral nutrition to improve agricultural production, yield, and amelioration of soil fertility. This book also addresses issues such as growth, production, yield and quality, which can be limited by sub-optimal conditions such as soil, salinity, poor drainage, water supply, fertilization programs, physical conditions affecting root growth and function or handling operations. The production of good quality food cannot be achieved without the strict control of the use of fertilizers and other pesticide sprays.

Nine chapters are included in the present volume, namely: Environmental Effects of Fertilizers; Crop Quality as Affected by Adverse Conditions: Importance of the Knowledge of Nutritional Status; The Importance of Boron in Apple Production Under Polish Conditions; Integrated Nutrient Management in Indian Soils for Sustainable Crop Production; Role of Phosporus on Carbon Uptake and Fruit Quality of Strawberry; Nutrition of Tropical Horticulture and Quality Product; Soil Fertility Management with Wood Ash; Role of the Grafting in the Horticultural Plants; and Domestication of Jujube Fruit Trees.

Our aim is to provide the scientific community with an accurate and up-to-date information in the field of fetrtilization and plant mineral nutrition. We believe that this book will stimulate researchers, students or extension workers' information to identify and solve problems related to plant growth, crop production and quality. Readers will be able to familiarize themselves with recent techniques to making a more efficient use of applied nutrients with a wide range of crops grown under different environmental conditions. More details and information can easily be found in www.world-food.net or www.food-faming.info.

Contents

General   v

Preface   vii

List of Contributors   xi

1. Environmental Effects of Fertilizers
    N. Sanjuán, G. Clemente and L. Úbeda   1

2. Crop Quality Under Adverse Conditions :
    Importance of Determining The Nutritional Status
    Gemma Villora, Diego A. Moreno and Luis Romero   55

3. The Importance of Boron in Apple Production
    Pawel Wójcik   77

4. Integrated Nutrient Management in Indian Soils for
    Sustainable Crop Production
    C.L. Acharya and J.K. Saha   93

5. Role of Phosphorus in Carbon Uptake, Fruit Yield
    and Quality in Strawberry
    Parmjit Singh, Zora Singh and M.H. Behboudian   115

6. Nutrition of Tropical Horticulture Crops and Quality
    Products
    B.C. Ghosh and S. Palit   133

7. Soil Fertility Management with Wood Ash
    J.C. Voundi Nkana   201

8. Role of Grafting in Horticultural Plants
    R.M. Rivero, J.M. Ruiz and Luis Romero   229

9. Domestication of Jujube Fruit Trees (*Zizyphus*
    mauritiana Lam.)
    A.M. Bâ, P. Danthu, R. Duponnois and P. Soloviev   255

Index   281

List of Contributors

C.L. Acharya

Division of Environmental Soil Science, Indian Institute of Soil Science, Nabibagh, Berasia Road, Bhopal, M.P. 462038, India
E-mail: jks@iissmp.mc.in

A.M. Bâ

Laboratoire de Biologie et Physiologie Végétales. UFR Sceinces Exactes et Naturelles, B.P. 592, Université des Antilles et de la Guyane, 97159 Pointe-à-Pitre, Guadeloupe

M.H. Behboudian

Institute of Natural Resources, Horticultural Science, Massey University, Private Bag 11 222, Palmerston North, New Zealand

Bijoy Chandra Ghosh

Department of Agricultural and Food Engineering, Indian Institute of Technology, Kharagpur, West Bengal 721 302, India

Gabriela Clemente

Food Technology Department, Universidad Politécnica de Valencia, Camino de Vera S/N 46022, Valencia, Spain

P. Danthu

CIRAD-Forêt, Progrmme "Arbes et Plantation", BP. 853, Antananarivo, Madagascar

R. Dupponois

IRD, UR IBIS "Interactions Bilogiques dans les Sols des systèmes Anthropis Tropicaux", 01 BP. 182. Ouagadougou. Burkina Faso

Diego A. Moreno

Biologia Vegetal, Facultad de Ciencias, Universidad de Granada, Fuentenueva, S/N E-18071, Granada, Spain

J.C. Voundi Nkana

Universität Bonn, Institut für Bodenkunde, Nussallee 13, D-53115 Bonn Germany

Soumen Palit

Department of Agricultural and Food Engineering Indian Institute of Technology, Kharagpur 721 302 India

R.M. Rivero

Department of Plant Biology, Faculty of Sciences, University of Granada, E-18071 Granada, Spain

Luis Romero

Department of Plant Biology, Faculty of Sciences, University of Granada, E-18071 Granada, Spain

Juan M. Ruiz

Department of Plant Biology, Faculty of Sciences, University of Granada, E-18071 Granada, Spain

J.K. Saha

Division of Environmental Soil Science, Indian Institute of Soil Science, Nabibagh, Berasia Road, Bhopal, M.P. 462038, India

Nieves Sanjuán

Food Technology Department, Universidad Politécnica de Valencia, Camino de Vera S/N 46022, Valencia, Spain

Parmjit Singh

Orange Agricultural Institute, New South Wales Agriculture, Orange, NSW 2800, Australia

Zora Singh

Department of Horticulture/Viticulture, Muresk Institute of Agriculture, Division of Resources and Environment, Curtin University of Technology, GPO Box U1987, Perth, Western Australia 6845, Australia
E-mail: 2.Singh@curtin.edu.au

P. Soloviev

Centre deFormation Professionnelle Horticole, BP. 3284, Dakar, Sénégal

Laura Úbeda

Food Technology Department, Universidad Politécnica de Valencia, Camino de Vera S/N 46022, Valencia, Spain

Gemma Villora

Biologia Vegetal, Facultad de Ciencias, Universidad de Granada, Fuentenueva, S/N E-18071, Granada, Spain

Pawel Wojcik

Department of Fruit Management and Plant Nutrition, Research Institute of Pomology and Floriculture, Pomologiczna 18, 96-100 Skierniewice, Poland

1

Environmental Effect of Fertilizers

N. Sanjuán, G. Clemente and L. Úbeda

Food Technology Department. Universidad Politécnica de Valencia.
Camino de Vera s/n. 46022 Valencia. SPAIN.

1. Introduction

Superior plants may contain up to 60 elements, of which only 16 are considered essential for the plants' development and reproduction. When one of the essential elements is missing, the plant cannot complete its vegetative cycle. The 16 essential elements are: carbon, oxygen, hydrogen, nitrogen, phosphorus, potassium, calcium, sulphur, magnesium, iron, boron, manganese, copper, zinc, molybdenum and chlorine. Of these elements carbon, oxygen and hydrogen are supplied by air and water and the rest by soil and can be classified as macro-elements (if plants need to absorb a relatively high quantity of these elements) and micro-elements (if plants need to absorb a relatively low quantity). Macro-elements can be further classified in two groups: primary elements (nitrogen, phosphorus and potassium), of which there are not enough present in soil; and secondary elements (calcium, sulphur and magnesium), which are sufficiently present in soil to cover plant necessities. The micro-elements referred to here comprise iron, boron, manganese, copper, zinc, molybdenum and chlorine.

Quantitatively, nitrogen, phosphorus and potassium are the most important elements for plants. The other macro-elements are also fundamental, but in practice, a lack of these elements does not necessarily have a limiting effect on the plant's growth (Davis and Haglund, 1999). Soil contains large amounts of the essential mineral elements but only a small fraction of these reserves are available to the plant (Dominguez, 1999) In order to be available to the plant, the mineral element must exist in either a plant-available form or a compound, which in all cases, except for boron, implies that the form or compound is an ion.

Since the beginning of agriculture, the importance of soil was understood. Man knew that he could not go on taking nourishment from the soil indefinitely without allowing it to recuperate. It was common knowledge that allowing the natural forest to recolonize the land made it fruitful again and that growing crops in certain sequences or rotations was helpful (Addiscott, *et al.*, 1991). One of the most interesting early examples of the beneficial effects of supplying nutrients for agriculture is the Nile Valley during the time of Egyptian Pharaohs (Park, 2001).

Fertilizers can be defined as organic or inorganic materials that are applied to soil and provide essential nutrients for growing plants, usually by absorption through the plant roots. They are used to supplement the natural supply of soil nutrients; to compensate for nutrients lost by the harvesting of crops, leaching or by gaseous exchange; and to maintain or improve the fertility of the soil (Park, 2001). Fertilizers can be classified according to their origin in organic fertilizers—such as manure—or mineral fertilizers. Mineral fertilizers provide available forms of the plant nutrition elements and are, therefore, not artificial to nature (Davis and Haglund, 1999). Organic fertilizers are usually derived from living organisms and do not contain petroleum-based fertilizers (Muchojev and Pakovsky, 1997).

The increase of human population over the past century has been closely associated to the growth in production of food and forest products (Dyson, 1996). Projected increases in the world population indicate that current production of food will need to be raised over the next few decades (Gregory and Ingram, 2000; Gregory *et al.*, 2002). At the same time, an increasing trend towards urbanization is expected, which will reduce the area of prime agricultural land, decrease the availability of rural labour and increase the pressures toward the intensification of crop production (Alexandratos, 1995). Consequently, the sustainability and potential environmental consequences (global change) of the current and future production systems must be questioned.

Global change is a primary environmental concern, not only for the scientific community, but also for the society. One area of particular concern is food and forestry production, where changes (e.g. use of fertilizer or application of a particular pesticide) usually occur at local scales but as the frequency of use increases, changes occur that become globally significant (Gregory and Ingram, 2000). The effect of agricultural methods on the quality of the environment has now been established as a factor behind many of the decisions being made that affect farming (Burton, 2001). In this sense, the fertility of soils and the use of fertilizers have played and continue playing an important role in world food production (Foth and Ellis, 1997).

Fertilization leads to changes in the properties of soil that can be either beneficial or deleterious. The emissions from agricultural management into water, soil and air cause environmental impacts. A major deleterious environmental aspect of intensive agriculture concerns the management of nutrients, with major consequences for climate forcing and water resources (Gregory *et al.*, 2002). Other emissions caused by the use of fertilizers contribute to global warming and to acidification. Moreover, the continuous application of large amounts of fertilizers can produce an accumulation of toxic levels of their trace element constituents which is potentially harmful to the environment (Raven and Loeppert, 1997). There is a need for more efficient use of fertilizers in order to reduce food production costs and environmental pollution.

The aim of this chapter is to review the global environmental impacts of fertilizers. In the first part, the effects of different fertilizers are considered: chemical fertilizers (namely nitrogen, phosphorus and potassium fertilizers) and organic fertilizers. In the second part, the application of Life Cycle Assessment as an integral tool used to evaluate the environmental impacts of agricultural practices will also be reviewed. Finally, an example of the use of nutrient balance and LCA applied to different fertilizers, strategies used in Valencia (Spain) for orange production is given.

2. Nitrogen Fertilizers

The first inorganic nitrogenous fertilizer was Chile nitrate, discovered in 1809 and imported into Europe and the USA from about 1830 onwards. The development of the town gas industry and coke for steel production produced ammonium sulphate as by-product, so it became available as fertilizer (Park, 2001).

A large-scale ammonia capacity development during the 1930s allowed for the production of ammonium nitrate and urea. At the beginning, nitrogen fertilizers (mainly ammonium sulphate, calcium cyanamide and calcium nitrate) had low concentrations, around 15–21% N, but over time, its concentration was increased. Around the 1940s, ammonium nitrate (34% N) and calcium ammonium nitrate (up to 27% N) were produced in increasing quantities. By the 1960s, these were the main nitrogen fertilizers. Nowadays, the cheapest and most widespread nitrogen fertilizer is urea (46% N). Nitrogen fertilizer consumption has grown in the last few years, and it is now the dominant factor for future consumption growth in the developing countries (Park, 2001). Production of nitrogen has also grown (Figure 1). Urea represents 50% of the world's nitrogen fertilizer production and about 44% of the total nitrogen fertilizer trade.

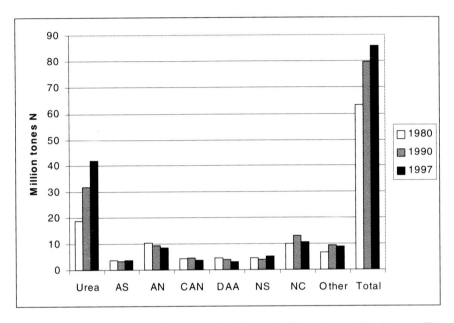

Fig. 1. World production of nitrogen fertilizers (million tonnes N). Source: IFA (International Fertilizer Industry Association). AS, Ammonium Sulphate; AN, Ammonium Nitrate; CAN, Calcium Ammonium Nitrate; DAA, Direct Applied Ammonia; NS, Nitrogen Solutions; NC, Nitrogen in Compounds (Data from Park, 2001).

Nitrogen is essential for plant growth. This macro-element is a part of all the essential constituents of cells: the chlorophyll (essential for photosynthesis); nucleic acids, DNA and RNA (the pattern for the plant's growth and development are encoded in them); proteins (including the enzymes, responsible for the catalyse of all biochemical processes) and the walls that hold the cell together (Addiscott *et al.*, 1991). Thus, nitrogen-containing compounds are involved in practically all the biochemistry of life.

Nitrogen fertilizer has to be applied to a crop in the correct amounts. The application of nitrogen fertilizers influences the yields of crops in four different ways: leaf area, crop development, crop quality and side effects (Addiscott *et al.*, 1991). It is also important to know the kind of fertilizer applied and when is it done. Nitrogen is absorbed in nitric form (NO_3^-) and ammonia form (NH_4^+). Both are metabolized by the plant. Nevertheless, the nitric form is absorbed with ease (Domínguez, 1997).

2.1. Nitrogen Balance

Nutrient balance is a useful tool when quantifying the flow of nutrients in agriculture production (Cederberg and Mattson, 2000) because it gives an indication of emissions derived from nutrient inputs. Concretely, nitrogen balance allows to compute the emissions caused by the application of N fertilizers in relation with the N cycle, one of the most important nutrient cycles in terrestrial ecosystems. It is a model that describes the movement of nitrogen in its many forms between the hydrosphere, lithosphere, atmosphere and biosphere. In Figure 2, a simplified nitrogen cycle is shown, focusing on the most important inputs and outputs of nitrogen.

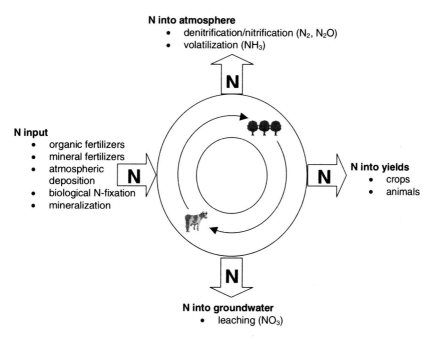

Fig. 2. Nitrogen cycle on a farm (Brentrup et al., 2000).

When the nitrogen balance is computed, the inputs and outputs of nitrogen must be quantified (Equation 1):

$$\text{N balance} = \Sigma \text{ input} - \Sigma \text{ output} \tag{1}$$

Nitrogen fertilizers (organic and mineral), biological N fixation, atmospheric N deposition and N mineralization are to be considered as nitrogen inputs. For quantifying N outputs, N removal with harvested crops, NH_3 N_2O, N_2 and NO emissions and N immobilization must be determined.

As an example, typically 20–60% of the N fertilizer applied is recovered by a wheat crop, while 20–60% remains in the soil and, on an average, 20% is lost from the soil/plant system (Pilbeam, 1996).

Due to human intervention in the global nitrogen cycle, more nitrogen is fixed in fertilizers and legumes in agricultural systems than by natural processes (Vitousek *et al.*, 1997). The enhanced losses of N from agricultural areas to adjacent lands can have major impacts on ecosystems (Matson *et al.*, 1997).

It is often difficult to calculate the exact rates of N released to the air and water, because emission rates vary, depending on the soil type, climatic conditions and agricultural managements' practices (Brentrup *et al.*, 2000). It is important to know how the loss of nitrogen occurs. Volatilization of ammonia, denitrification/nitrification and leaching can all be contributors in this occurrence. Another N emissions' pathway such as soil erosion or surface run-off is less important in quantification of N emissions. Each one of these losses produces a different impact and may need different control measures (Addiscott and Powlson, 1992).

2.2. Ammonia Volatilization

The contribution of agriculture to the anthropogenic NH_3 released annually to the atmosphere varies between 55% (Duxbury, 1994) and 95% or even more (Isermann, 1994). The main source of atmospheric NH_3 is the storage and application of livestock manure (Asman, 1992; ECETOC, 1994). Nevertheless, the application of synthetic fertilizers—mainly urea—contributes to this release too.

The rate of NH_3 volatilization varies among forms of fertilizers (McGinn and Janzen, 1998). Certain circumstances encourage ammonia volatilization (Fuentes, 1999):

- Excessive ammonification. If a surface soil presents a high organic matter content with high concentrations of N which is easily broken down in favourable conditions (high temperature and humidity), losses due ammonia volatilization are enhanced.
- Low nitrification. Low aeration and temperatures higher than 30°C inhibit nitrification and increase ammonia volatilization.
- Alkaline soil with low change capacity. Under these circumstances, ammonia salts produce ammonia that is partially volatilized because it is produced naturally.

Other factors that influence Ammonia volatilization as proposed by Bentrup and Küsters, 2000 are:

- Average air temperature. Temperatures increase ammonia emissions.

- Infiltration rates describe the capability of the soil to take up the NH_3/NH_4^+. The infiltration of NH_3/NH_4^+ into the soil reduces the volatilization rates because within the soil, NH_3/NH_4^+ either stays in a solution form and is plant available; it is biological oxidized (nitrification) or it is absorbed by clays and organic matter.
- The amount of volatilized ammonia depends on the time NH_3 is present on the soils surface. Rainfall reduces the volatilization of NH_3 considerably due to the increase in the solution of NH_3/NH_4^+ and infiltration into the soil.
- The incorporation of organic fertilizers also reduce NH_3 losses, because NH_3/NH_4^+ moves deeper into the soil.

As it can be seen, there are many circumstances which influence ammonia volatilization, and it is very important to take all of them into account in order to determine the quantity of ammonia volatilized. The ECETOC (1994) proposed a method to estimate the emissions for the different factors throughout Europe. First, three kind of countries with different regional sensitivities to NH_3 volatilization have been defined (Table 1). After that, based on a literature review, NH_3 emission factors for six groups of mineral fertilizers according to the regional differences have also been developed (Table 2).

Table 1. Grouping of European countries according to their NH_3 volatilization sensitivity (ECETOC, 1994)

Group	Countries	Calcarous soil	pH (usually)	Sensitivity
I	GR, E	common	> 7	high
II	I, F, UK, IRL, P, B, NL, L	partly existent	7	medium
III	N, S, FIN, DK, D, CH, A	rare	< 7	low

A high atmospheric ammonia concentration can produce acidification of land and water surfaces; cause plant damage (van der Eerden *et al.*, 1991; McGinn and Janzen, 1998); and reduce plant biodiversity in natural systems (Sutton *et al.* 1993; McGinn and Janzen, 1998). NH_3 also contributes to the greenhouse effect.

2.3. Denitrification/Nitrification

Around 80% of the N_2O emissions caused by agriculture are due to the use of mineral and organic fertilizers. The responsible agents for most of these emissions are two microbial processes: denitrification and nitrification. Nitrification occurs in aerobic soils with a pH of 6 or higher; NH_4^+ from mineralization is rapidly oxidized by specialized

Table 2. Emission factors (%NH₃-N loss of total applied mineral N) for different mineral fertilizers in Europe. (ECETOC, 1994)

Fertilizer type	Groups of European countries		
	Group I	Group II	Group III
Urea	20	15	15
Ammonium Nitrate, NP, NK, NPK, Calcium Ammonium, Nitrate	3	2	1
Ammonium Phosphate	5	5	5
Ammonium Sulphate	15	10	5
Anhydrous Ammonia	*	*	4
Urea Ammonium Nitrate solution	8	8	8

*fertilizer not common in the group country.

chemoautropic bacteria to nitrite as well as nitrate. A small proportion of ammonium is converted to N_2O and NO during nitrification, and the gas is also produced during denitrification of the nitrate (Granli and BÆckman, 1994). Nitrification is an environmentally-dependent process, depending not only on acidity, but also on temperature, water and O_2 supply. Thus, little nitrification occurs in wet and cold soils (Foth and Ellis, 1997).

Denitrification can be defined as the chemical reduction of nitrate and nitrite to a gaseous form, nitric oxide, nitrous oxide and dinitrogen (Foth and Ellis, 1997). For denitrification, anaerobic conditions are needed. There are two decisive factors: the available amount of nitrogen in the soil and the availability of degradable organic matter, because denitrifying microorganisms need organic carbon as an energy source (Brentrup et al., 2000). Other factors that influence this process (Fuentes, 1999) are:

- The presence of sulphur; some microorganisms such as *Thiobacillus* uses oxygen of nitrates for oxidizing sulphur to SO_4H_2, which results in denitrification with a loss of molecular nitrogen.
- Temperature; around 28-30°C is the optimum condition for denitrifying microorganisms.

Bouwman (1994, 1996) has observed from field experiments that the emissions from fertilized land are higher if the measurements have been conducted for longer periods. In experiments with a duration of longer than one year a linear relationship was observed between fertilizer N applied and N_2O emission (Equation 2).

N_2O emission (kg N_2O-N/ha) = 1 + 0.0125 * N application (Kg N/ha./year)

(2)

The complex interaction of soil, crop and environmental factors cause many variations in emission rates of N_2O (Smith *et al.*, 1997). It is clear that this emission increases with temperature, provided that other factors are not limiting (Granli and Bøckman, 1994; Smith, 1997). In Table 3, the influence of soil and climate factors and agricultural management in N_2O emissions are summarized.

Table 3. Parameters which influence N_2O emissions from agricultural soils (Brentrup et al., 2000)

Parameter	*Effect on N_2O emissions*
Soil aeration	• intermediate aeration → highest N_2O production
	• low aeration → high denitrification rate, but mainly N_2 production
Soil water content	• increasing soil water content → increasing N_2O emissions, but
	• under very wet conditions → decline
	• changing conditions (dry/wet) → highest N_2O production
Nitrogen availability	• increasing NO_3/NH_4 concentrations → increasing N_2O emissions
Soil texture	• from sand to clay → increasing N_2O emissions
Tillage practice	• ploughing → lower N_2O emissions
	• no/low-tillage → higher N_2O emissions
Compaction	• increasing compaction → increasing N_2O emissions
Soil pH	• where denitrification is the main source of N_2O emission: increasing pH results in decreasing N_2O emissions
	• where nitrification is main source of N_2O emission: increasing pH results in increasing N_2O emissions
Organic material	• increasing organic carbon content → increasing N_2O emissions
Crops and vegetation	• plants, but especially their residues and remaining roots after harvest increase N_2O emission
Temperature	• increasing temperature → increasing N_2O emission
Season	• wet summer → highest N_2O production
	• spring thaw → high N_2O production
	• winter → lowest N_2O emission

Natural and cultivated soils are the major global source of N_2O, accounting for 65% of all emissions (Prather *et al.*, 1995). The total emissions of NO due to the use of fertilizers are more uncertain than the ones of N_2O due to the sparse information available (Smith *et al.*, 1997). Skiba *et al.* (1997) reviewed 12 studies and found that the fraction of N applied reported to be lost as NO varied between 0.02 and 3.25% for bare soil, between 0.003 and 3.2% for grass swards and between 0.53 and 2.5% for arable crops.

The emissions that are caused by denitrification are benign if the ultimate gas produced is dinitrogen. The evolution of dinitrogen to nitrous oxide is dangerous because this gas contributes both to the greenhouse effect and to the depletion of ozone in the upper atmosphere (Addiscott and Powlson, 1992). Nitric oxide (NO) causes acid rain and takes part in reactions which produce the formation of ozone in the troposphere (Prather *et al.*, 1995).

2.4. Nitrate Leaching

In the soil, nitrate is the main mineral nitrogen form. As it is hardly adsorbed by soil particles, it is easily leached into the groundwater. The loss of nitrate by leaching and run-off to watercourses is serious in the short-term as also locally because it affects drinking water supplies and other ecosystems (Gregory and Ingram, 2000). For this reason water pollution, especially from nitrate leaching, has now become an issue of concern in all EU countries. The EU Nitrate Directive (91/676/EC), which sets a 50 mg L^{-1} limit for drinking water, has been adopted in many countries.

The general principle is that if nitrate supply is low, plants are effective in removing nitrate from the soil solution, but if the supply is high, the crops cannot absorb it all and a large amount is susceptible to loss in drainage water (Back, 1993). Leaching losses are very important when there is a large downward movement of water and scarce radical activity. Thus, leaching losses are more important in a bare soil rather than in cultivated ones and it can be significant during rainy winters (Fuentes, 1999). Nevertheless, the link between fertilizer use and leaching is usually indirect (Addiscott *et al.*, 1991).

The amount of leached nitrate mainly depends on three different factors, which are agriculture related, soil related and climate related (Brentrup *et al.*, 2000). The influence of these factors have been examined and are stated below.

Almost all nitrates in the soil are dissolved in water, thus the amount of NO_3^- which is in the soil at the beginning of the leaching periods in autumn, probably will be available for leaching. The quantity of nitrates that are leached can be calculated by Equation 3 (Brentrup *et al.*, 2000).

leached NO_3^-N (kg N/(ha· year)) =

$$NO_3^-N_{\text{in soil in autumn}} \text{ (kg N/ha·)} \cdot \text{exchange frequency/year} \qquad (3)$$

NO_3^-N in soil during autumn indicates the quantity of nitrates in the soil that can be leached in autumn, after the vegetation period. It is calculated by working out the balance of nitrogen, as stated before. In this calculation, the nitrate-leaching agricultural-related parameters are taken into account.

Direct losses of nitrates from fertilizers are not the main source of leached nitrates. Addiscott and Powlso (1992) showed that for wheat crops, some of the N fertilizer absorbed by the crop in spring or summer is remobilized by microbes from dead roots or other crop residues during the following autumn and leached along with any unused N fertilizer.

Exchange frequency of the drainage water is calculated by Equation 4. It reflects the share of nitrates lost by leaching. The maximum value used in Equation 3 is 1, although if it is one or higher, it can be considered that all of the nitrates will be leached (Brentrup *et al.*, 2000).

$$\text{exchange frequency/year} = W_{drain} \text{ (mm/year)}/FC_{RZe} \text{ (mm)} \qquad (4)$$

For the application of Equation 4, two more parameters are needed: the rate of drainage water (W_{drain}) and the field capacity in the effective rooting zone (FC_{RZe}). They are related to climate and soil conditions, respectively.

Due to the fact that nitrate dissolve in water, the quantity of water which can be drained from the soil has a great influence on nitrate leaching. This quantity is related to climatic conditions, mainly with precipitation. Thus, in order to calculate the rate of drainage water (W_{drain}), the precipitation rate (W_{precip}) and its distribution through the year (the quotient between the precipitation in summer and in winter) are needed. This rate can be calculated according to Equation 5 (DBG, 1992).

$$W_{drain} \text{ (mm)} = 0.86 \cdot W_{precip\_year} \text{ (mm)} - 11.6 \cdot (W_{precip\_summer}/ W_{precip\_winter}) \text{ (mm)} - 241.4 \qquad (5)$$

The last parameter needed to determine the quantity of leached nitrate is the field capacity in the effective rooting zone (FC_{RZe}). It describes the capacity of the soil to absorb water in the section of the soil in which the roots are able to take it in. For its calculation, the texture of the soil must be known as was stated before. FC_{RZe} is calculated as shown in Equation 6, where FCa is the available fields capacity and RZe is the effective rooting zone (Brentrup *et al.*, 2000):

$$FC_{RZe} \text{ (mm)} = FCa \text{ (mm/dm)} * RZe \text{ (dm)} \qquad (6)$$

Both indexes, FCa and RZe, depend on the soils texture. Tables 4 and 5 show different available field capacities and effective rooting zones, respectively proposed by the German Soil Science Association (DBG) (1992).

Table 4. Assignment of soil textures to available field capacity, medium soil density (DBG, 1992)

Class (evaluation)	Soil texture[a]	FCa (mm/dm) Range	average
1 (very low)	S	< 10	8
2 (low)	lT	10 – 14	12
3 (medium)	lS, tS, sL, tL, uT, T	14 – 18	16
4 (high)	uS, sU, uL	18 – 22	20
5 (very high)	lU, tU, U	> 22	24
6 (swamp)	Hh, Hn		60

(a) S = sand, s = sandy, U = silt, u = silty, T = clay, t = clayey, L = loam, l = loamy, H = swamp, h = swampy, n = half-swampy

Table 5. Assignment of soil textures to effective rooting zone, medium soil density (DBG, 1992)

Class (evaluation)	Soil texture[a]	RZe (mm/dm) Range	average
1 (very low)	Hn	< 3	2
2 (low)	S, Hn	3 – 5	4
3 (medium)	lS, uS	5 – 7	6
4 (high)	tS, lS	7 – 9	8
5 (very high)	U, sU, lU, tU, sL, uL, tL, lT, T	> 9	10

(a) S = sand, s = sandy, U = silt, u = silty, T = clay, t = clayey, L = loam, l = loamy, H = swamp, h = swampy, n = half-swampy

2.5. Nitrogen Fertilizer Recommendations

The variations in mineralization, denitrification and leaching rates from one soil to another and from one year to the next, have complicated the use of a soil test for making N fertilizer recommendations (Foth and Ellis, 1997). The basis for the assessment of the efficiency of N fertilizer must be different from that for P and K because only very small amounts of any excess of these two elements accumulate in soil to benefit subsequent crops; but there are adverse environmental impacts from excess nitrate or ammonia lost from the soil. The most accurate way for this assessment is the use of ^{15}N-labelled fertilizer in order to evaluate the amount of N absorbed by the crop. But its problem is that it is very expensive (Johnston, 2000).

Controlled-release fertilizers are the ones which supply nutrients to the soil solution and, hence, to the crop roots at a rate which more or less matches plant demand. The use of this kind of fertilizer has attracted considerable interest in order to improve fertilizer use and its efficiency. Considerable advances have been made in formulating them, particularly for the ones used in high value crops in which the extra costs over conventional materials is not so important (Smith *et al.*, 1997).

Nitrification inhibitors are another way to improve fertilizer efficiency and crop yields and minimize denitrification and/or leaching losses of NO_3^- by maintaining the applied N fertilizer in the soil as NH_4^+-N (Jonhnston, 2000, Bronson *et al.*, 1991, Smith *et al.*, 1989). Another consequence of keeping the N in the ammonium form is the reduction of NO and N_2O emissions from nitrification and N_2O emission from denitrification (Bronson *et al.*, 1992).

When irrigation is needed, the application of the required nutrients along with irrigation (fertirrigation) offers great promise to improve nutrient use efficiency, especially if the quantity of water is controlled because in this case, the nutrients are not leached below the zone of active root growth (Johnston, 2000).

Another alternative for increasing N use efficiency besides minimizing the adverse effects of fertilizers in the environment is the deep placement of nitrogenous fertilizer. For wetland rice, its application limits the loss of N due to surface run-off, leaching, volatilization and denitrification, leading to a reduction in applied N (Bautista *et al.*, 2001).

The main recommendation is that N fertilizer should be applied at an appropriate time and in the correct amount if it is to be used efficiently as estimated by the amounts of N mineral that have remained in the soil at harvest. Also, another very important issue is the development of non-destructive plant analysis methods used for monitoring the N status of the crop throughout the growing season, which will allow for more accurate decisions regarding the amounts of N fertilizers that must be applied (Johnston, 2000).

3. Phosphate Fertilizers

The use of phosphate fertilizers in western Europe started at the beginning of nineteenth century and the phosphorus source was bones from animals (Davis and Haglund, 1999). The production of phosphoric acid began in Europe in the 1870s. During the 1960s, ammonium phosphates were developed. Today, diammonium phosphate is the main product in the world phosphate trade (Park, 2001).

Nowadays, there are three main uses of phosphorus. In western Europe, 79% is used to make fertilizers for use in agriculture for food production; around 11% is used to make feed grade additives for animal feed fillers and approximately 7% is used to make detergents (Johnston and Steén, 2000). Figure 3 shows the world production of phosphate fertilizers during 1980, 1990 and 1997. Diammonium phosphate is the key product, having increased from 31% of the total world production during 1980 to 48% in 1997, while the production of all the other products has declined (Park, 2001).

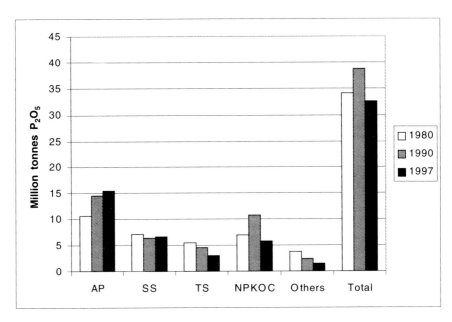

Fig. 3. World production of phosphate fertilizers (million tonnes P_2O_5). Source: IFA (International Fertilizer Industy Association). AP, Ammonium Phosphates; SS, Single Superphosphate; TS, Triple Superphosphate; NPKOC, NPK and Other Compounds; Others include mainly basic slag, ground phosphate rock for direct applications, fused magnesium phospate (Data from Park, 2001).

Phosphorus has a very important role in the building of nucleic acids and the storage and use of energy in the cells (Ozanne, 1980). Among the most important compounds, the following are emphasized: adenosin di and triphosphate (ADP, ATP), phospholipids, and nucleic acids (RNA, DNA). Thus, it is a remarkable fact that phosphorus is essential for photosynthesis (Johnston and Steén, 2000). The content of phosphorus in plants ranges between 0.1 and 1.2% from which, 80% is incorporated in organic compounds. Plants can absorb phosphorus in very low concentrations and the energy needed for this is derived from breathing (Domínguez, 1997).

3.1. Phosphorus Cycle

In soil solutions, phosphorus is either taken up by plant roots or goes into a readily available pool, where it is held weakly. From there,

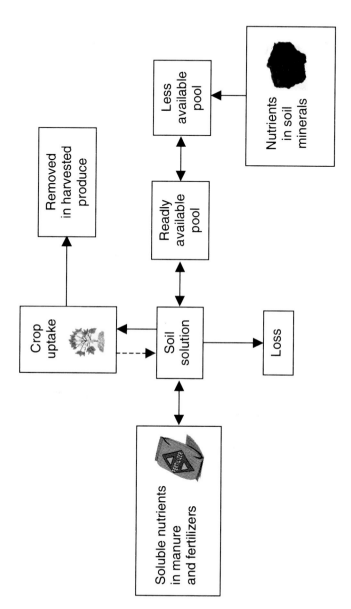

Fig. 4. Schematic representation of the P-cycle in the soil-plant system (adapted from Johnston and Steén, 2000).

phosphorus moves to a less readily available pool, where it is held more strongly (Johnston and Steén, 2000), as is showed in Figure 4.

Inputs of P to soils can occur by atmospheric deposition (pollen or dust) and by the application of fertilizers (mineral or organic) in agricultural systems (Chardon, 2000). After the application of inorganic P fertilizers to soil, several reactions can take place with the soil, as described by Chardon (2000):

- Adsorption, fast reaction with outside soil particles.
- Absorption, slow migration into the pores of soil aggregates.
- Immobilization, incorporation into soil organic matter.
- Precipitation in soils with high pH, binding with other chemical elements.

The difference between the amount of P applied and that exported is the P balance. If the P balance is negative, P in the soil becomes depleted, while if the balance is strongly positive, P in the soil can exceed crop needs and a risk for the environment exists.

3.2. Phosphate Fertilizers and Water Eutrophication

The main problem derived from the repeated application of phosphate fertilizers to soil is the loss of freshly-applied and soil-accumulated P from agricultural land to water, causing eutrophication. Eutrophication problems in freshwater are commonly related to P inputs (Eghball et al., 1996; Gibson, 1997; Foth and Ellis, 1997). P levels in surface waters higher than 10 ppb (10 mg P/L) have been associated with increased algae growth in streams and lakes (Foth and Ellis, 1997). Only very low concentrations of P are required for the appearance of eutrophication symptoms (Gibson, 1997).

Phosphorus can be transported from agricultural land to water by different pathways (Figure 5). If P is accumulated in soils because of the addition of fertilizers, the soil surface will be either very high in labile and total P and lead to a downward movement in the soil profile or will increase the P content of water and soil particles transported to surface waters through run-off and erosion (Foth and Ellis, 1997). In soils with shallow groundwater tables, the main pathway of water transport is via subsurface run-off through the unsaturated part of the soil (Eghball et al., 1996; Van der Molen et al., 1998). P movement through the soil profile is substantial under favourable conditions, especially in sandy and coarse-textured soils (Van der Molen et al., 1998).

Run-off loss of P has been a major environmental concern regarding water pollution. It usually occurs in suspended soil colloidal materials (particulate P) via erosion. Even where erosion is minimal, elevated soil P can sustain high P losses (Heathwaite, 2000). Van der Molen et al.

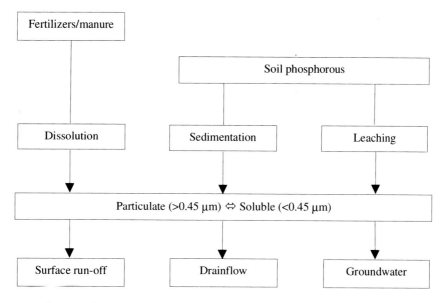

Fig. 5. Pathways of P loss from agricultural land to water (Withers, 2000).

(1998) after field experiments and model results, showed that in the Netherlands, the highest P emissions to surface waters are found in P-saturated sandy soils. Undesirable agricultural practices which accelerate P loss are the unnecessary accumulation of P, the application of P amendments at rates and times which cause direct run-off, and the adoption of land management practices, which increase erosion risk on unstable soils (Withers, 2000).

Three subsurface pathways are recognized as having potential for P transport (Heathwaite, 2000): surface lateral flow, due to higher soil P concentrations in upper soil horizons, preferential flow that enables rapid subsurface transport of mobile P through soil macropores, and artificial drainage that encourages rapid transit of water from land to stream. In sandy soils close to water, for example in the coastal plain of southwestern Australia, direct leaching of P can deliver enough P to estuarine channels to render them eutrophic (Gregory and Ingram, 2000). In Tables 6 and 7, geographical and management factors determining flows of phosphorus have been summarized.

The primary manner of diminishing environmental problems caused by P is improving the management of P fertilizers through integrated plant nutrient management in order to avoid over application and reduce the export of P from agricultural land (Van der Molen *et al.*, 1998; Weaver and Reed, 1998; Johnston, 2000). The addition of Fe

Table 6. Geographical factors determining flows of phosphorus (Cowell, 2000)

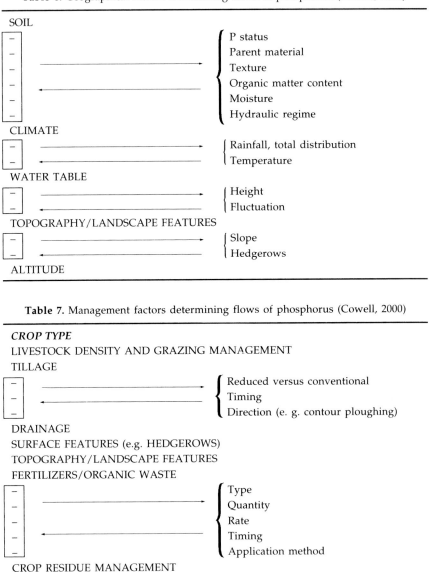

SOIL
- –
- –
- –
- –
- –
- –

{ P status
Parent material
Texture
Organic matter content
Moisture
Hydraulic regime

CLIMATE
- –
- –

{ Rainfall, total distribution
Temperature

WATER TABLE
- –
- –

{ Height
Fluctuation

TOPOGRAPHY/LANDSCAPE FEATURES
- –
- –

{ Slope
Hedgerows

ALTITUDE

Table 7. Management factors determining flows of phosphorus (Cowell, 2000)

CROP TYPE
LIVESTOCK DENSITY AND GRAZING MANAGEMENT
TILLAGE
- –
- –
- –

{ Reduced versus conventional
Timing
Direction (e. g. contour ploughing)

DRAINAGE
SURFACE FEATURES (e.g. HEDGEROWS)
TOPOGRAPHY/LANDSCAPE FEATURES
FERTILIZERS/ORGANIC WASTE
- –
- –
- –
- –
- –

{ Type
Quantity
Rate
Timing
Application method

CROP RESIDUE MANAGEMENT
APPLICATIONS OF OTHER NUTRIENTS, LIME AND PESTICIDES

oxides to the soil may raise the P adsorption capacity of P-saturated soils (Van der Molen, 1998). The treatment of selected parts of catchments with highly-adsorbing bauxite waste may also reduce P treatments to estuarine systems (Summers and Pech, 1997).

3.3. Other Environmental Problems

The use of phosphate fertilizers has other environmental concerns such as the exhaustive mining of rock phosphate reserves in developing countries and the air pollution by manufacturing inorganic P products; the accumulation of potentially-harmful metals in soils from the repeated application of rock phosphate and its products, particularly cadmium (Cd) (Whiters, 2000).

The raw material for most P fertilizers is apatite or rock phosphate. The mineral deposit is exposed generally by strip mining (Foth and Ellis, 1997). Approximately, five tonnes of ore must be mined and processed to produce one tonne of commercial phosphate rock with an average P_2O_5 content of 32% (Ullmann, vol. A 19). The mining, crushing and grinding of phosphate ore are extremely energy intensive unit operations in rock phosphate processing.

Phosphorus as such is not a non-renewable limited resource but rock phosphate is a limited resource. It is very difficult to judge the remaining reserves of phosphorus. In any case, even if it is not exactly known as to when this resource will cease, if the extraction continues at the present rate, the reserves will run out one day. Therefore, there is the need to create a system where P is recycled within a closed cycle (Davis and Haglund, 1999).

Fertilizers, together with atmospheric deposition, contribute to the build up of heavy metals in soil. On the other hand, heavy metals leave the field through leaching and erosion and in the cropped product (Audsley *et al.*, 1997). Raven and Loppert (1997) evaluated the trace element and heavy metal composition of a wide variety of fertilizers and soil amendments and came to the conclusion that rock phosphate, followed by sewage sludge and phosphorous fertilizers, had the highest concentrations.

Interactions of trace metals with the solid phase of soil include chemisorption of minerals, precipitation with different anions, co-precipitation in minerals, and complexation with organic matter (McBride, 1989). For instance, Almas and Singh (2001) found that low pH and organic matter content increase the solubility of Cd and Zn in soil, while organic matter addition lead to an increased mobility in soil and, consequently, their uptake by plants is also increased.

In order to diminish the impact caused by toxic elements in soil, it is once again advisable to review the P fertilizer rates applied in order to avoid not only over application of P, with the previously exposed consequences, but also the accumulation of heavy metals and other toxic elements.

4. Potash Fertilizers

Potassium is absorbed by plants in great quantities. It constitutes, together with lime, the main proportion of the mineral material of vegetables. For this reason, plant ashes contain a large quantity of this element, which plants absorb as potassium ions (K^+).

The use of potash as a modern fertilizer can probably be traced back to the first settlers in North America who noticed that the Native Americans applied wood ash to their crops (Park, 2001). The first commercial-scale production of K fertilizers began in Germany in 1861, and it dominated world trade until after the World War I (Foth and Ellis, 1997).

As it has been shown in Figure 6, the main product in the potash fertilizer industry is potassium chloride (more than 90%). Among other products, natural potassium sulphate and potassium magnesium sulphate are the most important. It can be seen that the production has not increased during the past two decades.

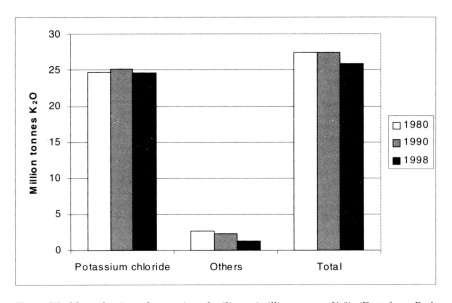

Fig. 6. World production of potassium fertilizers (million tonnes K_2O). (Data from Park, 2001)

The role of potassium in plants varies, although some aspects are not well known. Potassium is part of a large number of enzymes favouring the synthesis of hydrocarbons in photosynthesis as well as the movement of these compounds in the plant and its accumulation in the reservoir organs. For this reason, plants with significant carbohydrates reserves (potatoes, beets, grapes) respond very well to potassium contributions. Potassium takes part in the protein formation process. Hence, in order to obtain a good yield of nitrogen fertilizers, it is important to supply this element adequately. Potassium contributes to maintaining cell swelling, thus favoring water benefits for plants. It also has a favorable effect on plant resistance to cold and frost and increases plant resistance to salinity and parasites (Beringer and Nothdurft, 1985).

Potassium deficiency is manifested by underdevelopment, affecting mainly those parts that accumulate reservoir substances (fruits, seeds, tubers). When the deficiency is very acute, the leaves show yellow spots followed by necrosis on the tips and edges. Potassium deficiency also causes a decrease in harvest yield and an increase of the vegetative period, a delay in fruit ripeness and a decrease in cold and drought resistance (Huber and Arnay, 1985).

4.1. Potassium Forms in Soil

Potassium in soil is in both organic and inorganic form. Inorganic potassium (1.5% average concentration in soil) is present mainly in silica minerals and is released by the weathering of these minerals. In general, soils have large quantities of potassium, mainly the ones consisting of clay. Organic potassium comes from vegetable decomposition and animal wastes, attributing a small part of the total potassium content in soil. An important part is immediately dissolved in soil solutions, while the rest requires the action of microorganisms. Binded potassium is recovered when the microorganism wastes are incorporated to the soil. Compared to nitrogen, potassium is not complexed or bound up into organic matter. Essentially, all of it is associated with the mineral fraction (Foth and Ellis, 1997).

From the point of view of its benefit to plants, potassium can be classified as (Dominguez, 1999):

- Non-available potassium, which constitutes most of the potassium in soil. It is a part of the structure of soil parent materials, mainly feldspars and micas. These minerals are weathered and K ions released are exchangeable and exist as K adsorbed and K solution.
- Quickly-available potassium. Potassium dissolved in a soil solution and potassium that is exchangeable and adsorbed into soil colloid is quickly available. In a normal soil it represents 1-2% of the total K content. From this, only 10% is in the soil solution, while the remaining 90% is adsorbed to soil colloids.

  Potassium ion concentration in soil remains constant. When plants absorb ions from soil solution, its concentration decreases and the clay-humic complex quickly releases ions that go into the solution in order to maintain the equilibrium. On the other hand, when a potash fertilizer is applied, the ion content of the solution increases. In order to maintain equilibrium, the surplus of ions in the solution is adsorbed by the complex.

- Slowly-available potassium. In some circumstances, K^+ from soil solutions is fixed by some clay types, impeding its absorption by plants. This potassium can be released with time (regeneration process) passing again to the soil solution. Fixation and regeneration processes play an important role in agriculture. One part of soluble potassium added with chemical fertilizers is absorbed by plants, while a larger part is adsorbed by colloids and the rest is not available temporarily, because it is being retained by clays. This retained potassium in a non exchangeable way is not lost by lixiviation. When a potash fertilizer is applied to the soil, it is quickly dissolved in water and a part is adsorbed by colloids. The quantity of adsorbed potassium depends on the clay content of the soil, the kind of clay that prevails and the calcium content.

4.2. Potassium Balance

Potassium balance in the soil—in available forms for plants—is the result of some processes with inputs and outputs. The inputs are produced by the following processes:

- Mineralization of organic matter.
- Meteorization of soil minerals.
- Chemical fertilizers.

The outputs are produced by the following processes:

- Extraction by plants and microorganisms. This potassium is temporarily immobilized until a part of it is given back to the soil in the form of organic waste.
- Fixation in the inner surface of some kinds of clay. This potassium is recovered after some time.
- Leaching.

4.3. Potash as a Non-Renewable Resource

There is no evidence of environmental problems caused by the use of potash fertilizers. Due to its low concentration and the tampon activity of the soil, potassium is not toxic or aggressive to the soil ecosystem. The most relevant impact is the depletion of potash salts reserves. Although potash, like phosphate, is a non-renewable resource, the known resources are much larger than for phosphate. Nevertheless, over the next 50 years, some potash producers will be obliged to mine lower grade ores, deeper layers or more distant regions (Park, 2001). In this sense, an efficient use of K fertilizers must be applied, considering the availability of nutrients from all other sources on each farm (Johnston, 2000). Once the plants' critical need of this nutrient is defined, the quantity of exchangeable K in the soil should be determined by the use of nutrient balance.

5. Manure and Other Biological Waste and By-Products

Organic wastes were the first fertilizers to be used in agriculture (manure, meat flour, agricultural and forest wastes, etc.). Afterwards, with the development of the chemical fertilizer industry, biological waste was replaced by non-organic fertilizers because these are cheaper and easier to transport and to apply. Nowadays, the large volume of waste generated by our society poses an environmental problem and its use in agriculture has resulted in a more interesting application than disposal. Furthermore, the search for more environment-friendly production systems is increasing the use of manure and other organic wastes in agriculture.

The current legal framework in Europe shapes the general trend with respect to the management and treatment of biological wastes. In this sense, the Directive 91/156/EEC establishes the basis for waste management. The keywords for this are reduction in origin, reuse and recycling. The Directive 99/37/EEC establishes the objectives for a gradual and progressive reduction of the disposal of biodegradable wastes in landfills (Burton, 2001).

It is easier to add the organic residues directly to the soil. This is a common technique in agriculture with crop residues. It can also be heaped and the natural fermentation process will take place, leading to anaerobic degradation of the material. According to Canet (2000), the use of fresh organic residues is a non-advisable alternative. Its use results in a drastic increase of the microbial activity of the soil, which in turn, causes a decrease in the oxygen available, the production of phytotoxic elements (ammonia and different organic acids such as acetic, propionic, butyric, etc.) as well as an acceleration of the mineralization of native soil's organic matter and possibly, a rise in soil temperature that could damage the growing seeds or plants. As N is a limiting factor of the soil's biological activity, an excess of soil activity could cause its immobilization and consequently, the plants would suffer. In addition to all of the previous factors, if the soil has not reached the high temperatures required for the composting process, undesirable seeds as well pathogenic organisms or its reproductive forms may be present in fresh organic residues. Sequi (1996) also points out that the advantage of using composted materials is that it is easier to handle and diminishes the resulting odour.

The anaerobic decomposition process has been studied and is used in controlled conditions as a waste treatment method. However, the composting process has been developed as a treatment of the organic portion of municipal solid waste much more than anaerobic digestion, particularly due to economic reasons.

Manures also need to be stabilized. Manure comprises animal excrement mixed with different wastes used as livestock bedding (straw, sawdust, rice shell, etc.). This mixture is constituted by nitrogen-rich materials (urine and excrements) and carbon-rich products, with a high content of cellulose and lignin.

Manure and other residual products such as sludge generated in wastewater treatment plants and agricultural by-products are rich in organic matter. Therefore, these complex fertilizers are beneficial in improving the edaphic characteristics of soil. Organic by-products and wastes have a significantly higher carbon content (approximately 50%) than inorganic fertilizers (often 1% or less). Organic amendments help to stabilize soil aggregates, prevent erosion and improve the structure of soil, promoting good tilth, good moisture retention in areas prone to drought and, paradoxically often good drainage in wet areas (Addiscot et al., 1992).

The N, P, and S present in organic residues are often covalently bound to C. Microbial decomposition releases ammonia, phosphates, or sulphates from the sources to be made available for the plant. Organic fertilizers also supply the soil with many other nutrients. The nutrient content and nutrient release rate of manure and other organic fertilizers varies with each organic source as also with other factors like application methods and soil characteristics. Evaluation of nutrient concentration is important in selecting an organic fertilizer, because this concentration governs nutrient availability, the bulk of material that must be applied, and other possible benefits that the fertilizer may have as a soil amendment (Huntley et al., 1997; Palm et al., 2001). In this sense, development of a rapid, accurate assay for predicting potential nutrient release rates would promote better waste utilization and nutrient management (Mikkelsen, 1997).

The range of problems associated with animal manure and its disposal is summarized in Figure 7 (Burton, 2001). In the past, odour nuisance was the main problem and so research and legislation was focused mainly on this aspect. Nowadays, nutrient leaching causing water pollution, more concretely eutrophication and, as mentioned before, a significant problem derived mainly from agricultural activity. Surface run-off (overland flow) transporting slurry with a high BOD and containing nitrogenous and phosphorous components is another cause of water pollution. Volatilized ammonia, which causes acid rain and eutrofication, as well as emissions of nitrous oxide, a greenhouse gas, also contributes also to air pollution.

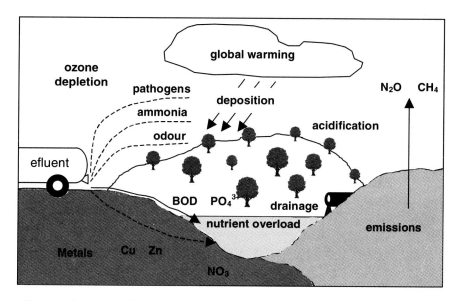

Fig. 7. A Summary of potential impacts on the environment deriving from livestock waste (Burton, 2001)

Concern over the environmental damage done by large amounts of ammonia emissions has resulted in legislation in Denmark and Holland that requires slurries to be covered. Nutrient surpluses in terms of nitrogen, phosphorus and organic material have been recognized as a common factor behind many of the pollution problems, especially in Northern Europe. On the other hand, in southern countries, including Greece and Portugal, soils are often deficient in organic matter. Therefore, redistribution of manure, especially as a composted product, thus received some interest (Burton, 2001).

5.1. Manure and Leaching of Nutrients

Manure is an excellent source of nutrients that should not be wasted; the main problem in using it is the uncertainty about its composition. Whatever the nutrient source used, a proper nutrient management based on nutrient balances must be a challenging goal in agriculture.

Nitrogen availability depends on the source of the manure but generally, most of it becomes available to plants in the first year, while the rest will be attainable in the following years. In the case of poultry manure, for instance, studies have shown that it contains large percentages of rapidly mineralizing organic N and that as much as 20-30% of the total N can be found in ammoniacal forms that are easily lost by volatilization (Sims, 1997). This reflects the large content of mineral nitrogen and urea present in fresh manure. Although a small amount of nitrogen becomes available in the following 4 or 5 years after application, the manure has a persistent effect (Addiscot et al., 1992). Consequently, the application of manure increases not only the fertility of the soil, but also the risk of serious nitrate loss in surface run-off water or by leaching, and these factors must be considered in the selection of an agronomically-and environmentally-effective manure.

The extent of N mineralization will depend not only on the properties of the manure, but also on the soil environment, such as the manure application method, soil temperature, moisture, and pH level. Some studies suggest that depending on the application method, approximately 30 to 90% of the total N in swine manure will be available for plant uptake during the season of application, with a smaller fraction released in subsequent years (Mikkelsen, 1997).

A greater loss of nitrate leaching can be expected in autumn and winter, where manure is ploughed into warm soils. Spreading during winter months is prohibited in a number of European countries including Sweden, Denmark, Germany and the Netherlands, although it is permitted, subject to some restrictions, in the UK (Burton 2001). Ploughing or spreading manure in spring rather than in autumn or winter is not always a solution, because apart from the problems of storing manure over winter, there is a risk of soil compression because of the traffic of farm vehicles (Addiscot et al., 1991). Gangbazo et al. (1998) concluded that pre-planting and post-emergence application of hog manure, at rates compatible with corn N needs, did not contaminate water more than chemical fertilizers. They also concluded that spring run-off and drainage accounted for 65% of the total annual water loss, and 40 to 70% of the annual nutrient loads.

The ratio of essential plant nutrients contained in manure is generally not well balanced with plant requirements. Amounts of P and K are proportionally greater than N concentration, since it is volatilizable. Manure (and other organic fertilizer) application rates are usually based on the N content, and consequently, can result in an accumulation of P and K in manured soils. As mentioned for P fertilizers, an increase in the amount of P in these soils corresponds to an increase in the potential for P loss via run-off to surface waters

(Eghball *et al.*, 1996; Gale *et al.*, 2000). In the same way, P leaching can occur in those areas that have received excessive P loading (Sharpley *et al.*, 1996; Kim *et al.*, 2001), especially in sandy soils (Mikkelsen, 1997; Hens and Merckx, 2001). Kim *et al.* (2001) observed in a study performed in plastic greenhouses in Korea that P had accumulated 1.6 times more in organic farming than in conventional farming soils, due to repeated and excessive applications of manure compost with a low N/P ratio. One important concern is the need to regulate manure application (mainly poultry wastes) based not only on N, but also on P.

Eghball *et al.* (1996) concluded that phosphorus from manure moved deeper into the soil than that from chemical fertilizer, even at similar P loading rates, this being probably due to the combination of P with organic compounds from manure. More research is needed to determine the fate and transport of P in soils considered 'excessive' due to long-term application of animal wastes (Sims, 1997). Another aspect to be noted in order to avoid this surplus of P in soil is to eliminate dietary excesses, where they exist. It has been well documented that cow farmers overuse P in feed (Van Horn and Hall, 1997; Kuipers and Mandersloot, 1999). In pig farming, a considerable reduction of P-losses is achieved by using enzyme phytase in concentrated feeds (Kuipers and Mandersloot, 1999).

Some management system treatments such as anaerobic lagoons (Mikkelsen, 1997), ash lagoons (Lau and Wong, 2001) or tangential flow separator (Westerman and Bicudo, 2000) have been proposed as effective methods to decrease nutrient composition in swine manure (80-90% P and 22% N removal in tangential flow separator). Anaerobic lagoons are especially popular for treatment in those areas where suitable land for effluent application is limited, since it requires significantly less land for manure management (Mikkelsen, 1997).

Due to the different characteristics of organic fertilizers and blended fertilizers (mixtures of different plant-, animal-, rock- or mineral-derived materials) which have relatively high nutrient, contents have been designed to mimic synthetic fertilizers with rapid nutrient availability (Huntley *et al.*, 1997). This option has other beneficial aspects, because as stated before, it improves soil properties and provides a pool of long-term attainable nutrients.

5.2. Manure and Atmospheric Emissions

Another problem with manure is the loss into the atmosphere, mainly as ammonia, but also as nitrous oxide, particularly in dry sandy soils. The major source of ammonia is urea from urine, or uric acid in the case of birds, which is converted to ammonia by bacterial urease, commonly available in soil. The volatilization of ammonia from livestock manure has been the focus of many studies because it causes a loss of nitrogen used in crop production and has an adverse impact on the environment. Odours resulting from the storage and land application of manure are also a serious concern for intensive livestock operations. Since NH_3 concentration is correlated with manure odours, it can be used to gauge the effectiveness of management practices on reducing odours (McGinn and Janzen, 1998).

The main release of atmospheric NH_3 occurs during the production, storage and application of livestock manure. Nearly 96% of the ammonia emissions in Europe are related to agriculture (Jol and Kiedman, 1997). According to Asman (1992), in Europe, 83% of NH_3 emissions come from livestock manure, while the remainder is from synthetic fertilizers.

In the previous section dealing with nitrogen fertilizers, the factors influencing NH_3 volatilization have been explained in detail. Other important factors are the spreading equipment and other management practices. A good method is to plough manure into the soil in order to reduce NH_3 loss but in doing so, this ensures that more N is in the soil and, therefore, is at the mercy of soil microorganisms. Regarding the use of slurry, a farm with a large slurry accumulation enables spreading to be timed to coincide with a high rate of uptake by growing plants, whereas a small accumulation requires extensive spreading in winter, which inevitably increases losses in various categories (McGechan and Wu, 1998). Slurry applications with a shallow injector reduce ammonia volatilization losses to a low level in comparison to surface spreading (McGechan and Wu, 1998; Mikkelsen, 1997).

Some studies have shown that chemical additives such as aluminium sulphate can significantly reduce NH_3 losses from poultry waste by volatilization, improving animal health and increasing the fertilizer N value of these wastes (Sims, 1997) although it can increase the aluminium content in soils. Recent studies have shown a very strong connection between high intensity protein feeding and high NH_3 losses (Smits et al., 1995).

As has been pointed out, lagoons are often used as a management system for cattle slurry. Studies dealing with cattle, poultry manure or slurry applications show little information concerning NH_3 and N_2O emissions associated with the application of waste lagoon effluent. Sharpe and Harper (2001) have studied this case and observed that an important part of ammonia was lost through drift volatilization during irrigation and also within 24 hours of application. Nitrous oxide emissions were also measured, but their contribution to N enrichment in the environment was considerably lowered.

5.3. Environmental Concerns with Other Constituents

Husbandry is mainly concerned with the profitable production of meat products. For this reason, fodders contain additives that promote animal growth and prevent diseases. Among these additives are growth hormones, antibiotics, insecticides, and feed additives containing trace elements such as arsenic (As), copper (Cu), and zinc (Zn). Sewage sludge and other composted materials such as municipal solid waste also present trace elements. Many of these products pass to animal faeces or urine and are present in manure applied to agricultural land. In the case of poultry waste, particular concerns exist with As, which has been reported to be present in broiler waste at concentrations that exceed regulatory limits established for land application, and with the concentration of pathogenic microorganisms (Sims, 1997).

When toxic metal concentrations of manure are high, metal leachability is of concern. Studies have shown that aspects such as soil pH and associated dissolution of organic matter from animal waste and sludge products may modify metal leachability more than the total concentration. High leachability of some metals (Cd, Cu, Mo, Ni and Zn) from alkaline stabilized sludge results from a greater concentration of dissolved organic matter, forming complexed metals into solution (McBride, 1998; Hsu and Lo, 2000; Almas and Singh 2001). Hsu and Lo (2000) assessed that although swine composts may contain high amounts of Cu, Mn and Zn, the results show that the potential leachability of these elements are most likely to be low. However, organic rich applications may influence the leachability of these elements, indicating that the application of swine manure may not be a risk-free management practice. More research needs to be done in this field. Nevertheless, at the moment, there is little evidence of the problems caused by fodder additives or pathogenic microorganisms.

6. LCA as a Tool to Evaluate Environmental Effect of Fertilizers

Since the late 1980s, the demand for a more ecological and sustainable life style has lead to further development in research methods in order to analyse and assess the environmental impact of products and systems. The environmental impact of food products depends on different factors such as agricultural practices (for instance, organic or intensive agriculture), packaging, the techniques applied for its processing in the industry, the destination and the means of transport used as well as the manner in which it is consumed. It is not easy for consumers or even for experts to account for these impacts, because it involves the aspects of emissions related to the use of energy, emissions of nutrients causing eutrophication, or pesticides and greenhouse effect emissions (Jungbluth et al., 2000).

Research work in food products that have taken simple indicators into account, such as energy use or greenhouse gas emissions, have been performed (Hülsbergen et al., 2001; Kramer et al., 1999). Nevertheless, these indicators cannot completely describe the global environmental impacts of agricultural production.

The Life Cycle Assessment (LCA) allows the evaluation of resource consumption and the environmental burdens associated to a product, process or activity from cradle to grave. In ISO 14040, the LCA is defined as the 'compilation and evaluation of the inputs, outputs and potential environmental impacts of a product's system through its life cycle, (Figure 8). The LCA process implies the identification and quantification of energy and material resources used in the system to produce the functional unit, together with the emissions into the environment during all the stages of the life cycle; the impact assessment of these mass and energy consumptions and environmental emissions; the assessment and implementation of methods to carry out form environmental improvements.

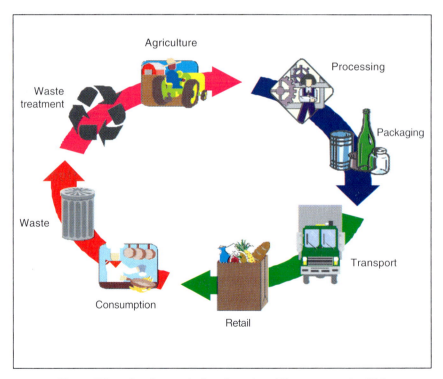

Fig. 8. Life cycle of an agricultural product (Clementre et al., 1999)

The environmental burdens cover all kinds of impacts such as the consumption of energy and other resources, land use and the effects of different hazardous substances on the ecosystems. Table 8 shows the main impact categories considered in an LCA. The main advantages of LCA are that it is product orientated, an integrative tool and is scientific and quantitative (UNEP, 1966).

Table 8. Examples of impact categories and some suggested end points (Andersson et al., 1994)

	Impact categories	*Suggested end points*
Consumption of resources	**Energy** (renewable/non-renewable)	MJ
	Raw materials (renewable/non-renewable)	kg
	Land	km^2
Global effects	Global warming	CO_2 equivalents
	Ozone depletion	CFC-11 equivalents
Regional effects	Acidification	H^+ equivalents loss of cations
	Eutrophication	O_2 consumption (aquatic) N (terrestrial)
	Photo-oxidant formation	NO_x Photochemical ozone creation
	Toxicity (human)	
	Ecotoxicity	Acute toxicity Potential bioconcentration Acute toxicity non-degradable
Local effects	Nuisance (noise, smell)	
	Working conditions	
	Effects of hazardous waste	
	Effects of solid waste	

In the previous years, the LCA has won general acceptance as a tool of great application in fields such as environmental labelling, product improvement and eco-design, that is to say, the design of more environmentally-friendly products, and the evaluation of environmental politics.

According to the ISO 14040, the main steps for performing an LCA are (UNEP, 1996):

- Analysis:
  1. Goal definition and scoping. The product(s) to be assessed is/are defined, a functional basis for comparison is chosen and the required level of detail is specified.

2. Inventory analysis. In this phase, the energy and raw material used and the emissions into the atmosphere, water and land are quantified for each process and then combined in the process flow chart.

- Assessment:
  3. Impact assessment. The effects of the resources used and emissions generated are grouped and quantified into a limited number of impact categories which may be then weighed by importance.
  4. Improvement analysis. The results are reported in the most informative way possible. The need and opportunities to reduce the impact of the product(s) on the environment are systematically evaluated.

When the LCA for agricultural products have been carried out, special demands on methodology are detected with respect to the specific features of agricultural production. Consequently, in order to satisfy these special demands, specific reports have been published in the past years (Weidema et al., 1995; Wegener Sleeswijk, 1996; Audsley, et al., 1997; Olsson, 2000; Weidema and Meeusen, 2000).

As illustrated in the previous points, some of the most important ecological effects of modern agriculture are related to non-functioning nutrient cycles, which entail losses of nitrogen, phosphorus and the associated trace elements. Nutrient balance is a useful tool in quantifying the flow of nutrients in food production systems (Cederberg and Mattsson, 2000). The nutrient cycle (N and P) is among these special demands when applying LCA to agriculture because emission rates can greatly vary, depending on soil types, climatic conditions and agricultural management practices (Brentrup et al., 2000). In order to know which emissions have originated by the use of fertilizers (ammonia, nitrate, phosphate...), structured estimation methods/models can be used. These estimation methods present some advantages, such as easy performance and less effort compared to measurements or values derived from literature (Bentrup and Küsters, 2000). Nevertheless, checking the estimated data against literature data is recommended.

There are many uncertainties associated with the nutrient cycle as also the effects that result from an accumulation of these elements in the system. These uncertainties are not associated with the basic knowledge of the different processes, but rather, with the complexity of the interactions among them. Consequently, when performing an LCA, the environmental effect of N and P should be contrasted with the status of the region, i.e. with the total N and P in the area (Erisman, 2000).

Another important aspect when performing an LCA is the inventory of data used in the evaluation (system inputs and outputs). With reference to this, Davis and Haglund (1999) performed a Life Cycle Inventory of fertilizer production in Europe and Sweden, where the emissions and use of resources in the production are thoroughly detailed to form a basic data source for LCA of food production systems. Here, environmental aspects associated with these processes have also been discussed.

When applying LCA methodology, one of the typical problems arises when a process fulfils more than one function in the system that is being studied. The problem is to decide what share of the environmental burdens of the activity should be allocated to the system that is being investigated (Ekvall and Finnveden, 2001). In this sense, when considering the use of manure in agriculture, the fact that manure is livestock waste must be taken into account. Therefore, its use in agriculture avoids the breeder to take responsibility of the impacts that originate by its disposal. Audsley et al. (1997) analysed different possibilities of allocating the environmental burdens between the manure producer system and the manure reception system. Nevertheless, they did not arrive at an agreement because each case implied different factors and it was difficult to find a universal method.

In the case of crop rotation, the problem lies in the manner of performing the allocation of impacts among the different crops. Arable farming may be seen as a multi-output process that yields a package of agricultural products and in terms of LCA, this implies that allocation of environmental impacts may be needed. Van Zeijts et al. (1999) proposed the following allocation system for the impacts of the different fertilizers in a rotation:

- Chemical nitrogen fertilizers should be allocated to the crop of application because farmers usually apply these fertilizers separately for each crop, due to the fact that the surplus of nitrogen will be lost to air and groundwater.
- Phosphate and potash fertilizers should be allocated to the different crops, according to the uptake and its efficiency. Phosphate does not easily leach to groundwater, thus over-fertilizing one crop may be beneficial to the following crops. The behaviour of potassium in clay soils is comparable to that of phosphate.
- With respect to animal manure, in the first year, available nitrogen should be allocated to the crop of application, and in the following years, available organic nitrogen should be allocated to all crops in the crop rotation system. For phosphate and potassium, the same principles as for chemical fertilizers should

be applied. The organic matter contribution from manure should be allocated to all crops, considering that it benefits soil structure and soil life.

When applying LCA to systems involving agricultural production, the soil quality and biodiversity are other areas which require more attention, and the development of assessment methods for soil quality and physical components of inhabitats affecting biodiversity is one of the goals for future research in this field (Cowell and Clift, 1997, 2000).

6.1. LCA and A Comparison of Farming Systems

In the past years, problems related to food production systems such as mad cow disease, and fears about nitrates and pesticides in our food and water have led to an increasing number of people changing from intensive farming to organic or traditional methods of farming. However, the resultant question is whether these methods are really safer for the environment.

In order to study the impacts of agricultural practices in food production systems, life cycle assessment studies have been performed on different products, comparing and evaluating different agricultural systems. LCA offers a degree of objectivity that enables comparisons to be made, based on global rather than solely on local impacts (Milà i Canals et al., 2001). Some of these works compare organic and conventional farming systems (Cederberg and Mattson, 2000; Mattson, 1999a; Nicoletti et al., 2001; Haas et al., 2001; Milà i Canals et al., 2001). Other works have focused on the agricultural practices in different countries (Stadig, 1997; Carlsson-Kanyama, 1998) or on intensive greenhouse production (Woerden, 2001). Brentrup et al. (2001) performed an LCA study on sugar beet production that took into account different forms of nitrogen fertilizers with the aim to study the suitability of the LCA methodology to analyse the environmental impact of agricultural production.

Several definitions for organic farming have used, ranging from the system that utilizes natural materials and practices to develop and protect biological systems for optimum soil health and crop ecology to a set of conditions that producers must meet in order to qualify their products as organic foods by the CEE (1991). Addiscot et al. (1992) stated that a convenient description from the point of view of fertilizer use is that organic farmers 'feed the soil not the plant'.

When comparing the environmental effects of the use of organic and synthetic fertilizers, different impact categories have to be considered. With respect to the manufacturing phase of fertilizers, it is obvious that organic farming implies a recycling of N and P nutrients from livestock production, while conventional or integrated production implies a depletion of phosphate ore and potassium salt mines for fertilizer manufacturing. The same can be stated with respect to the use of primary energy for fertilizer manufacturing. Regarding energy consumption, Cowell (1998) stated that over 90% of the energy used in fertilizer manufacturing is for nitrogen fertilizer production. Davis and Haglund (1999), in a life cycle inventory for fertilizer production, observed that the production of ammonia was found to be the step that required high energy consumption. Approximately, 90% of the total energy needed to produce a multi-nutrient fertilizer is connected with the production of ammonia, which is also true for emissions of CO_2 that are responsible for global warming.

Other impacts related to fertilizer production are the emissions that contribute to acidification (nitrogen and sulphur oxides) and eutrophication (nitrogen oxides and NH_3 to air; and NO_3^- and PO_4^{3-} to water). Davis and Haglund (1999) stated that for N_2O, the production of nitric acid alone generates a significant part of this emission. However, for nitrogen oxides (NO_x), the situation is different, since these emissions originate from several production steps. Emissions of sulphur oxides (SO_x) are caused mainly by production of sulphuric acid. Emissions of N and P containing compounds that end up in water occur most often in the final production steps. The fact that these emission probably have less importance when the entire life cycle of the product is considered, implies that the application of fertilizers, together with other agricultural practices, has to be taken into account (Davis and Haglund, 1999). Regarding the emissions from fertilizer production and transportation, Carlsson-Kanyama (1998) in an assessment of greenhouse emissions during the life-cycle of the carrots and tomatoes consumed in Sweden, concluded that they accounted for 4 to 10% of the total emissions.

The application of fertilizers into the soil has been shown to be related to serious global impacts such as eutrophication and acidification. Both organic and chemical fertilizers have significant results in eutrophication and acidification. Ammonia, being an acidifying and a nutrifying compound, is closely connected to the handling of farmyard manure (Cederberg and Mattsson; 2000). For carrot production in Sweden, Mattson (1999) detected a much higher potential for eutrophication and acidification in organic rather than in the integrated farming system. Nevertheless, when comparing different impacts on grassland farms in Germany, three different farming intensities were analysed (intensive, extensive, and organic). Haas et al. (2001) found that for both N and P balances, the intensive farms accounted for the highest inputs and the organic farming systems the lowest. When comparing the application of different synthetic fertilizers (calcium ammonium nitrate, urea ammonium nitrate solution and urea) for sugar beet production, Brentrup et al., (2001) found that they all contributed to acidification and eutrophication, but urea had the strongest environmental impact and calcium ammonium nitrate the lowest. The differences were mainly due to different ammonia volatilization after the application of the fertilizers.

Global warming emissions of N_2O, which are connected to synthetic fertilizer production and agricultural activity, play an important role with respect to the amount of total emissions contributing to this effect; hence, the potential negative effect on this impact category is more apparent in a conventional system due to the higher fertilizer rates (Cederberg et Mattsson, 2000).

Another aspect to consider when applying synthetic P fertilizers is the heavy metals originating from phosphate ore. Mattson (1999a), in a study that compared carrot purée from organic and integrated productions, observed that the amount of heavy metals was three times higher in integrated agricultural production.

All of these studies agree that LCA is a tool that allows distinction among the systems compared and consequently, the results of an LCA can help to choose the agricultural practices which will improve the nutrients use and their efficiency, such as the appropriate combination of fertilizers, manure, nutrient management and rates and the timing of fertilizer applications. Research shows that the effects of fertilizers have strong influences regarding the total impact on farming systems; thus the balance of the nutrients should be reviewed in order to attain more sustainable production systems. Finally, although organic farming allows for the use of wastes, that in other cases would have to be disposed of, the source of the manure that is used along with the timing and application rates should be reviewed in order to improve these types of agricultural systems.

7. Application to Orange Crop in la Comunidad Valenciana

Spain is the fourth largest orange producer in the world after Brasil, the USA and China, with 5.3 million tons produced in 1998/99. Of this, nearly 3.5 million tons were produced in the Comunidad Valenciana. Almost 70% of the orange production is exported. The main market for Spanish oranges is western Europe, importing around 85% of the citric exportations (Llorens 1999).

As a consequence of the increasing environmental concern regarding, citrus production following the early 1980s the Comunidad Valenciana´s citrus production started the Integrated Production system. Since then, it has increased in Spain, especially in the

Comunidad Valenciana which in the year 2000 was responsible of the 65% of this production (Coscollá *et al.*, 2000). The main purpose of Integrated Production is that agricultural practices are controlled in order to optimize the use of resources, and that the available knowledge is being adhered to.

7.1. Inventory Data

The agricultural practices regarding fertilization systems carried out in the Comunidad Valenciana integrated production of oranges have been assessed in this example by means of a modular LCA. Only the agricultural practices were examined. The data regarding the agricultural practices included in this study have been obtained from annual reports of the producers. The modules considered are fertilizer production, pesticide production, production of energy to be used directly on the farm (machinery and watering) and finally, the agricultural phase that includes the farm practices (Figure 9). The functional unit used was 1 kg of oranges leaving the farm.

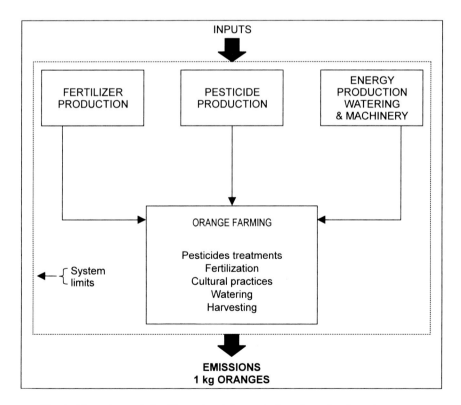

Fig. 9. Flow chart of the life cycle of the agricultural production of oranges.

Irrigation. Two irrigation systems have been considered—gravity and drip. In the first case, it is considered an average annual water volume of 6000 m^3/ha. and 65% irrigation efficiency. For drip irrigation, the average annual water volume of 5000 m^3/ha. and 85% irrigation efficiency has been considered. Four scenarios have been studied, depending on the origin of water, from a well or from a river, and on the fertilization system, fertirrigation or solid fertilization. Table 9 summarizes the scenarios.

Table 9. Scenarios that were examined in the study

Scenario	Water	Fertilization
A	Well	Solid
B	Well	Fertirrigation
C	River	Solid
D	River	Fertirrigation

Fertilization. When the gravity irrigation system is examined, a solid fertilization is carried out applied in three fractions. For the first one, 600 kg/ha. of 15-15-15 complex is applied in March, together with 3600 kg/ha. of sheep manure. The second fertilizer application is made between May and June; 300 kg/ha. of ammonium nitrate (33.5% N) is applied. In the third application, between July and August, the previous application is repeated. All the fertilizers that are used are applied by hand. The manure that is used is not buried in the soil due to the high cost it would imply.

For drip irrigation, the following solid soluble fertilizers are applied: ammonium nitrate (33.5% de N), phosphoric acid (54% P_2O_5) and potash sulphate (13% N y 46% K_2O). The applied rates are 782, 120 and 293 kg/ha, respectively. These fertilizers are applied together with water. Table 10 shows their distribution throughout the year.

Magnesium fertilization has not been examined in this study because this element is not environmentally problematic. Nevertheless, high concentrations of this element have been detected in irrigation water, both from river and well water. This fact should be considered when carrying out the fertilization plan.

7.2. Nutrient Balance

In order to estimate the emission data from fertilization practices, nutrient balances (N, P, K) for the different scenarios have been performed. It is important to emphasize the difficulty that can be found when estimating these emissions due to the complexity of the nutrients'

Table 10. Nutrient distribution (N P K) applied in fertirrigation (% per month)

Month	N (%)	P (%)	K (%)
January	0	0	0
February	5	25	0
March	10	0	10
April	15	0	15
May	15	25	15
June	20	0	20
July	20	0	20
August	0	25	20
September	0	0	0
October	15	0	0
November	0	25	0
December	0	0	0

behaviour in nature, which also depends on the application rate (Weidema and Meeusen, 2000). It must also be pointed out that the soil's nutrient stock has not been considered, due to the impossibility to analyse soil samples at the beginning and the end of the study.

The average nitrate, potassium and magnesium content in water— used for both irrigation systems—have been obtained from public organizations data bases. In order to compute the nutrients that were applied to the water, the irrigation efficiency, together with the annual water volume, have all been examined. The elements that are provided by rain water have been taken from Carratala *et al.* (1998) in consideration of an average annual rainfall in the zone of 500 mm.

It has been presumed that there is an average 1.5% of organic matter in soil and also that the organic matter content is stable, i.e. organic matter contributions compensate mineralization. Organic matter mineralization has been computed using the method proposed by Henin and Dupuis (1945). The mineralization coefficient has been taken from the data from Tamés (1975) in Pomares (2001) for irrigation farming in Eastern Spain, which had a value of 0.02. The analysis used soil samples from a depth of up to 15 cm. The C/N rate in soil humus has been concluded as 10 and 58% carbon content in soil (Brady and Weil, 1999; White, 1997; Rowell, 1994).

The results of nutrient extraction by the crop were obtained from Legaz and Primo Millo (1998) for the cultivation density examined.

Nitrate leaching has been calculated according to the results obtained by Ramos et al. (2002) for orange cultivation in the Comunidad Valenciana. The authors state that the average nitrate leaching is approximately 33% of the total NO_3^--N applied.

In order to compute immobilized N by the microorganisms, the ratio C/N has been concluded as 6 and its efficiency factor as 1/3 (Brady and Weil, 1999; White, 1997). The organic matter inputs are sheep manure, prunes and fruits and leaves that have fallen from the trees. It has been considered that the latter does not cause N immobilization due to its high N content. According to the manure composition provided by suppliers, it has also been determined that it does not cause N immobilization. Prunes do, however, cause N immobilization, as has been calculated according to data assessed by Moreno (2001).

Nitrogen outputs by the process of volatilization and denitrification have been calculated according to Bentrup and Küsters (2000). Table 11 shows the results of the nutrient balance.

Table 11. Nutrient balance for agricultural practices in the system under study (kg/ha.)

Inputs (kg/ha.)		N	P	K	Outputs (kg/ha.)	N	P	K
Mineratization (Organic					Extraction	226,5	22	123
matter soil)		31,32			Leaching			
(MO suelo)					A	124,58		
Fertirrigation		291,59	28,32	111,87	B	126,34		
Fertilization		291,00	39,33	74,70	C	110,87		
Rain		2,52	0,00	0,00	D	111,39		
Irrigation								
(ground water)	A	52,68	*	9,69	Immobilization	44,63		
	B	57,41	*	10,56	Volatilization	4,97		
Irrigation								
(surface water)	C	11,12	*	16,72	Denitrification N_2O	4,53		
	D	12,11	*	18,22	Denitrification N_2	32,18		
TOTAL A		377,52	39,33	84,39	TOTAL A	437,39	22,00	123,00
TOTAL B		382,84	28,32	122,42	TOTAL B	439,14	22,00	123,00
TOTAL C		335,96	39,33	91,42	TOTAL C	423,67	22,00	123,00
TOTAL D		337,55	28,32	130,09	TOTAL D	424,20	22,00	123,00
BALANCE (kg/ha.)					N	P		K
A					– 59,87	17,33		-38,61
B					– 56,30	6,32		-0,58
C					– 87,71	17,33		-31,58
D					– 86,65	6,32		7,09

Factors that were not considered in the nutrient balance are integrated in the difference between inputs and outputs. The figures obtained in this balance agree with those obtained by Lidón (1994) for orange crops in the Comunidad Valenciana.

7.3. Impact Assessment

The impact assessment has been calculated using the TEAM 3.0 (Ecobilan, 1999 France) software tool with the data calculated in the inventory analysis phase. In Figure 10, the relative contribution of each module to the total dimension of each impact category is shown for scenario B.

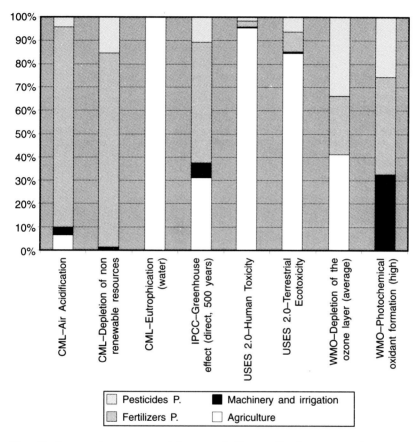

Fig. 10. Impact Assessment of Integrated Orange Production in the Comunidad Valenciana (scenario B).

As can be observed, the major impact on the categories of atmospheric acidification and the depletion of non-renewable resources is mainly caused by fertilizer productions. The agriculture module is also a main impact contributor in the eutrophication, as well as the human and terrestrial toxicity categories. In the greenhouse effect, it is the fertilizer production module that is of more relative importance, although agriculture also has an important effect on this impact category. The depletion of the ozone layer and photochemical oxidant formation are caused mainly by the fertilizer and pesticide productions and machinery and irrigation.

The results for each impact category have been subsequently displayed for each impact category affected, which demonstrates the contribution of each scenario to the total impact and the relative importance of the modules. Scenario B, considered as the reference scenario in this study, has the most impact in all of the categories. According to Mattsson (1999b), a difference in the impact produced by one scenario compared with the scenario B is considered to be equally significant if it constitutes more than 20% of its dimensions.

Figure 11 shows the results of depletion in the non-renewable resources impact category as a fraction of the reserve. As it is conceded, scenarios with fertirrigation deplete nearly double than those with solid fertilization. However, this could produce uncertainty because of the different data sources used for the different fertilizer production processes.

Figure 12 shows the results obtained for the atmospheric acidification impact category. As can be observed, the scenarios with fertirrigation present a greater impact than those with solid

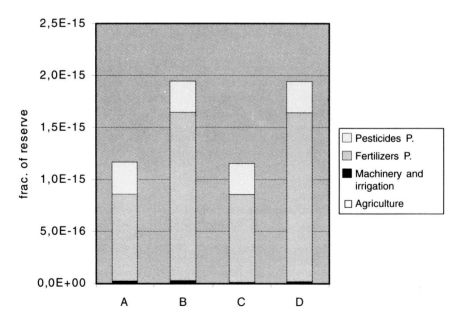

Fig. 11. Depletion of non-renewable resources in the different scenarios studied.

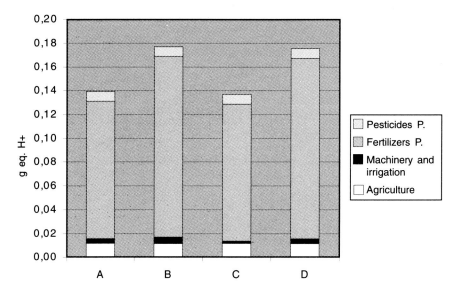

Fig. 12. Atmospheric acidification in the different scenarios studied.

fertilization. These differences are significant. The chief contributor to this is the fertilizer productions module.

Figure 13 shows the results for the water eutrophication impact category. No significant differences can be observed. The agricultural

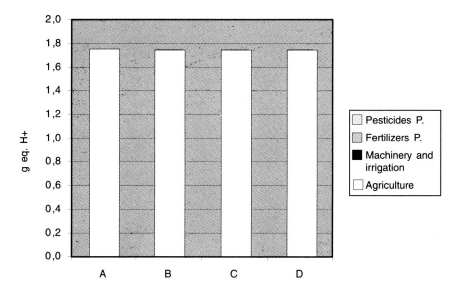

Fig. 13. Eutrophication to water in the different scenarios studied.

phase, concretely due to the NO_3^- in water, is the main contributor in this impact category.

Figure 14 shows the results obtained for the greenhouse effect impact category. As can be observed, all of the scenarios with fertirrigation have a stronger impact than those with solid fertilization. Nevertheless, these differences are not significant. The principal emissions that contribute to this impact are CO_2 produced specifically during ammonium nitrate production, followed by potassium nitrate production; and N_2O produced by the agricultural module during the denitrification process.

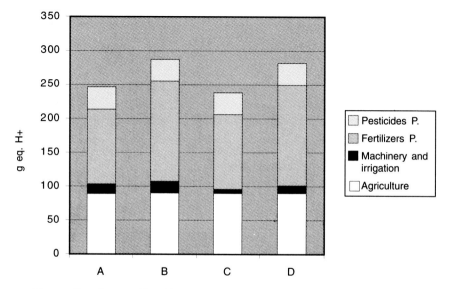

Fig. 14. Greenhouse effect (direct, 500 years) in the different scenarios studied.

Figure 15 shows the results obtained for the human toxicity impact category. Again, no significant differences can be observed between the studied scenarios. The impact category is mainly affected by the agriculture module, especially (97% of the impact contribution) by the use of Cu as a fungicide.

Figure 16 shows the results for the terrestrial ecotoxicity impact category with no significant differences between the studied scenarios. Again, the impact category was mainly affected by the agriculture module, especially by the use of Cu as a fungicide (68% of the impact contribution).

Figure 17 shows the results for the depletion of the ozone layer impact category. Differences between all of the scenarios can be

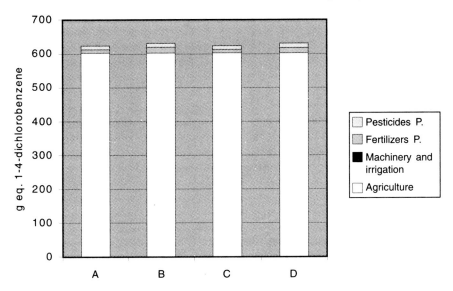

Fig. 15. Human toxicity in the different scenarios studied.

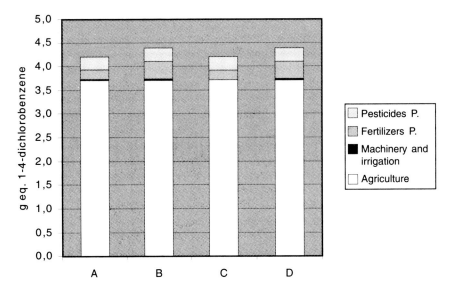

Fig. 16. Terrestrial ecotoxicity in the different scenarios studied.

observed. However, these differences are only significant in scenario C. The unique emission that is recognized as a contributor to this impact category is the gas Halon 1301 (CF3Br), mainly produced from diesel combustion in the machinery and irrigation module and during herbicide production.

Figure 18 shows the results for the photochemical oxidant formation impact category. Differences between all of the scenarios can be observed, although these variations are only significant in scenario C.

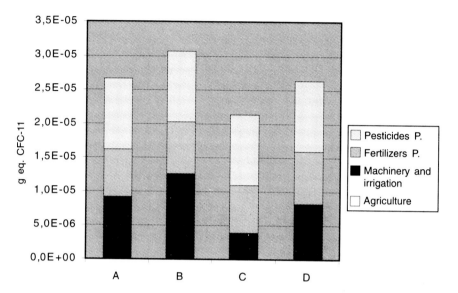

Fig. 17. Depletion of the ozone layer in the different scenario studied.

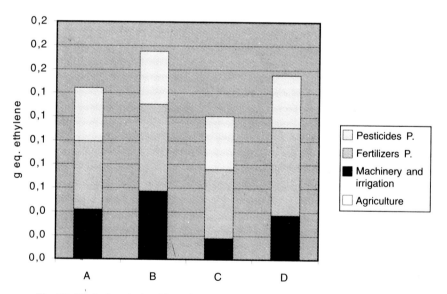

Fig. 18. Photochemical oxidant formation in the different scenarios studied.

The emission that contributes the most to this impact category is the one denominated by the used software as 'hydrocarbons', which includes hydrocarbons produced mainly during fossil fuel combustion. These emissions are produced first and foremost in the machinery and irrigation module and during ammonium nitrate production.

In summarizing the given information, the origin of water used in irrigation had little influence on the impact results, as compared to the other modules that were analysed in the system. This slight effect was related to the quantity of energy needed to propel the water in each case that was considered, and the subsequent emissions derived from the use of energy. In the aspect of the production of fertilizers, it can be stated that here the ones that were used for fertirrigation had a greater impact on the environment. However, these results depend mainly on the type of fertilizer used in the agricultural system, and this is subject to many possibilities. The source of the information from the production processes of the different fertilizers that were used is also a crucial point, which produces a degree of uncertainty when the sources are different. No differences have been found among the impacts that were caused by the use of different types of fertilizers. This can be attributed to the fact that almost the fertilization units were applied same in the different scenarios analysed.

It can be concluded from these results that fertilizers exert an impact on the environment and that LCA is a tool that can help us to assess the environmental effects of different fertilizers. The impact on the environment is not only due fertilizer applications in agriculture but also due to their production processes. The production of chemical fertilizers implies a depletion of non-renewable resources and use of energy, and also the emissions which derive from the manufacturing process have a significant impact on air acidification and the greenhouse effect, and a lesser effect on photochemical oxidant formation and the depletion of the ozone layer. The emissions that originate from the use of fertilizers contribute to eutrophication and the greenhouse effect. For this reason, fertilization rates need to be reviewed.

8. References

Addiscott, T. M., Whitmore, A. P. and Powlson, D. S. 1991. *Farming, Fertilizers and the Nitrate Problem*. CAB International, Wallingford, UK.

Addiscott, T. M. and D. S. Powlson. 1992. Partitioning losses of nitrogen fertilizer between leaching and denitrification. *Journal of Agricultural Science* 118, 101 – 107.

Alexandratos, N. 1995. *World Agriculture: Towards 2010—An FAO Study*. Wiley, Chichester, UK.

Almas, A. R., Singh, B. R. 2001. Plant uptake of Cadmium-109 and Zinc-65 at different temperature and organic matter levels. *Journal of Environmental Quality* 30, 869-877.

Andersson, K., Ohlsson, T. and Olsson, P. 1994. Life Cycle Assessment (LCA) of Food Production Systems. *Trends in Food Science & Technology* 5, 134-138.

Asman, W. A. H. 1992. Ammonia emission in Europe: Update emission and emissions variations. Report 228471008. National Institute of Public health and Environmental protection, Bilthoven, The Netherlands.

Audsley, E., Alber, S., Clift, R., Cowell, S., Crettaz, P., Gaillard, G., Hausheer, J., Jolliet, O., Kleijn, R., Mortersen, B., Pearce, D., Roger, E., Teulon, H., Weidema, B., van Zeijts, H. 1997. Harmonization of environmental life cycle assessment for agriculture. Final Report Concerted Action AIR3-CT94-2028. European Comission DG VI Agriculture. England, UK.

Bautista, E. U., Koike, M. And Suministrado, D. C. 2001. Mechanical Deep Placement of Nitrogen in Wetland Rice. *Journal of Agricultural Engeneering Research* 78 (4), 333-346.

Beringer, H. and Nothdurft, F. 1985. Effects of potassium on plant and cellular structures. *In: Potassium in Agriculture*. Munson, R. D. (ed.). ASA-CSSA-SSSA. Wisconsin, USA, pp. 352-368.

Brentrup, F. and Küsters, J. 2000. Methods to estimate potential N emissions related to crop production. In: *Agricultural Data for Life Cycle Assessments*. Weidema, B. P. and Meeusen, M.J.G. (eds). Agricultural Economics Research Institute (LEI), The Hague, Wageningen, Holland, pp. 104-113.

Brentrup, F., Küsters, J., Lammel, J. Kuhlmann, H. 2000. Methods to estimate on-field nitrogen emissions from crop production as an input to LCA studies in the agricultural sector. *International Journal of LCA* 5(6), 349-357.

Brentrup, F., Küsters, J., Kuhlmann, H., Lammel, J. 2001. Application of Life Cycle Assessment methodology to agricultural production: an example of sugar beet production with different forms of nitrogen fertilizers. *European Journal of Agronomy* 14, 221-223.

Black, C. A. 1993. *Soil Fertility Evaluation And Control*. Lewis Publishers, Florida, USA.

Brady, N. C. and Weil, R. R. 1999. *The Nature and Properties of Soils*. Prentice-Hall International.

Bouwman, A. F. 1994. Method to estimate direct nitrous oxide emissions from agricultural soils. National Institute of Public Health and Environmental Protection, Report 773004004, Bilthoven, The Netherlands.

Bouwman, A. F. 1996. Direct emissions of nitrous oxide from agricultural soils. *Nutrient Cycling in Agroecosystems* 46, 53-70.

Bronson, K. F., A. R. Mosier and S. R. Bishnoi. 1992. Nitrous oxide emissions in irrigated corn as affected by nitrification inhibitors. *Soil Science Society of America Journal* 55, 130-135.

Burton, H.C. 2001. Dealing with livestock manures: management strategies. In: *Aplicación agrícola de residuos orgánicos*, Boixadera, J. and Teira, R. (eds). Edicions de la Universitat de Lleida. Lleida, Spain.

Carlsson-Kanyama, A. 1998. Food consumption patterns and their influence on climate change. *Ambio* 27(7), 528-534.

Carratalá A, Gómez A and Bellot J. 1998. Mapping Rain Composition en the East of Spain by Applying Kriging. *Water, Air and Soil Pollution* 104, 9-27.

Cederberg, C. and Mattsson, B. 2000. Life cycle assessment of milk production – a comparison of conventional and organic farming. *Journal of Cleaner Production* 8(1), 49-60.

CEE. 1991. REGLAMENTO (CEE) N° 2092/91, DEL CONSEJO de 24 de junio de 1991 sobre la producción agrícola ecológica y su indicación en los productos agrarios y alimenticios. (DO L 198, 22.7.1991, p.1).

Chardon, W. J. 2000. The role of the soil in phosphorus cycling. In: Weidema, B. P. and Meeusen, M. J. G. (eds). Agricultural Data for Life Cycle Assessments. Agricultural Economics Research Institute (L. E. I.), Report 2.00.01, The Hague, Wageningen, Holland, pp. 8-12.

Clemente, G., Sanjuán, N., Cárcel, J. A., Benedito, J. y Bon, J. 1999. Análisis de Ciclo de Vida en Alimentos. *Alimentación, Equipos y Tecnología* 9, 103-108.

Coscollá, R., Malagín, J. y Fabado, F. 2000. La Producción Integrada de Cítricos en la Comunidad Valenciana. *Fruticultura Profesional* 70, 66-71.

Cowell, S.J. 1998. Environmental Life Cycle Assessment of Agricultural Systems: integration into decision-making. Ph.D. thesis. Centre for Environmental Strategy, University of Surrey, Guilford, U.K.

Davis, J. and Haglund, C. 1999. Life Cycle Inventory (LCI) of fertilizer production. Fertilizer products used in Sweden and Western Europe. SIK, the Swedish Institute for Food and Biotechnology. SIK-Report 654, Gothenburg, Sweden.

DBG (German Soil Science Association). 1992. Strategies to reduce nitrate loads in Ground water tied to the place and due to the use. German Soil Science Association (DBG), Geβen, Germany.

Dominguez, A. 1999. *Tratado de fertilización*. Mundi-Prensa, Madrid, Spain.

Duxbury, J. M. 1994. The significance of agricultural sources of greenhouse gases. *Fertilization Research* 38, 151 – 163.

Dyson, T. 1996. *Population and Food*. Routledge, London.

ECETOC (European Centre for Ecotoxicology and Toxicology of Chemicals). 1994. Ammonia emissions to air in Western Europe. Technical Report No. 62. European Chemical Industry Ecology & Toxicology Centre (ECETOC), Brussels, Belgium.

ECETOC. 1998. Nitrate and drinking water. Technical Report No. 27. European Chemical Industry Ecology & Toxicology Centre (ECETOC), Brussels, Belgium.

Eghball, B., Binford, G. D., Baltensperger, D. D. 1996. Phosphorous movement and adsorption in a soil receiving long term manure and fertilizer application. *Journal of Environmental Quality* 25, 1339-1343.

Ekvall, T., Finnveden, G. 2001. Allocation in ISO 14041-a critical review. *Journal of Cleaner Production* 9, 197-208.

Erisman, J. W. 2000. Key factors necessary to determine the impact on the nitrogen cycle and the resulting environmental effects as part of LCA for agricultural products. In: *Agricultural Data for Life Cycle Assessments*. Weidema, B. P. and Meeusen, M. J. G. (eds.). Agricultural Economics Research Institute (LEI), The Hague, Wageningen, Holland, pp. 104-113.

Foth, H. D. and B. G. Ellis. 1997. *Soil Fertility*. Lewis Publishers, Michigan, USA.

Fuentes, J. L. 1999. El Suelo y los Fertilizantes. Ministerio de Agricultura, Pesca y Alimentación (MAPA). Mundi-Prensa, Madrid, Spain.

Gale, P. M., Mullen, M. D., Cieslik, C., Tyler, D. D., Duck, B. N., Kirchner, M., McClure, J. 2000. Phosphorous distribution and availability in response to dairy manure applications. *Communications in Soil Science and Plant Analysis* 31(5), 553-565.

Gangbazo, G., Pesant, A. R., Côté, D., Barnett, G. M., Cluis, D. 1998. Spring runoff and drainage N and P losses from hog-manured corn. Journal of the American *Water Resources Association* 33(2), 405-411.

Gibson, C. E. 1997. The dynamics of phosphorus in freshwater and marine environments. In: *Phosphorus Loss from Soil to Water*. Tunney, H., Carton, O. T., Brookes, P. C., Johnston, A. E. (eds). CAB International, Wallingford, UK.

Granli, T. and Bøckman, O. C. 1994. Nitrous oxide from agriculture. *Norwegian Journal of Agricultural Science*, Supplement No. 12, 1994. Agricultural University of Norway, Ås, Norway.

Gregory, P. J. and Ingram, J. S. I. 2000. Global change and food and forest production: future scientific challenges. *Agriculture, Ecosystems and Environment* 82(1-2), 3-14.

Gregory, P. J., Ingram, J. S. I., Andersson, R., Betts, R. A., Brovkin, V., Chase, T. N., Grace, P. R., Gray, A. J., Hamilton, N., Hardy, T. B., Howden, S. M., Jenkins, A., Meybeck, M., Olsson, M., Ortiz-Monasterio, I., Palm, C.A., Payn, T.W., Rummukainen, M., Schulze, R. E., Thiem, M., Valentin, C., Wilkinson, M. J. 2002. Environmental consequences of alternative practices for intensifying crop productions. *Agriculture, Ecosystems and Environment* 88, 279-290.

Haas, G., Wetterich, F., Köpke, U. 2001. Comparing intensive, extensive and organic grassland farming in southern Germany by process life cycle assessment. *Agriculture, Ecosystems and Environment* 83(1-2), 43-53.

Heathwaite, L. 2000. Flows of phosphorous in the environment: identifying pathways of loss from agricultural land. In: *Agricultural Data for Life Cycle Assessments*. Weidema, B. P. and Meeusen, M. J. G. (eds) Agricultural Economics Research Institute (L. E. I.), Report 2.00.01, The Hague, Wageningen, Holland, pp 25-38.

Henin, S., Dupuis, M. 1945. Essai de bilan de la matière organique des sols. *Annales Agronomiques*. 15 (1) 161-172.

Hens, M., Merckx, R. 2001. Functional characterization of colloidal phosphorous species in the soil solution of sandy soils. *Environmental Science & Technology* 35, 493-500.

Hsu, J. H. and Lo, S. L. 2000. Characterization and extraction of copper, manganese, and zinc in swine manure composts. *Journal of Environmental Quality* 29, 447-453.

Huber, D. M. and Arny, D. C. 1985. Interactions of potassium with plant disease. In: *Potassium in Agriculture*. Munson, R. D. (ed.). ASA-CSSA-SSSA. Wisconsin, pp. 467-488.

Huntley, E. E:, Barker, A.V., Stratton, M. L. 1997. Composition and uses of organic fertilizers. In: *Agricultural Uses of By-products and Wastes*. Rechcigl, J. E. and MacKinnon, H. C. (eds). American Chemical Society, Washington DC, USA, pp. 120-139.

Isermann, K. 1994. Agriculture's share in the emission of trace gases affecting the climate and some cause-oriented proposals for sufficiently reducing this share. *Environmental Pollution* 83, 95-111.

Johnston, A. E. 2000. Efficient use of nutrients in agricultural production systems. *Communications in Soil Science and Plant Analysis* 31 (11), 1599-1620.

Johnston, A. E. and I. Steén. 2000. Understanding phosphorus and its use in agriculture. European Fertilizer Manufacturers Association (EFMA), Brussels, Belgium.

Jol, A., Kielland (ed.). 1997. *Air Pollution in Europe 1997*. European Environmental Agency (EEA), Copenhagen, Denmark.

Jungbluth, N., Tiejte, O. y Scholz, R.W. 2000. Food purchases: impact from the consumers' point of view investigated with a modular LCA. *International Journal of LCA* 5(3), 134-142.

Kim, P. J., Chung, D.Y., Malo, D. 2001. Characteristics of phosphorus accumulation in soils under organic and conventional farming in plastic houses in Korea. *Soil Science and Plant Nutrition* 47(2), 281-289.

Kramer, K. J., Moll, H. C.and Nonhebel, S. 1999. Greenhouse emissions related to Dutch food consumption. *Energy Policy* 27(4), 203-216.

Kuipers, A., Mandersloot, F. 1999. Reducing nutrient losses on dairy farms in The Netherlands. *Livestock Production Science* 61, 139-144.

Lau, S. S. S. and Wong, J. W. C. 2001. Toxicity evaluation of weathered coal fly ash-amended manure compost. *Water, Air, and Soil Pollution* 128(3-4), 243-254.

Legaz, F., Primo Millo, E. 1988. Normas para la Fertilización de los Agrios. Serie: "Fullets Divulgació" n° 5-88, Direcció General de Innovació y Tecnologia Agrària. IVIA, Valencia, Spain.

Lidón, A. 1994. Lixiviación de nitratos en huertos de cítricos bajo diferentes tratamientos de abonado nitrogenado. Tesis Doctoral. Departamento de Química, Universidad Politécnica de Valencia (UPV), Valencia, Spain.

Llorens Climent, S. M. 1999. La Producción Integrada en Cítricos. *Levante Agrícola* 346, 40-48.

Mattson, B. 1999a. Life Cycle Assessment of carrot purée: a case study of organic and integrated production. SIK, the Swedish Institute for Food and Biotechnology. SIK-Report 642. Gothenburg, Sweden.

Mattsson, B. 1999b. Environmental Life Cycle Assessment (LCA) of Agricultural Food Production. Doctoral Thesis. Ph. D. Thesis. Swedish University of Agricultural Sciences, Alnarp, Sweden.

Matson, P. A., W. J. Parton, A. G. Power and Swift. M. J. 1997. Agricultural intensification and ecosystems properties. *Science* 277, 504 – 509.

McBride, M. B. 1989. Reactions controlling heavy metal solubility in soils. *Advances in Soil Science* 10, 1-56.

McBride, M. B. 1998. Soluble trace metals in alkaline stabilized sludge products. *Journal of Environmental Quality* 22, 857-863.

McGechan, M. B. and Wu, L. 1998. Environmental and economic implications of some slurry management options. *Journal of Agricultural Engineering Research* 71, 273-283.

McGinn, S. M. and Janzen, H. H. 1998. Ammonia sources in agriculture and their measurement. *Canadian Journal of Soil Science* 78(1): 139-148.

Mikkelsen, R. L. 1997. Agricultural and environmental issues in the management of swine waste. In: *Agricultural Uses of By-Products and Wastes*. Rechcigl, J. E. and MacKinnon, H. C. (eds). American Chemical Society, Washington DC, USA, pp. 110-119.

Milà i Canals, L., Burnip, G. M., Suckling, D. M., Cowell, S. J. 2001. Comparative LCA of apple production in New Zealand under two production schemes: integrated and organic fruit production. In: *Proceedings of International Conference on LCA in Foods*. Gerrken, T., Mattsson, B., Olsson, P. and Johansson, E. (eds). SIK-Dokument 143. Gothenburg, Sweden.

Moreno, R. 2001. Caracterización analítica de residuos orgánicos para su posterior compostaje y aprovechamiento agrícola. TFC realizado en el Instituto Valenciano de Investigaciones Agrarias (IVIA). Valencia, Spain.

Muchojev, R. M. C. and Pacovsky, R. S. 1997. Future directions of by-products and wastes in agriculture. In: *Agricultural Uses of By-products and Wastes*. Rechcigl, J. E. and MacKinnon, H. C. (eds). American Chemical Society, Washington DC, USA, pp. 1-19.

Nicoletti, G. M., Notarnicola, B. Tassielli, G. 2001. Comparison of conventional and organic wine. In: Proceedings of International Conference on LCA in Foods. Gerrken, T., Mattsson, B., Olsson, P. and Johansson, E. (eds). SIK-Dokument 143. Gothenburg, Sweden, pp. 48-52.

Ozanne, P. G. 1980. Phosphate Nutrition of Plants – A General Treatise. In: *The Role of Phosphorus in Agriculture*, Khasawneh, F. E., Sample, E. C. and Kamprath, E. J. (eds). ASA-CSSA-SSSA. Wisconsin, pp. 559-589

Palm, C. A., Gachengo, C. N., Delve, R. J., Cadish, G., Giller, K. E. 2001. Organic inputs for soil fertility management in tropical agroecosystems: application of a resource database. *Agriculture, Ecosystems and Environment* 83, 27-42.

Park, M. 2001. *The Fertilizer Industry*. Woodhead Publishing Limited. Cambridge, England.

Pilbeam, C. J. 1996. Effect of climate on the recovery in crop and soil of 15N-labelled fertilizer applied to wheat. *Fertilization Research* 45, 209-215.

Pomares, F. 2001. La fertilización Orgánica. XIII Curso de Producción Integrada. Federación de Cooperativas Agrarias Valenciana (FECOAV). Valencia, Spain.

Prather, M., R. Derwent, D. Ehhalt, P. Fraser, E. Sanhueza and X. Zhou. 1995. Other trace gases and atmospheric chemistry. In: *Climate Change 1994: Radiative Forcing of Climate Change and An Evaluation of the IPCC IS92 Emission Scenarios*. Houghton J. T. et al., (eds). Cambridge University Press, Cambridge, UK, pp.73-126.

Ramos, C., Agut, A., Lidón, A. L. 2002. Nitrate Leaching in Important Crops of the Valencian Community Region (Spain). *Environmental Pollution* 118, 215-223.

Raven, K. P. and Loeppert, R. H. 1997. Heavy metals in the environment. *Journal of Environmental Quality* 26: 551-557.

Rowell, D. L. 1994. *Soil Science: Methods and Applications*. Longman Scientific and Technical, Harlow, England.

Sharpe, R. R., Harper, L. A. 1997. Ammonia and nitrous oxide emissions from sprinkler irrigation applications of swine effluent. *Journal of Environmental Quality* 26, 1703-1706.

Sims, J. T. 1997. Agricultural and environmental issues in the management of poultry wastes: recent innovations and long-term challenges. In: *Agricultural Uses of By-products and Wastes*. Rechcigl, J. E. and MacKinnon, H. C. (eds). American Chemical Society, Washington DC, USA.

Smith, K. A., Crichton, I. J., McTaggart, I. P. and Lang, R. W. 1989. Inhibition of nitrification by dicyandiamide in cool temperature conditions. In: *Nitrogen in Organic Wastes Applied to Soils*. Hansen, J. A. and Henriksen, K. (eds). Academic Press, London, UK, pp. 289-303.

Smith, K. A. 1997. The potential for feedback effects induced by global warming on emissions of nitrous oxide by soils. *Global Change Biology* 3, 327 – 338.

Smith, K. A., McTaggart, I. P. and Tsuruta, H. 1997. Emissions of N_2O and NO associated with nitrogen fertilization in intensive agriculture and the potential for mitigation. *Soil Use and Management* 13, 296 – 304.

Summers, R. N. and Pech, J. D. 1997. Nutrient and metal content of water, sediment and soils amended with bauxite residue in the catchment of the Peel Inlet and Harvest Estuary, Western Australia. *Agriculture, Ecosystems and Environment* 64, 219 – 232.

Stadig, M. 1997. Life cycle assessment of apple production – case studies for Sweden, New Zealand and France. SIK, the Swedish Institute for Food and Biotechnology. SIK-Report 630. Gothenburg, Sweden.

Sutton, M. A., Pitcairn, C. E. R. and Fowler, D. 1993. The exchange of ammonia between the atmosphere and plant communities. *Advanced Ecological Research* 24, 301-391.

Ullmann's Encyclopedia of Industrial Chemistry. 1991. Volume A 19, 448-455.

UNEP (United Nations Environmental Programme). 1996. *Life Cycle Assessment: What To Do and How To Do It*. UNEP. New York, USA.

van der Eerden, L. J., Dueck, T. H. A., Berdowski, J. J. M., Greven, H. and van Dobben, H. F. 1991. Influence of NH_3 and $(NH_4)_2SO_4$ on heathland vegetation. *Acta Botanica Neerl.* 40, 281-296.

van der Molen, D. T., A. Breeuwsma and P. C. M. Boers. 1998. Agricultural Nutrient losses to surface water in the Netherlands: impact, strategies and perspectives. *Journal of Environmental Quality* 27, 4 – 11.

van Horn, H. H. and Hall, M. B. 1997. Agricultural and environmental issues in the management of cattle manure. In: *Agricultural Uses of By-products and Wastes*. Rechcigl, J. E. and MacKinnon, H. C. (eds). American Chemical Society, Washington DC, USA, pp. 91-109.

van Zeijts, H., Leneman, H., Wegener Sleeswijk, A. 1999. Fitting fertilization in LCA: allocation to crops in a cropping plan. *Journal of Cleaner production* 7, 69-74.

Vitousek, P. M., J. D. Aber, R. W. Howarth, G. E. Likens, P. A. Matson, D. W. Schindler, W. H. Schlesinger and D. Tilman. 1997. Human alteration of the global nitrogen cycle: sources and consequences. *Ecological Applications* 7, 737-750.

Weaver, D. and A. E. G. Reed. 1998. Patterns of nutrients status and fertilizers practice on soils of the south coast of Western Australia. *Agriculture, Ecosystems and Environment* 76, 37-53.

Wegener Sleeswijk, A., Kleijn, R., Meeusen-van Onna, M. J. G., Leneman, H., Sengers H. H. W. J. M., van Zeijts, H., Reus, J. A. W. A. 1996. *Application of LCA to Agricultural Products. Part 1: Core Methodological Issues; Part 2: Supplement to the LCA Guide; Part 3: Methodological Background.* CML Report 130. Leiden: CML/The Hague: LEIDLO/ Utrech: CLM, Wageningen, Holland.

Weidema, B. P. and Meeusen, M. J. G. (eds.). 2000. Agricultural data for Life Cycle Assessments: Concerted Action PL-97-3079 of the EU Food Agricultural programme (FAIR) Volume 1. Agricultural Economics Research Institute (LEI). Wageningen, Holland.

Westerman, P. W. and Bicudo J. R. 2000. Tangential flow separation and chemical enhancement to recover swine manure solids, nutrients and metals. *Bioresource Technology* 73(1), 1-11.

White, R. E. 1997. *Introduction to the Principles and Practice of Soil Science.* Blackwell.

Withers, P. J. A. 2000. Cycling and sources of phosphorus in agricultural systems and to the wider environment: a UK perspective. In: *Agricultural Data for Life Cycle Assessments.* Weidema, B. P. and Meeusen, M. J. G. (eds). Agricultural Economics Research Institute (L. E. I.), Report 2.00.01, The Hague, Wageningen, Holland.

Woerden, S. C. 2001. The application of Life Cycle Analysis in glasshouse horticulture. In: *Proceedings of International Conference on LCA in Foods.* Gerrken, T., Mattsson, B., Olsson, P. and Johansson, E. (eds). SIK-Dokument 143. Gothenburg, Sweden. pp. 136-140.

2

Crop Quality under Adverse Conditions: Importance of Determining the Nutritional Status

Gemma Villora, Diego A. Moreno, and Luis Romero

Fisiología Vegetal, Facultad de Ciencias, Universidad de Granada, Fuentenueva s/n E-18071, Granada, Spain. E-mail: lromero@ugr.es

1. Introduction

During plant growth, environmental changes affect the nutrient uptake and plant development. Moisture supply, temperature, light and soil properties influence element availability as well as nutrient uptake, concentration and accumulation in the plant. Different plant species vary not only in the rate at which they absorb an available nutrient, but also in the manner by which they spatially distribute the element to different organs in the same plant. However, all the nutrients present in the soil or applied in a soil system are not available to the plant. The importance of nutrient uptake to crop productivity is assessed not only from the viewpoint of the accumulation of dry matter but also in terms of the economic return and the environmental pollution by way of nutrient leaching. Interaction between nutrients can occur either on the root surface or within the plant. In crop plants, such interactions are generally measured as growth response, either positive or negative. When nutrient combinations prompt a growth response greater than the sum of their individual effects, the interaction is positive and the nutrients are synergistic, whereas when the combined effect is less, the interaction is negative and the nutrients are antagonistic (Fageria et al., 1997a).

Interactions among nutrients, both in the growth medium of the plant as well as within plants, can lead to nutritional imbalances and, therefore, to deficiency or toxicity in elements needed for good development, decreasing plant growth and crop yield. The deficiency of an element, as well as its toxicity not only affects plant growth, but also induces morphological changes that can resemble effects caused by pathogens. Morphological changes, interactions among nutrients, in addition to deficiency and toxicity problems alter plant growth and development, particularly with age, depending on the nutrient levels in the different organs and tissues of the plant. Usually, with plant age, the nutrient concentrations—expressed on a dry-weight basis—decrease (Mozafar et al., 1993), perhaps due to a dilution process during growth (Jarrel and Beverly, 1981). Concentrations also depend on whether the elements are mobile (N, K,...) or immobile (Ca, Mg), since in older plants, the foliar levels of N, P, K,... tend to fall, as long as the levels of Ca, Mg, etc., rise, although such trends are not uniform along the entire leaf section (Mills and Jones, 1996).

Besides these considerations, which are dependent primarily on external factors, nutrient uptake and plant growth depend on genetic factors. In fact, differences in nutrient uptake and use between plant crops and even between cultivars within the same species are well established (Siddiqui et al., 1987). Thus, in recent years, it has been demostrated that the most important approach is to identify the genetic specificity of the plant mineral nutrition (Saric, 1987). Genetic variability in relation to a yield is defined as the hereditary characteristics of the species or cultivars that cause differences in the yield despite favorable or unfavorable environmental conditions (Fageria et al, 1994b). These genetic factors inevitably determine the productivity of the species or cultivar, and thus, together with fertilizer application, determine the crop yield. The quantitative evaluation of these factors and their interactions can help in developing systems to optimize the production. Therefore, we can alter not only the growing media and the cultivation practices but also the physiological characteristics of the plant in order to achieve an optimal yield in a given environment.

Since crop yield in experiments under controlled conditions is usually higher than the same under field conditions, we undertook a series of experiments designed to explore the influence of different crop conditions on different species and cultivars. Conducting controlled versus field experiments, using horticulturally and economically-important species that respond differently under adverse conditions, we tested variations in fertilizer application, fungicides, bioregulators, different environmental conditions, genetic variability, and grafting. For each situation, we here conduct a nutritional diagnosis study as well as analysis of response of yield and quality parameters of the crops.

2. Growing Conditions that Alter the Nutritional Status, Yield and Crop Quality

To ascertain the nutritional status of a plant or a crop, we need to know the appropriate conditions for optimum growth and development—that is, maximum yield and quality in agricultural plants with certain economic impact. Since environmental conditions vary among countries, regions and even in the same cultivation area, as also in different seasons, field experiments on plant nutrition are not always applicable to different geographical areas. Thus, crop experimentation at different sites provides the researcher with a wide range of fully-controlled environmental factors (i.e. growth-chamber versus greenhouse or the open field). In this way, we can establish the limits of growth, development and yield of a crop, thereby gaining a clearer idea of the nutritional status in plants of economic interest under specific conditions.

2.1. Growth-Chamber Experiments

The use of growth-chambers allows the control and maintenance of certain environmental conditions, and thus enables the quantification reaction to a determined stress, biotic or abiotc, without existing external interferences.

2.1.1. B-Ca Interaction

When we provide an economically-important crop plant such as tobacco with appropriate growth conditions and applied $CaCl_2$ in rising levels, the highest Ca dose (5 mM) lowered foliar levels of P and cause P accumulation in the roots (Ruiz and Romero, 1998a). As a bioindicator of quality and uptake of P, the level of non-structural carbohydrates also responded to increments of Ca in the rhizosphere (Hepler and Wayne, 1985). Other authors have observed an accumulation of sugars, particularly starch in the presence of high Ca concentrations (Wei and Sung, 1993). In tobacco, the application of increasing $CaCl_2$ dosages showed that the highest concentrations of glucose, sucrose, fructose and starch in roots were induced by the lowest dosage. In contrast, in the leaves (Table 1) the highest levels in these sugars resulted from the highest Ca dose (Ruiz and Romero, 1998). Thus, the effect of Ca on sugar levels was possibly the most important fact concerning the impact of the Ca on the P translocation. As opposed to the levels of sugars, the dry biomass yield remained stable in the leaves at the highest $CaCl_2$ dosage (5mM), whereas in roots, the dry weight increased with Ca dosages. Both processes were related to the effect of the Ca on the levels of P, which were higher than necessary for optimal growth (Table 1;

Ruiz and Romero, 1998a). In these experiments conducted on tobacco, the development under growth-chamber conditions reveals the true effect of the $CaCl_2$ on sugar levels and dry-matter production. However, $CaCl_2$, when applied to field-grown apple trees, boosted fruit quality and reduced physiological disorders caused by Ca deficiency (Malakouti et al., 1999). It should be noted, however, that these trees were grown on soil with high levels of $CaCO_3$, a condition which may have partly covered the Ca needs of this crop. Both examples indicate the manner in which different environmental conditions—absence or excess of Ca in the substrate, plus $CaCl_2$ supplementation induce similar reactions—enhance crop quality.

Table 1. Effect of Ca ($CaCl_2$) application on carbohydrates (mg/g f.w.) and dry matter (g) on root and leaves of tobacco plants

	Ca Treatment	Glucose	Fructose	Sucrose	Starch	Dry Matter
Root	2.5 mM	8.4	5.3	6.3	14.1	0.78
	5 mM	4.4	2.1	2.9	20.3	1.13
Leaf	2.5 mM	13.4	8.1	9.7	21.6	0.88
	5 mM	22.2	12.7	16.7	34.4	0.85

Data extracted from Ruiz and Romero (1998a)

2.1.2. Pesticides

In both examples of the previous section, the application of an external agent—in this case $CaCl_2$—increased the crop quality in species as different as apple and tobacco. Another very common external agent used in cultivation is pesticide application. In the field, these substances are degraded by water, temperature, light, etc., making it difficult to know the full effects on the metabolism of the plant. When we studied the application of the fungicide *Carbendazim* on tobacco plants, we found an increased dry-matter accumulation. This effect was stronger when the foliar application of *Carbendazim* was combined with boron (B) at 8 mM, since this micronutrient can strengthen the plant's resistance to pathogens. In this combination, the recommended dosage of fungicide can be reduced without sacrificing its effectiveness (Ruiz et al., 1999). As opposed to tobacco, in the apple tree, B treatments indicated that only the foliar application of B increased the yield, but not root application. In addition, the harvested apples from trees treated before flowering presented high incidences of bitter-pit, internal breakdown and *Gloeosporium* rot, during storage, probably related to accelerated ripening, a sympton apparently related to B application. The physiological utilization of B in the apple tree depends on the particular cultivar (Wójcik et al., 1999).

2.2. Greenhouse Experiments

The rigorous control of environmental conditions in the growth chamber, being expensive, is not profitable for commercial production. Consequently, despite higher yield than with other techniques, it becomes necessary to use less stringent methods. Thus, greenhouse cultivation has spread tremendously in the last 20 years, and information is needed concerning the response of the crops to the changes under such agricultural, environmental and physiological conditions. Studies needed include the evaluation of the fertilizer application, the irrigation-water quality, and environmental factors. In contrast to the absolute control in the growth chambers, the conditions used in the greenhouse experiments depend largely on local environmental, sociological and economic variables and even on the building materials of the greenhouse (See Víllora et al., 1998, 1999, and references therein).

Greenhouse-grown plants are usually vegetables, ornamental species and fruits, the developmental cycle of which could be altered by local environmental factors (temperature, humidity, illumination, soils, etc.) outside the greenhouse. Greenhouse cultivation is widespread throughout Mediterranean Europe, where the climate and soils favor early yields, resulting in attractive economic benefits for the farmers despite the heavy expenses as compared with the cost of other cultivation techniques.

2.2.1. Rootstocks

While the greenhouse themselves encourage greater yield, specific cropping techniques inside the greenhouses are used to correct problems and boost the yield even more. One such technique also employed in field crops is the use of rootstocks to induce resistance against soil pathogens (Ashworth, 1985), salinity (Behboudian et al., 1986; Picchioni et al., 1990; Walker et al., 1987), and chlorosis caused by calcareous soils (Bavaresco et al., 1991; Romera et al., 1991; Shi et al., 1993; Sudahono and Rouse, 1994). Rootstocks are also used to alter the foliar content of certain nutrients (Ruiz et al., 1997). Grafting (rootstocks) can improve the plant-substrate relationship. For example, pumpkin rootstocks (*Cucurbita maxima* × *Cucurbita moschata*) of the varieties Shintoza, RS-841 and Kamel were grafted to melon varieties Yuma, Gallicum, Resisto and Arava, using as control non-grafted melon plants (*Cucumis melo* L.) without implanting of the same varieties. Yield in kg/plant (Table 2) proved higher than in the control, although differences appeared between rootstock varieties with the same scion, but not between different scions. Therefore, the fruit production was significantly affected by the interaction between rootstock and scions, while the scion showed no effect on the yield (Ruiz et al., 1997). These results agree with those of Neilsen and Kappel (1996), who indicated that the rootstock increases the yield of the grafted plants by altering the foliar content of different nutrients (Neilsen and Kappel, 1996).

Table 2. Fruit yield (Kg/plant) in rootstocks of melon varieties (scions) grafted onto pumpkin (rootstocks)

Grafts	Kg/plant	Grafts	Kg/plant
Resisto	6.05	Arava	5.16
Shintoza × Resisto	10.20	Shintoza × Arava	8.46
RS-841 × Resisto	9.67	RS-841 × Arava	9.62
Kamel × Resisto	7.86	Kamel × Arava	6.60
Yuma	5.07	Gallicum	5.16
Shintoza × Yuma	9.63	Shintoza × Gallicum	9.65
RS-841 × Yuma	8.44	RS-841 × Gallicum	8.45
Kamel × Yuma	11.30	Kamel × Gallicum	11.4

Extracted from Ruiz et al., (1997)

2.2.2. N-P-K Fertilization

A major factor in greenhouse cultivation is the use of inorganic fertilizers, often exceeding the requirements of a crop for optimal growth. Excess fertilizers, besides increasing salt contamination of the soil and groundwater, reduce the yield, lower the crop quality, and shorten the storage time after harvest, resulting in short-term as well as long-term economic losses. Currently, research is focused on determining the maximum levels that different crops can tolerate without yield and/or quality reduction and with maximum economic return. For example, greenhouse-grown cucumber (*Cucumis sativus* L. cv. Brunex F1) with increasing applications of $N-NO_3$, increased in yield up to a dosage of 20 g/m² (Table 3), above which the yield fell by 30% (Ruiz and Romero, 1998b). In fact, there is evidence of the negative influence by which N stimulates vegetative growth to the detriment of the marketable fruit yield (Davenport, 1996). However, fruit quality (sugars, soluble solids) increased with the N applications, becoming highest at 40 g/m² (Ruiz and Romero, 1998b).

Table 3. Effect of NxK interaction on biological production (Kg/Ha.) of pepper plants.

		Nitrogen (g/m^2)			
		6	12	18	24
Potassium (g/m^2)	4	42.836	49.927	51.129	47.709
	8	41.024	47.475	48.202	44.882
	12	42.964	45.457	45.378	50.685

Extracted from Valenzuela and Romero (1996)

If potassium is applied in addition to N, the plant's reaction undergoes a transformations. That is, pepper fruit yield is highest at 18 g/m² of N, but with a minimum application of K of 4 g/m² (Table 3). Thus, the N rate can be reduced when applying K fertilizer, and the values of commercial yield will surpass those reached with N application alone (Valenzuela and Romero, 1996). In addition, excess N-P-K applied to eggplant (*Solanum melongena* L. cv. Bonica), although tolerated, depressed yield at N dosages higher than 4 g/m² (Table 4), a level substantially lower than that needed by the pepper under similar growth conditions (Valenzuela and Romero, 1996; Víllora et al., 1998). Meanwhile, the P dosage that induced the highest yield was the 26 g/m² whereas higher application rates reduced yield up to 24% (Table 4; Víllora et al., 1998). As with N, the application of K with greater than 5 g/m² diminished yield up to 45% at 20 g/m², a result similar to that reported by Martin et al., (1994) for tomato. However, the interaction between fertilizer elements—in this case N and P—applied at low dosages, can be positive.

Table 4. Effect of N-P-K fertilization on the biological yield (Kg/Ha.) of eggplant

		Kg/Ha.		*Kg/Ha.*		*Kg/Ha.*		*Kg/Ha.*
Control	17,531		4	23,942	13	19,351	5	25,199
	N g/m²				P g/m²		K g/m²	
			8	16,056	26	20,199	10	20,057
			12	15,085	52	15,428	20	13,942
							30	15,999

Extracted from Víllora et al., (1998)

In fact, increased N fertilization that boosts the yield also enhances the response to the P, thereby producing a positive net interaction. Thus, 15 g/m² of N and 24 g/m² of P gave the highest yield in the eggplant crop (Table 5) when the irrigation water presented a great content of dissolved ions (López-Cantarero et al., 1993).

Table 5. Effect of NxP interaction on fruit production (fruit number/Ha.) of eggplants

	Nitrogen (g/m²)		
Phosphorus (g/m²)	*15*	*22.5*	*30*
24	285,502	281,270	266,878
36	253,334	274,285	268,676

Extracted from López-Cantarero et al., (1993)

2.2.3. Salinity Stress

Not only is the Mediterranean environment harsh for many horticultural crops, but there are additional problems of salinity in aquifers due to sea water intrusion and the leaching of fertilizers into the groundwater. This situation translates into increasing salts in the irrigation water, mainly of Cl and Na. As a consequence, the growth, yield and quality are altered in certain crops generally considered sensitive to the salinity. However, not all crops are affected by high levels of Cl and Na in the irrigation water (classified as dangerous water because of its electric conductivity; C.E.; USSL, 1954). For example, zucchini (*Cucurbita pepo* L. cv. Moschata or *Cucurbita moschata*) was experimentally grown with increasing levels of NaCl in addition to that already present in the irrigation water (Table 6). In this case, we found that 1 g/L of NaCl increased the crop yield, with respect to the lower dosage, and improved firmness, diameter, length and concentration of soluble solids in fruits (Víllora et al., 1999). Similar improvements were reported in tomato (Satti and López, 1994; Pulupol et al., 1996), and cucumber (Ruiz and Nuez, 1997). When the NaCl levels applied to the zucchini exceed 1 g/L, the yield falls and morphological damages appear in the plants. Total destruction of the crop occurs when the NaCl dosage reaches 9 g/L (Greifenberg et al., 1996). For field experiments, Mitchell et al., (1991) reported that saline irrigation water could significantly improve the fruit quality by increasing the total soluble-solids concentration without depressing marketable yield. That is, the NaCl applied in the irrigation water in Mediterranean agricultural areas seriously affects economically important crops. Nevertheless, the use of tolerant or semi-tolerant cultivars can partly alleviate yield loss, thereby providing economic benefits.

Table 2.6. Quality parameters on Zucchini fruit under salinity conditions

NaCl (g/L)	Yield Kg/Plant	Firmness (Kg/100g)	SSC (%)	Diameter (cm)	Length (cm)
Control	1.04	2.40	4.31	2.55	13.42
0.25	0.91	3.10	4.65	2.77	13.46
0.50	1.02	3.15	4.77	2.70	13.75
1.00	1.45	3.25	5.30	2.79	14.54

Extracted from Víllora et al., (1999)

2.2.4. Bioregulators

The use of bioregulators is a common horticultural practice to improve crop yield (Latimer, 1992). Bioregulators can act on rooting, flowering, fruiting and fruit growth, leaf or fruit abscission, senescence, metabolic processes, and stress resistance (Nickell, 1988). Nickell (1982) found that an antiauxin (toluepthalamic acid) extended the spring and winter greenhouse production seasons of many plant species, including tomato, potato and pepper, by promoting fruit set and development. Berkowitz and Rabin (1988), reduced transplant shock and increased plant yield by applying abscisic acid to bell pepper seedlings immediately before transplanting, while Csizinsky (1990) reported that the application of a mixture of growth regulators and nutrients augmented pepper yield and nutrient availability in the fruit. Different commercial bioregulators applied under greenhouse conditions induced different responses independently of the cultivar used. For instance, the application of NAA (naphthalenacetic acid) to grape boosted yield (Reynolds, 1988), while the application of Biozyme (Gibberellic acid + Indole-3-acetic Acid) increased the dry-weight production of bean and corn (Campos et al., 1994).

In different pepper cultivars, Biozyme increased the yield (Elsayed, 1995) and improved parameters of quality. In the Lamuyo pepper (Table 7), different bioregulators induced varying yield results; the highest yield resulted from the application of NAA and Biozyme, while quality parameters proved highest with the application of GA_3 (gibberellic acid; Table 7). However, the effects of bioregulators are not exclusively due to the chemical composition of the substance, but also to application frequency and dosage (Belakbir et al., 1998).

Table 7. Effect of bioregulator applications on quality parameters of fruit in pepper

	Total Yield *(t/Ha.)*	*Firmness* *(Kg)*	*SSC* *(%)*
Control	44	0.60	0.805
CCC	49	0.57	1.000
NAA	55	0.59	0.880
GA_3	47	0.63	0.980
Biozyme	53	0.56	1.020

CCC: chlormequat chloride; NAA: naphthaleneacetic acid; GA_3: gibberellic acid; Biozyme: GA_3+IAA (indole-3-acetic acid)+zeatine+micronutrients.
Extracted from Belakbir et al., (1998)

2.2.5. Genetic Variability

As mentioned in the Introduction to this chapter, different cultivars vary in their response to the use of any nutrient under varying conditions of light, temperature and humidity. For example, genetic variability in nitrogen-use efficiency has been recognized for many years (Smith, 1934). This variability has been partitioned into differences in the uptake and use of N (Pollmer et al., 1979; Reinink et al., 1987). Teyker et al. (1989) by quantifiying genetic variation in N-use efficiency in corn, demonstrated that the selection for increased efficiency was possible. The presence of genotypic variation related to N accumulation and its use has also been demonstrated in wheat (Dhugga and Waines, 1989). Thus, the potential exists for developing superior N-efficient cultivars in some crops.

In greenhouse-grown tomato, genotype influenced the marketable, non-marketable, and total yield (Table 8). The highest non-marketable yields occurred in G9, G10, and G12, exceeding the levels of G2, G3, G6, and G8 by up to 205%. The G6 plants registered the highest levels of marketable yield, surpassing that of G10 by 75%. The highest biological yield was harvested from G6 and the lowest from G5 and G10 (Ruiz and Romero, 1998c). Confirming that N utilization can be a determining factor for yield (Mattson et al., 1991; MacDonald et al., 1996), the highest marketable yields were recorded for G2, G3, G6, and G8 (Table 8), the genotypes defined above as having intermediate efficiency in N-utilization (NO_3). In contrast, genotypes that were highly efficient in N-utilization (G7, G9, G11, and G12) had low marketable and high non-marketable yields. High N utilization of these genotypes can encourage excessive vegetative growth and less fruiting (Davenport, 1996). Finally, the genotypes defined as having low efficiency in N utilization (G1, G4, G5, and G10) behaved similarly to those of high efficiency, falling substantially in marketable and total yields. In addition, López-Cantarero et al. (1997) found that heavier N fertilization and, therefore,

Table 8. Genotypic variability in relation to the tomato fruit yield (Kg/Plant)

	Genotype	Non-marketable	Marketable	Biological
G_1	Bufalo	0.93	3.43	4.36
G_2	Corindon	0.56	4.21	4.77
G_3	Dombelo	0.57	4.14	4.73
G_4	GC 773	0.95	3.19	4.14
G_5	GC 775	0.67	3.01	3.68
G_6	Nancy	0.48	4.62	5.10
G_7	Noa	0.94	3.54	4.48
G_8	Sarky	0.53	4.31	4.84
G_9	Yunke	1.11	3.68	4.79
G_{10}	Volcani	1.04	2.66	3.70
G_{11}	617/83	0.81	3.69	4.50
G_{12}	2084/81	1.46	3.61	5.07

Extracted from Ruiz and Romero (1998c)

increased N utilization by eggplant (*Solanum melongena*) boosted non-marketable yield. These facts indicate the close relationship between the yield and N metabolism (Ruiz and Romero, 1998a, 1998b).

One of the parameters which best defines fruit quality is its carbohydrate concentration (Ho, 1996). The major sugar reservoir is found in fruit but, depending on the photosynthetic capacity of the leaf, this reservoir or sink can be enriched, as carbohydrates are easily conveyed to the fruit. For example, in experiments using different watermelon cultivars (*Citrullus lanatus* [Trumb.] Mansfeld), and similar crop conditions, the foliar accumulation of sugars and pectins differed substantially (Table 9), highlighting among the used varieties, 'Perla Negra' plants, in which the levels of both quality parameters surpass those found in the other cultivars (Vargas et al., 1990). In order to apply the carbohydrate metabolism characteristics in a crop-improvement programme, an adequate initial range of variation of the character is required, particularly in the direction in which improvement is sought as also necessary to confirm to what degree the character is inheritable, and the mode of genetic action (i.e. whether it is a dominat or recessive trait, simple or multigenic, additive or non-additive). The rate of photoassimilate export from source leaves depends upon plant age, leaf age and position, and the rate of carbon fixation (Sánchez et al., 1990). Genotypic variation in the assimilate export has been reported in soybean (Egli and Crafts-Brandner, 1996), tomato (Hewitt et al., 1982), and mango (Tandon and Kalra, 1983). The rate of assimilate transport to the fruit depends upon the fruit's developmental stage, and recent work has demonstrated that the fruits themselves control the rate of import of

Table 9. Genotypic variability in relation to the carbohydrate concentration (mg/g d.w.) in watermelon leaves

Genotype	Sucrose	Fructose	Glucose	Starch
Sugar Bell	25.0	6.6	9.0	14.3
Panonia	28.6	7.3	9.6	13.4
Perla-Negra	34.5	10.0	12.9	18.0
Rocio	32.6	8.4	12.2	19.1
Tolerant	31.2	8.5	11.9	22.3
Candida	30.7	8.3	10.9	26.3
Fabiola	22.1	4.8	7.3	22.3
Early-Star	26.9	6.5	9.3	19.0
Carmit	33.9	8.1	12.1	26.6
Resistent	24.2	4.5	7.7	18.5

Extracted from Vargas et al. (1990)

assimilates. In melon (*Cucumis melo*, L.), the different genotypic varieties studied show a broad variability in yield and sugar levels. In this way, the highest yield was reached in the genotype Galia, while the highest foliar concentrations of sugars were reached in the genotypes Gallicum and Gold-King (Sánchez et al., 1990), consistently under the same greenhouse conditions.

2.3. Field Experiments

By experiments conducted under field conditions, scientists seek to approach the real situations faced by farmers and, thereby, achieve a higher and better yield in adverse environments and, finally, improve economic benefits. This implies developing new techniques within established crop practices while protecting the environment.

2.3.1. Mulch

Semi-forcing techniques using plastic covers on the soil surface (mulching, mulches) benefit the soil thermal regime, raising soil temperature and moisture, conserving the soil water, avoiding soil erosion, enhancing root growth and bioavailability of nutrients, and ultimately, improving the crop yield and quality (Decoteau and Friend, 1990; Schmidt and Worthington, 1998). These covers could be made of plastic (e.g. different polyethylene types) or natural materials (e.g. wheat straw or pine straw). The aim is to insulate plants from unfavorable external conditions, encourage early development of the crops either over the entire cycle of development or only during certain developmental phases, for example, to avoid autopolinization or pest attack (Hanna et al., 1997).

The effect of different colors of plastic material has been studied in different crops, such as strawberry (Albregts and Chandler, 1993) and tomato (Decoteau et al, 1989), and the manner in which these materials alter the aerial and root-zone temperatures (Ham et al., 1993) and the soil-water content (Mbagwu, 1991). By contrast, organic covers exert opposite effects, as biological decomposition of the material raises ambient temperatures. The control of air and root temperatures is one of the main reasons for using plastic mulches. In watermelon cultivation under black or white polyethylene, in contrast to open-air cultivation, white polyethylene reduced the heat accumulation on the surface of the plastic and in the soil at 10 cm deep, with respect to black polyethylene cover from April until July. Thus, clearer plastics can be used to reduce heat accumulation, thereby allowing a longer transplant period for heat-sensitive crops, while the darker plastics would be more appropriate for crops with higher heat requirements (Schmidt and Worthington, 1998).

For example, a crop like potato, which is sensitive to high temperatures, is also affected by the color of the polyethylene mulch used. As in the previous cases, any plastic increases root temperatures with respect to open-air plots. However, black polyethylene induces a maximum temperature of 30°C, reducing tuber production by 4% (Table 10), compared with the yield under coextruded cover (mixing black and white plastic), or white cover (Moreno et al., 1999). When these results are compared with those reported for tomato grown under black plastic and the same crop grown under white-black covers, a similar conclusion is reached¾that is, black plastic reduces yield both in Ton/ Ha. and in g/fruit with regard to the white-black mulch (Hanna et al., 1997), although root temperatures under both plastics remain lower (± 1°C) than in the case of the potato (± 3°C). While these plastic mulches increased yield with respect to the open field, transparent polyethylene reduced yield to below that of the open-air plots (Moreno et al., 1999). This decrease was possibly induced by lower temperatures, that in turn, depressed the plant growth and metabolism (Atkinson and Porter, 1996). This contrasts with the results of Ghawi and Battikhi (1998), who attributed lower root temperatures to the shade-effect by the vigorous vegetative growth under the clear cover.

Table 10. Root zone temperature and potato tuber yield (Kg/Ha.) under plastic mulches

Mulch Type	Temperature (°C)	Potato Tuber Yield
Control (uncovered)	16	41,110
Transparent polyethylene	20	40,020
White polyethylene	23	44,870
Black and White polyethylene	27	45,720
Black polyethylene	30	43,940

Extracted from Moreno et al., (1999)

2.3.2. Floating Row Covers

Plastics covers in agriculture are also used to enclose the complete plant, by fixing polyethylene sheets over hoops anchored in the soil. This method is called floating row covers. Due mainly to their ability to trap heat, polyethylene row covers are often used for the production of early vegetables in several regions of the world (Dalrymple, 1973). Slitted or perforated polyethylene row covers eliminate the need for manual ventilation required by solid row covers (Guttormsen, 1972). Such techniques have increased day and night soil and air temperatures in several studies, and have usually increased early crop yield, although total yields have varied (Taber, 1983; Jenni et al., 1998).

In watermelon, high-density polyethylene row covers inflicted physical damages (puncturing and abrasion) when the covers were removed, causing a reduction in the yield. This problem could be minimized by the use of low-density polyethylene cover, without incurring any yield losses. The combined application of mulches and rowcovers, regardless of the characteristics of the rowcovers has also been known to reduce the yield. In conclusion, the type of row cover should be appropriate to the crop in order to increase the yield without inflicting mechanical damage (Baker et al., 1998).

An example of the application of this technique is the Chinese cabbage (*Brassica pekinensis* (Lour) Rupr.) production under Mediterranean agricultural conditions. This plant is native to areas with high humidity and warmer climatic conditions than in Mediterranean areas, and therefore, these types of environments require protection in the row covers. In fact, floating row covers increased production—both biological and commercial—of this plant with regard to a open-air crop (Table 11). Perforated polyethylene and non-woven fleece polypropylene (agrotextile) resulted in upto fivefold yield increase (Table 11). These data agree with the finding of Loy and Wells (1982), who observed that the polypropylene (agrotextyle) provided lower air and root temperatures than did perforated polyethylene, possibly for the high porosity of the material (Technique Interprofessionnel Centers des Fruits et Légumes, 1987), but significantly surpassing the open-field yields.

Table 11. Influence of thermal regime under row-covers on marketable yield (Kg/Ha.) in Chinese cabbage

Treatments	Air Temperature (°C)	Soil Temperature (°C)	Yield
Control (open air)	14.9	20.1	12,797
Polyethylene	20.5	22.9	79,880
Polypropylene	19.2	22.6	74,959

Extracted from Pulgar et al., (1998)

Reductions in night-time temperatures under the covers, with regard to the external temperatures, are reported mainly when the wind is scarce or null and when relative humidity is low (Goldsworthy and Shulman, 1984). Higher temperatures under these covers can reach 30°C in summer (Jenni et al., 1998). Thus, perforated polyethylene considerably increases production, especially in melon plants (Motsenbocker and Bonano, 1989) and Chinese cabbage (Pulgar et al., 1999), compared with either agrotextile covers or no cover at all.

2.3.3. Deciduous Fruit Trees
The applicability of different techniques depends fundamentally on the growth habit of the species to be used. That is, not all crops can be

covered with organic or plastics materials, as in the case of the fruit trees, although in some areas, even these are grown in greenhouses. Therefore, trees must be studied mainly in the open-field in order to improve parameters governing the yield and fruit quality (Moreno et al., 1998). Foliar analysis is used to evaluate the nutritional status of fruit trees to identify the causes of altered yield, in relation to visual symptoms of nutritional imbalances (Tagliavini et al., 1992). In fruit trees, these kinds of analysis have revealed, on the one hand, yield and fruit quality, and, on the other, the relationships between the foliar levels of several nutrients (Fallahi and Simons, 1996). However, the foliar analysis of trees presents certain disadvantages, since accurate results require the standardized collection of samples and the appropriate time of sampling during the biological cycle. In addition, it is necessary to know the agricultural practices in the crop area. In southern Europe, the diversity in species of deciduous fruit trees gives rise to a variety of responses to the same agricultural practice such as the application of identical fertilization levels. In this sense, for 4 consecutive years, we applied Ca-superphosphate (250 kg/Ha.), potassium chloride (150 kg/Ha.), ammonium nitrate (300 kg/Ha.) and 8000 Kg/Ha. of organic manure to almond, apple, pear, pomegranate, hazelnut, persimmon and fig trees (Table 12), drip irrigated every 15 days. The average yield in almond, with a 1 tree/15m^2 plant density, was 2800 kg/Ha., although the second year yield reached 3200 kg/Ha. In the apple tree (1 tree/12 m^2, plant density), higher yield occurred in the second and the fourth year, in both cases exceeding mean value. The pear tree (1 tree/12 m^2, plant density) registered maximum production in the second year, with 4400 kg/Ha. Meanwhile, the hazelnut tree (1 tree/2 m^2 plots), averaged 6500 kg/Ha. over the four years, and yield was the highest during the second year, surpassing 16% on an average. A characteristic Mediterranean tree, the pomegranate, at 1 tree/11 m^2, yielded a 4-year average of 14575 kg/Ha., this value being surpassed the second year.

Table 12. Fruit yield (Kg/Ha.) of different field grown deciduous fruit trees under mediterranean conditions during four consecutive years.

Fruit Tree	Species	1992	1993	1994	1995	Average yield
Almond	*Amygdalus communis* L.	2500	3200	2500	3000	2800
Apple	*Malus communis* Poir.	2900	3600	2800	3500	3200
Pear	*Pyrus communis* L.	3600	4400	3500	4200	3925
Pomegranate	*Punica granatum* L.	13200	16500	12900	15700	14575
Hazelnut	*Coryllus avellana* L	5900	7700	7100	5400	6525
Persimmon	*Diospyros kaki* L.	45700	36400	36100	44700	40725
Fig	*Ficus carica* L.	26000	27800	17300	32000	25775

Another common fruit tree in the Mediterranean area, the persimmon, reached the highest yield during the first year, 11% higher than the average. Lastly, the fig tree, of tropical origin but widely cultivated in the Mediterranean basin, yielded 32000 kg/Ha. during the last year, this value being far higher that under optimum rates established for maximum yield (Moreno et al., 1998). In summary, the second year predominantly gave the highest experimental yields, except for the persimmon (the first year), and the fig tree (the fourth year). These findings confirm, as indicated at the beginning of the chapter, that the species has a decisive influence on yield under similar crop conditions. These examples illustrate that under field conditions, characteristic plants of the agricultural area, or plants well adapted to those environmental conditions are indispensable, especially when they present growth habits that prevent the use of cover materials or greenhouse systems. As shown here, fruit-tree cultivation in a Mediterranean climate does not necessarily require cover techniques, but fertilization applications should be reduced to achieve high yields with better quality, and to reduce the soil pollution as well as costs of agricultural management.

3. Present and Future Efforts

In view of the results presented from experiments carried out in growth chambers, greenhouses or the open field, we can deduce that the genetic variability within a species as well as between different species is one of the main variables faced by researchers, particularly in tree crops. It is equally evident that the use of controlled atmospheres for the commercial crops is economically profitable—that is, these techniques can enable species or cultivars to grow either in otherwise forbidding areas or even during inhospitable seasons. In this manner, market demand can be partially regulated, and the agricultural usefulness of an area can be expanded. Next in importance to increasing yield by means of new crop techniques, improving the storage conditions of the products constitutes another focus of the present study. The goal is to maintain as much quality over the storage period, since post-harvest losses can greatly reduce the benefits. Such improvements are challenging because not only due to occurrence of nutritional alterations, but also morphological problems. Therefore, the efforts should be directed towards preventing not only the nutritional disorders caused by cultivation practices or environmental factors during the cultivation, but also some agricultural handling that deteriorates produce quality during harvest, transport and storage. Current efforts in this sense include research in fungicide application to reduce losses during the storage. Also, to avoid chilling injury and its consequent decay, citrus fruit, for example, are intermittently warmed (Wang, 1993) and dipped in hot water (Rodov et al., 1995). In grapefruit, treatment with methyl-jasmonate have also been reported to reduce chilling injury (Meir et al., 1996). Jasmonate appears naturally in the plants as growth regulators in grape, and defends the potato and the tomato from attack by Phytophtora and protects barley from mildew (Droby et al., 1999). The success of these techniques varies not only according to the crop species, but also with the post-harvest period of each crop. Besides jasmonate application, other techniques employed with effectiveness are storage at low temperatures in atmospheres with high levels of CO_2 or low ethylene concentrations. Such treatments seem to be effective for the strawberry storage (Ku et al., 1999), while Ca applications to the harvested fruits discourage fungal decay (Faust, 1989; Janisiewicz et al., 1998). These examples reflect that although the crop processes are vital to good quality and quantity production, the maintenance of this quality through non-aggressive methods for the consumer is one of the current priorities in post-harvest vegetable nutrition.

4. References

Albregts, E.E. and C.K. Chandler. 1993. Effects of polyethylene mulch color on the fruiting response of strawberry. *Proceedings of Florida Soil and Crop Science Society* 52: 40-43.

Ashworth, J. 1985 Verticillium resistant rootstock research. Annual Report of Californian Pistachio Industry. Fresno, CA. pp: 54-56.

Atkinson, D. and J.R. Porter. 1996. Temperature, plant development and crop yields. *Trends in Plant Science,* 1: 105-132.

Baker, J.T.; D.R. Earhart; M.L. Baker; F.J. Dainello and V.A. Haby. 1998. Interactions of poultry litter, polyethylene mulch, and floating row covers on triploid watermelon. *HortScience,* 33: 810-813.

Bavaresco, M.; M. Fregoni and P. Fraschini. 1991. Investigations of ion uptake and reduction by excised roots of different grapevine rootstocks and a *Vitis vinifera* cultivar. *Plant and Soil,* 130: 109-113.

Behboudian, N.M.; R.R. Walker and E. Torokflvy. 1986. Effects of water stress and salinity on photosynthesis of Pistachio. *Scientia Horticulturae,* 29: 251-261.

Belakbir, A.; Ruiz, J.M. and L. Romero. 1998. Yield and fruit quality of pepper (*Capsicum annuum* L.) in response to bioregulators. *HortScience,* 33: 85-87.

Berkowitz, G.A. and J. Rabin. 1988. Antitranspirant associated abscisic acid effects on the water relations and yield of transplanted bell peppers. *Plant Physiology,* 96: 329-331.

Campos, C.A.; D.C. Scheuring and J.C. Miller, Jr. 1994. The effect of Biozyme on emergence of bean (*Phaseolus vulgaris* L.) and sweet corn (*Zea mays* L.) seedlings under suboptimal field conditions. *HortScience,* 29: 734.

Centre Technique Interprofessionel des Fruits et Légumes. 1987. Cultures légumières sous bâches. Paris.

Csizinszky, A.A. 1990. Response of two bell pepper (*Capsicum annuum* L.) cultivars to foliar and soil applied biostimulants. *Soil and Crop Science Society Florida Proceedings,* 49: 199-203.

Dalrymple, D.G. 1973. Controlled environment agriculture: A global review of greenhouse food production. U.S. Dept. Agr. Econ. Serv. Rpt. 89. Washington, D.C.

Davenport, J.R. 1996. The effect of nitrogen fertilizer rates and timing on cranberry yield and fruit quality. *Journal of American Society for Horticultural Science*, 121: 1089-1094.

Decoteau, D.R. and H.H. Friend. 1990. Seasonal mulch color transition. *Proc. 22nd Natl. Agr. Plast. Congr. pp:* 13-18.

Decoteau, D.R.; M.J. Kasperbauer and P.G. Hunt. 1989. Mulch surface color affects yield of fresh-market tomatoes. *Journal of American Society for Horticultural Science*, 114: 216-219.

Dhugga, K.S. and J.G. Waines. 1989. Analysis of nitrogen accumulation and use in bread and durum wheat. *Crop Science*, 29: 1232-1239.

Droby, S.; R. Porat; L. Cohen; B. Weiss; B. Shapiro; S. Philosoph-Hadas and S. Meir. 1999. Suppressing green mold decay in grapefruit with postharvest jasmonate application. *Journal of American Society for Horticultural Science*, 124: 184-188.

Egli, D.B. and S.J. Crafts-Brandner. 1996. Soybean. In: *Photoassimilate Distribution in Plants and Crops.* Source-Sink Relationships. (Eds) E. Zamski and A.A. Schaffer. Marcel Dekker, Inc. N.Y. pp. 595-623.

Elsayed, S.F. 1995. Response of 3 sweet pepper cultivars to Biozyme^{TM} under unheated plastic house conditions. *Scientia Horticulturae*, 61: 285-290.

Fageria, N.K.; V.C. Baligar and C.A. Jones. 1994a. Diagnostic techniques for nutritional disorders. In: *Growth and Mineral Nutrition of Field Crops.* (Eds) N.K. Fageria; V.C. Baligar; C.A. Jones. Marcel Dekker, Inc. N.Y. pp: 83-134.

Fageria, N.K.; V.C. Baligar and C.A. Jones. 1994b. Factors affecting production of field crops. In: *Growth and Mineral Nutrition of Field Crops.* (Eds) N.K. Fageria; V.C. Baligar; C.A. Jones. Marcel Dekker, Inc. N.Y. pp: 11-59.

Fallahi, E. and B.R. Simons. 1996. Interrelations among leaf and fruit mineral nutrients and fruit quality in Delicious apples. *Journal of Tree Fruit Production*, 1: 15-25.

Faust, M. 1989. *Physiology of Temperate Zone Fruit Trees.* Wiley, New York. pp.: 53-132.

Ghawi, I. and A.M. Battikhi. 1988. Effects of plastic mulch on squash (*Cucurbita pepo* L.): Germination, root distribution, and soil temperature under trickle inrrigation in the Jordan Valley. *Journal of Agronomy Crop Science*, 160: 208-215.

Goldsworthy, W.J. and M.D. Shulman. 1984. A statistical evaluation of near-ground frost processes. *Agriculture, Forestry and Meteorology*, 31: 59-68.

Graifenberg, A.; L. Botrini; L. Giustiniani and M. Lipucci de Paola. 1996. Yield, growth and elemental content of zucchini squash grown under saline-sodic conditions. *Journal of Horticultural Science*, 71: 305-311.

Guttormsen, G. 1972. The effect of perforation on temperature conditions in plastic tunnels. *Journal of Agricultural Engineering and Pesquises*, 17: 172-177.

Ham, J.M.; G.J. Kluitenberg and W.J. Lamont. 1993. Optical properties of plastic mulches affect the field temperature regime. *Journal of American for Society Horticultural Science*, 118: 188-193.

Hanna, H.Y.; E.P. Millhollon; J.K. Herrick and C.L. Fletcher. 1997. Increased yield of heat-tolerant tomatoes with deep transplanting, morning irrigation, and white mulch. *HortScience*, 32: 224-226.

Hepler, P.K. and R.O Wayne. 1985. Calcium and plant development. *Annual Review of Plant Physiology*, 36: 397-439.

Hewitt, J.D.; M. Dinar and M.A. Stevens. 1982. Sink strength of fruits of two tomato genotypes differing in total fruit solids content. *Journal of American Society for Horticultural Science*, 107: 896-900.

Ho, L.C. 1996. Tomato. In: *Photoassimilate distribution in plants and crops. Source-Sink Relationships.* (Eds) E. Zamski and A.A. Schaffer. Marcel Dekker, Inc. N.Y. pp. 709-728.

Janisiewicz, W.J.; W.S. Conway; D.M. Glenn and C.E. Sams. 1998. Integrating biological control and calcium treatment for controlling postharvest decay of apples. *HortScience,* 33: 105-109.

Jarrell, W.M. and R.B. Beverly. 1981. The dilution effect in plant nutrition studies. Advances in Agronomy, 34: 197-224.

Jenni, S.; K.A. Stewart; D.C. Cloutier and G. Bourgeois. 1998. Chilling injury and yield of muskmelon grown with plastic mulches, row covers, and thermal water tubes. *HortScience,* 33: 215-221.

Ku, V.V.V.; R.B.H. Wills and S. Ben-Yehoshua. 1999. 1-Methylcyclopropene can differentially affect the postharvest life of strawberries exposed to ethylene. *HortScience,* 34: 119-120.

Latimer, L.G. 1992. Drought, paclobutrazol, abscisic acid, and gibberellic acid as alternatives to daminozide in tomato transplant production. *Journal of American for Society Horticultural Science,* 117: 243-247.

López-Canterero, I.; A. del Río; A. Sánchez; J.L. Valenzuela and L. Romero. 1993. The influence of fertilized irrigation with brackish water on the number of fruits produced by *Solanum melongena* L. *Acta Horticulturae,* 335: 121-129.

López-Cantarero, I.; J.M. Ruiz; J. Hernández and L. Romero. 1997. Nitrogen metabolism and yield response to increases in nitrogen-phosphorus fertilization: Improvement in greenhouse cultivation of eggplant (*Solanum melongena* L. cv. Bonica). *Journal of Agriculture and Food Chemistry,* 45: 4227-4231.

Loy, J.B. and O.S. Wells. 1982. A comparison of slitted polyethylene and spunbonded polyester for plant row cowers. *HortScience,* 17: 405-407.

Malakouti, M.J.; S.J. Tabatabaei; A. Shahabil and E. Fallahi. 1999. Effects of calcium chloride on apple fruit quality of trees grown in calcareous soil. *Journal of Plant Nutrition,* 22: 1451-1456.

Martin, H.W. and W.C. Liebhardt. 1994. Tomato response to long-term potassium and lime application on a sandy ultisol high in non-exchangeable potassium. *Journal of Plant Nutrition,* 17: 1751-1768.

Mattson, M.; T. Lundborg; M. Larsson and C.M. Larsson. 1991. Nitrogen utilization in N-limited barley during vegetative and generative growth. I. Growth and nitrate uptake kinetics in vegetative cultures grown at different relative addition rates of nitrate-N. *Journal of Experimental Botany,* 43: 15-23.

Mbagwa, J.S.C. 1991. Influence of different mulch materials on soil temperature, soil water content and yield of three cassava cultivars. *Journal of Science and Food Agriculture,* 54: 569-577.

McDonald, A.J.; T. Ericsson and C. Larsson. 1996. Plant nutrition dry matter gain and partitioning at the whole plant level. *Journal of Experimental Botany,* 47: 1245-1253.

Meir, S.; S. Philosoph-Hadas; S. Lurie; S. Droby; M. Akerman; G. Zanberman; B. Shapiro; E. Cohen and Y. Fuchs. 1996. Reduction of chiling injury in stored avocado, grapefruit, and bell pepper by methyl jasmonate. *Canadian Journal of Botany,* 74: 870-874.

Mills, H.A. and J.B. Jones Jr. 1996. *Plant Analysis Handbook II.* MicroMAcro Publishing, Inc. Georgia, USA.

Mitchell, J.P.; C. Shennan; S.R. Grattan and D.M. May. 1991. Tomato yields and quality under water deficit and salinity. *Journal of American for Society Horticultural Science,* 116: 215-221.

74   *Gemma Víllora et al.*

Moreno, D.A.; G. Pulgar; G. Víllora and L. Romero. 1998. Nutritional diagnosis of fig tree leaves. *Journal of Plant Nutrition,* 21: 2579-2588.

Moreno, D.A.; L. Ragala; J. Hernández; N. Castilla and L. Romero. 1999. Optimum range in leaves of potato grown under plastic mulches: I. Macronutrients. *International Journal of Experimental Botany* (Phyton), 64: 67-72.

Motsenbocker, C.E. and A.R. Bonano. 1989. Row covers effects on air and soil temperatures and yield of muskmelon. *HortScience,* 24: 601-603.

Mozafar, A.; P. Schreider and J.J. Oertli. 1993. Photoperiod and root-zone temperature: Interacting effects on growth and mineral nutrients of maize. *Plant and Soil,* 15: 71-78.

Neilsen, G. and F. Kappel. 1996. "Bing" sweet cherry leaf nutrition is affected by rootstock. *HortScience,* 31: 1169-1172.

Nickell, L.G. 1982. *Plant Growth Regulator.* Agricultural uses. Springer-Verlag, Berlin-Heidelberg.

Nickell, L.G. 1988. Plant growth regulator use in cane and sugar production. Update. *Sugar Journal,* 50: 7-11.

Picchioni, G.A.; S. Miyamoto and J.B. Storey. 1990. Salt affects on growth and ion uptake of pistachio rootstock seedlings. *Journal of American Horticultural Science,* 115: 647-653.

Pollmer, W.G.; D. Eberhard; D. Klein and B.S. Dhillon. 1979. Genetic control in nitrogen uptake and translocation in maize. *Crop Science,* 19: 82-86.

Pulgar, G.; D.A. Moreno; J. Hernández; N. Castilla and L. Romero. 1999. Semi-forcing with floating mulch increases yield of chinesse cabbage (*Brassica pekinensis* [Lour.] Rupr. cv. Nagaoka 50). *International Journal of Experimental Botany* (Phyton), 64: 19-21.

Pulupol, L.U.; M.H. Behboudian and K.J. Fisher. 1996. Growth, yield and postharvest attributes of glasshouse tomatoes produced under deficit irrigation. *HortScience,* 31: 926-929.

Reinink, K.; R. Groenwold and Bootsma. 1987. Genotypical differences in nitrate content in *Lactuca sativa* L. and related species and correlation with dry matter content. *Euphytica,* 36: 11-18.

Reynolds, A.G. 1988. Effectiveness of NAA and paclobutrazol for control of regrowth of trunk suckers on "Okanagan Riesling" grapevines. *Journal of American Society for Horticultural Science,* 113: 484-488.

Rodov, V.; S. Ben-Yehoshua; R. Albagli and D.Q. Fang. 1995. Reducing chilling injury and decay of stored citrus fruit by hot water dips. *Postharvest Biology and Technolgy,* 5: 119-127.

Romera, F.J.; E. Alcántara and M.D. de la Guardia. 1991. Characterization of the tolerance to iron in different peach rootstocks grown in nutrient solution: I. Effect of bicarbonate and phosphate. *Plant and Soil,* 130: 115-119.

Ruiz, J.J. and F. Nuez. 1997. The pepino (*Solanum muricatum* Ait.): An alternative crop for areas affected by moderate salinity. *HortScience,* 32: 649-652.

Ruiz, J.M.; A. Belakbir; I. López-Cantarero and L. Romero. 1997. Leaf-macronutrient content and yield in grafted melon plants. A model to evaluate the influence of rootstock genotype. *Scientia Horticulturae,* 71: 227-234.

Ruiz, J.M. and L. Romero. 1998a. Calcium impact on phosphorus and its main bioindicators: response in the roots and leaves of tobacco. *Journal of Plant Nutrition,* 21: 2273-2285.

Ruiz, J.M. and L. Romero. 1998b. Commercial yield and quality of fruits of cucumber plants cultivated under greenhouse conditions: response to increase in nitrogen fertilization. *Journal of Agriculture and Food Chemistry,* 46: 4171-4173.

Ruiz, J.M. and L. Romero. 1998c. Tomato genotype in relation to nitrogen utilization and yield. *Journal of Agriculture and Food Chemistry*, 46: 4420-4422.

Ruiz, J.M.; P.C. García; R.M. Rivero and L. Romero. 1999. Response of phenolic metabolism to the application of carbendazim plus boron in tobacco. *Physiologia Plantarum*, 106: 151-157.

Sánchez, A.; F.A. Lorente;A. del Río; J.L. Valenzuela and L. Romero. 1990. Production and transport of carbohydrates in some cultivars of muskmelon. *Acta Horticulturae*, 287: 485-493.

Saric, M.R. 1987. Progress since the first international symposium: "Genetic aspects of plant mineral nutrition" Belgrade, 1982 and perspectives of future research. *Plant and Soil*, 99: 197-209.

Satti, S.M.E. and M. López. 1994. Effect of increasing potassium levels for alleviating sodium chloride stress on the growth and yield of tomato. *Communications in Soil Science and Plant Analysis*, 25: 2807-2823.

Schmidt, J.R. and J.W. Worthington. 1998. Modifying heat unit accumulation with contrasting colors of polyethylene mulch. *HortScience*, 33: 210-214.

Shi, Y.; D.H. Byrne; D.W. Reed and R.H: Loeppert. 1993. Iron chlorosis development and growth response of peach rootstocks to bicarbonate. *Journal of Plant Nutrition*, 16: 1039-1046.

Siddiqui, M.Y.; A.D.M. Glass; A.I. Hsiao and A.N. Minjas. 1987. Genetic differences among wild oat lines in potassium uptake and growth in relation to potassium supply. *Plant and Soil*, 99: 93-105.

Smith, M. 1934. Response of inbred lines and crosses in maize to variation of nitrogen and phosphorus supplied as nutrients. *Journal of American Society of Agronomy*, 26: 785-804.

Sudahono, D.H.B. and R.E. Rouse. 1994. Greenhouse screening of citrus rootstocks for tolerance to bicarbonate-induced iron chlorosis. *Horticultural Science*, 29: 113-116.

Taber, H.G. 1983. Effect of plastic soil and plant covers on Iowa tomato and muskmelon production. *Proceedings of National Agricultural Plastics Conference* 17: 37-45.

Tagliavini, M.; D. Scudellari; B. Marangoni; A. Bastianel; F. Franzin and M. Zamborlini. 1992. Leaf mineral composition of apple tree: Sampling date and effects of cultivar and rootstock. *Journal of Plant Nutrition*, 15: 605-619.

Tandon, D.K. and S.K. Kalra. 1983. Changes in sugars, starch and amylase activity during development of mango fruit cv. Dashehjart. *Journal of Horticultural Science*, 58: 449-453.

Teyker, R.H.; R.H. Moll and W.A. Jackson. 1989. Divergent selection among maize seedling for nitrate uptake. *Crop Science*, 29: 879-884.

U.S.S.L. 1954. *Diagnosis and Improvement of Saline and Alkaline Soils. U.S.D.A., Handbook 60*, Washington D.C.

Vargas, L.; F.A. Lorente; A. Sánchez; J.L. Valenzuela and L. Romero. 1990. Phosphorus, calcium, pectin and carbohydrate fractions in varieties of watermelon. *Acta Horticulturae*, 287: 469-476.

Valenzuela, J.L. and L. Romero. 1996. Yield and optimum nutrient range in capsicum plants (*Capsicum annuum* L. cv. Lamuyo). *International Journal of Experimental Botany* (Phyton), 58: 63-75.

Víllora, G.; G. Pulgar; D.A. Moreno and L. Romero. 1998. Eggplant yield response to increasing rates of N-P-K fertilization. *International Journal of Experimental Botany* (Phyton), 63: 87-91.

Víllora, G.; D.A. Moreno; G. Pulgar and L. Romero. 1999. Zucchini growth, yield and fruit quality in response to sodium chloride stress. *Journal of Plant Nutrition*, 22: 855-861.

Walker, R.R.; E. Torokflvy and N.M. Behboudian. 1987. Uptake and distribution of chloride, sodium and potassium ions and growth of salt-treated Pistachio plant. *Australian Journal of Agricultural* Research, 38: 383-394.

Wang, C.Y. 1993. Approaches to reduce chilling injury of fruits and vegetables. *Horticultural Review*, 15: 83-95.

Wei, M.L. and J.M. Sung. 1993 Carbohydrate metabolism enzymes in developing grains of rice cultured in solution with calcium supplement. *Crop Science*, 33: 174-177.

Wójcik, P.; G. Cieslinski and A. Mika. 1999. Yield and fruit quality as influenced by boron application. *Journal of Plant Nutrition*, 22: 1365-1377.

3

The Importance of Boron in Apple Production

Pawel Wójcik

Research Institute of Pomology and Floriculture
ul. Pomologiczna 18, 96-100 Skierniewice, Poland

1. Introduction

Apple production in Poland is of great importance. In the previous decade, it amounted to 1500 to 2000 thousand tons, comprising approximately 60-80% of total fruit production. Moreover, apple production in Poland shows a tendency to increase. This is due to planting new apple cultivars ('Jonagold', 'Szampion', 'Elstar', 'Gala') on semi-dwarf rootstocks (M.26, P.14, P.2) at high densities (1500-3000 trees ha^{-1}).

Boron is an essential element necessary for optimal growth and development of higher plants (Marschner, 1995). Boron deficiency is one of the most widespread of all micronutrients (Sparr, 1970). Apple (*Malus domestica* Borkh.) trees have been known to have high requirements for B (Shorrocks, 1997). Despite the fact that plants have lower B requirement compared to N, P, K, Mg and Ca, a shortage of this trace element often reduces cropping and deteriorates the yield quality. Boron deficiency occurs mostly on course-textured soils with low pH and organic matter (Berger and Truog, 1945). In some harvests, plant B deficiency may take place on heavy-textured soils with high pH because under these conditions, B is strongly adsorbed by soil particles (mainly clay minerals). Low soil moisture and/or low air temperatures in spring may also induce plant B deficiency.

In Polish agriculture, sandy soils with low levels of the available B are predominant. More than 70% of agricultural soils have B deficiency.

In central Poland, the percentage of soils of this quality in the total agricultural area is even higher and ranges from 80 to 100%. Central Poland is the largest fruit-producing region, with approximately 20% of the total fruit production. It is estimated that about 50% of apple orchards in this region show symptoms of B deficiency.

In this chapter, the issues under study are symptoms of deficiency and B toxicity in apple trees, factors influencing uptake and B distribution in plant, diagnosis of B nutritional status, and the efficiency of B fertilization in controlling plant B deficiency.

2. Deficiency and B Toxicity in Apple Tree

Low B status in apple tree results in poor fruit set and, consequently, low cropping (Peryea, 1994). Low fruit set is due to the fact that B plays a key role in pollen germination and pollen tube growth. Apples from trees with low B status are small; a result of the reducing cell division in fruitlets, decreasing photosynthesis rate and limiting movement of carbohydrates to fruit tissues. Apples with low B levels are also deformed and corked (cork spot); often have low concentrations of acids and soluble solids; are sensitive to cracking and russeting; have yellow skin with a poor blush; ripe and drop; prematurely, and have a short storage life because of high susceptibility to breakdown. Severe B deficiency in apple tree causes symptoms on the vegetative parts of plants, such as internal bark necrosis (bark measles), dead terminal buds and shoot dieback, shortened internodes and dwarfed, thick and brittle leaves.

Boron toxicity, not as widespread as B deficiency, can occur under three main conditions: (i) in soils inherently rich in B or when B has accumulated naturally; (ii) as a result of over fertilization with minerals high in B; and (iii) through the use of irrigation water rich in B. In Poland, symptoms of B toxicity in apple orchards are rarely observed and are mainly related to over fertilization with B. Boron toxicity in plants can be readily induced due to the fact that the limit between deficiency and B toxicity is very narrow. On light-textured soils with low buffering capacity for B, the risk of B toxicity as a result of application of this element is particularly high.

Excessive content of B in apple trees has a negative effect on growth, cropping and fruit quality. High B status in the apple trees results in reduction of fruit set and fruit yield, premature ripening and fruit dropping, increasing fruit sensitivity to internal breakdown and water core, dead terminal buds and shoot dieback, marginal leaf chlorosis and necrosis, and defoliation.

3. Concentration and B Forms in Soil

Total B concentration in Polish soil ranges from 4 to 200 mg kg^{-1} and depends on composition and intensity of parent material weathering, contents of clay minerals and organic matter, and climatic conditions (Shorrocks, 1997). In Poland, total B concentrations in soils are 5 to 14 mg kg^{-1} in mineral soils and 8 to 25 mg kg^{-1} in organic soils (Musierowicz, 1960).

Soil parent materials too differ widely in B concentration. The lowest B levels occur in soils derived from igneous rocks (gabbro, bazalt, diorite, granite, rhyolite); total B concentrations ranges from 5 to 30 mg kg^{-1}. Soils derived from metamorphic rocks have slightly higher B concentrations (25-40 mg B kg^{-1}) than those from igneous rocks. The highest B levels (40-200 mg kg^{-1}) are found in soils derived from sedimentary rocks (shale, sandstone, dolomite). The highly-weathered soils have generally low B concentrations because B released from primary minerals is leached down to the soil profile. Boron leaching is especially intensive on the course-textured soils with low water retention. Under Polish conditions, B leaching from top soil ranges from 90 to 260 g ha^{-1}. per year^{-1}. The lower B leaching takes place in soils with high contents of clay minerals and organic matter. Therefore, the fine-textured soils have generally higher B concentrations than sandy soils. Soils in humid regions have generally low B concentrations because of intensive B leaching down the soil profile.

Boron exists in four forms in the soil: (i) within the structure of primary minerals; (ii) adsorbed on clay minerals, hydrous oxides of Al, Fe and Mn and on Ca $(CO_3)_2$; (iii) associated with organic matter; and (iv) in soil solution.

The most common primary mineral containing B is tourmaline, closely followed biotyte, micas and muskovite. However, all these primary minerals are resistant to weathering and, therefore, they do not make B source for plants.

Boron adsorbed on clay minerals, $Ca(CO_3)_2$, and oxides of Fe, Mn and Al has a great importance in plant nutrition because there is dynamic equilibrium between B in soil solution and B adsorbed on the above-mentioned soil compounds (Goldberg, 1997). In case of decline of B concentration in soil solution as a result of B uptake by plants or by leaching, B desorption from surface of clay minerals and oxides takes place. This phenomenon maintains B in soil solution, limiting B deficiency. As B concentration in soil solution increases as a result of soil B application, some amounts of B added are adsorbed on clay minerals and oxides. In this way, B toxicity is reduced. Thus, clay minerals and oxides of Al, Fe and Mn are important compounds influencing soil buffering capacity for B. However, the rate of B adsorption on clay minerals and oxides of Fe, Al and Mn is influenced by pH. With increasing pH values B adsorption enhances with maxima at pH 8 to 10. Thus, the soil liming having a high content of clay minerals and Fe, Al and Mn oxides may induce plant B deficiency. This effect is due to not only the increase of soil solution pH but also B retention by $Ca(CO_3)_2$.

Organic matter is an important soil constituent affecting the availability of B (Elrashidi and O'Connor, 1982). Soils with high organic matter have generally high concentrations of both total and available B. Soil organic matter adsorbs more B than mineral constituents on the basis of weight. Adsorption of B on soil organic matter enhances with increasing pH values up to a maximum near pH 9. Ligand exchange is a possible mechanism for B sorption by organic matter. Boron-diol complexes forming with the breakdown products of soil organic matter are stable and make an important reserve of B in soil. Thus, the risk of B deficiency in soil rich in organic matter is negligible.

Soil solution B is the most important form since plants take up B directly from this source. Undissociated molecule of boric acid [$B(OH)_3$] in soil solution is predominant. Borate ions [$B(OH)_4$] also occur in soil solution but its contribution to the total amount of B in the soil solution is low (1 to 10%). Boron concentration in soil solution ranges from 5 mg to 30 mg L^{-1} (Barber, 1995). The level of B in soil solution below 0.05 mg L^{-1} is believed to be too low for most plant species. The level of B in soil solution undergoes significant changes among years and also during growing season. A great fluctuation of B in soil solution occurs mainly on sandy soil with low buffering capacity for B.

4. Uptake of B by Plants

The uptake process of B by plants can be divided into three stages: (i) the movement in soil solution to the root surface; (ii) the adsorption by the cell wall of rhizodermis; and (iii) the movement across plasma membrane.

Boron moves within the soil solution down to the root surface mainly by mass flow. The amount of B supplied by mass flow is influenced by plant transpiration rate and B concentration in soil solution. With increasing transpiration rates and B concentrations in soil solution, there is an increase in the B movement to root surface. In the case of water deficiency and low B concentration in soil solution, B can be also transported in the soil via diffusion, wherein thermal motion induces a net movement of B along a concentration gradient. Diffusion coefficient for B is influenced mainly by soil density and soil moisture and ranges from 0.94 to 4.63×10^{-6} cm s^{-1}. Thus, B movement in soil solution via diffusion has generally little importance.

Cell walls have negative charges forming as a result of dissociation of carboxylic groups (R$\bullet$ COO$^-$) of polygalacturonic acid. However, negative charges of cell walls has no effect on B adsorption because it is mainly absorbed by roots is the form of undissociated molecule of B(OH)$_3$. Despite the fact that B is not sorbed exchangeably, B level in cell wall is 60 to 90% of the total cell B. Such a high level of B results from high B ability to form complexes with diols and polyols.

Free B(OH)$_3$ molecules are transported in apoplast to the surface of plasma membrane via diffusion. The same part of B forms complexes with compounds of plasma membrane, maintaining its integrity and permeability, while the other molecules of B(OH)$_3$ move across the plasma membrane. In existing literature, there are contradictory results explaining mechanism of the transport of B across plasma membrane. Bingham et al. (1970) showed that the uptake of B by barley (*Hordeum vulgare* L.) increased with enhancing concentrations of B in a nutrient medium. They also indicated that the decrease of temperature and application of respiration inhibitors had no effect on the uptake rate of B. Moreover, after transfer of plants to a nutrient solution without B, leakage of B from root tissues to solution was noted. These results indicate that uptake of B by plants was a passive process. The authors of this experiment also proved that only B as B(OH)$_3$ was taken up by plants. However, Shu et al. (1991) reported that uptake of B by peach (*Prunus persica* L.) trees was an active process. In their experiment, uptake of B increased with increasing temperatures (from 2°C to 40°C) and also when the plants were treated with adenosine triphosphate (ATP). Wildes and Neales (1971) claim that in a sufficient amount of B(OH)$_3$ in nutrient solution, B absorption by plants is passive, whereas at low level of B(OH)$_3$ or high concentration of B(OH)$_4^-$ active absorption is predominant.

4.1. Soil and Biological Factors Influencing B Uptake

4.1.1. Soil Moisture

Under field conditions, water shortage in soil is a critical factor reducing B uptake by plants. Therefore, plant B deficiency often occurs in years with low precipitations. Reduction of B uptake by plants in water-deficient soil is related mainly to limiting B movement in soil solution to root surface. Additionally, under conditions of water deficiency in soil, the total concentration of mineral nutrients in a soil solution is high, reducing the uptake of water and, consequently, of B by roots. Also, polimerization of B(OH)$_3$ molecules in soil solution in water deficient soil can reduce B uptake. Negative effect of water deficiency in soil on B uptake by plants is especially pronounced at high B concentrations in soil solution.

4.1.2. Soil pH

One of the most important soil factors affecting the availability of B to plants is pH. In general, with increasing pH values, B becomes less available to plants. Therefore, application of lime can result in plant B deficiency. Reduction of B uptake by plants at high soil pH is due mainly to strong B adsorption on clay minerals, organic matter, and oxides of Al, Fe and Mn. Insufficient B uptake by plants at high soil pH can also result from a transfer of $B(OH)_3$ molecules in soil solution to $B(OH)_4^-$ ions, which are less available to plants. Barber (1995) suggests that the decline of B uptake, by plants with increasing soil pH is caused by lowering kinetics of B uptake induced by changes in permeability of plasma membrane.

At soil pH > 9, there is a dramatic increase in the uptake of B by plants. This phenomenon is related to the displacement of $B(OH)_3$ by OH^- from surface of clay minerals and oxides of Al, Fe and Mn, which consequently increases the amount of B in soil solution and its uptake by the plants.

4.1.3. Boron Interactions with Other Nutrients

Uptake of B by plants can be markedly affected by the presence of some nutrients in the soil solution. According to Eck and Campbell (1962), the highest effect on B absorption have Ca ions. These scientists have reported that liming soil having high B concentration in soil solution reduced B uptake by plants. Fox (1968) found that cotton (*Gossypium hirsutum* L.) and white lupine (*Lupinus albus* L.) grown on soil rich in Ca were more tolerant to B toxicity than those on soil with low Ca level. Tanaka (1967) reported that B absorption by radish (*Raphanus sativus* L.) decreased with increasing Ca concentration in solution culture. Gupta (1972) separated pH and Ca effects and found that Ca:B ratios of 10 : 45 were toxic to barley (*Hordeum vulgare* L.), a ration of 180 was optimal, and rations greater than 697 produced B deficiency. Sinha (1961) reported that high rates of K applied on soil having low B level induced plant B deficiency. Also, Patel (1967) showed that B deficiency symptoms in plants were more severe with increasing K:B ratio in soil. In contrast, Cutcliffe and Gupta (1980) found that B concentration in leaves of cauliflower (*Brassica oleracea* L.) was not affected by soil K status. In existing literature on this subject, there is little information on the effect of N on B availability. Gupta et al. (1976) reported that N fertilization had no effect on B uptake by wheat (*Triticum aestivum* L.) and barley (*Hordeum vulgare* L.).

4.1.4. Rootstock

Many experiments have shown that B uptake is species- and cultivar-dependent (Bagheri et al., 1992; Blatt, 1976; Brooks, 1991; Francois and Clark, 1979; Jamjod, 1996; Picchioni and Miyamoto, 1991; Yau et al., 1995). Therefore, the breeders have tried to obtain genotypes with high ability to B uptake at low concentrations of this element in soil (El-Motaium at al. 1994; Schuman, 1969). According to Atkinson and Wilson (1980), Fallahi et al., (1984a,b), Olszewski et al., (1994) and Smith et al. (1987) B absorption by apple trees depends mainly on rootstock. Fallahi et al. (1984a) found that leaves and fruit of 'Starkspur Golden Delicious' apple trees grafted on OAR-1 rootstock had higher B concentrations as compared to those on M.26, M.7, MM.106 and M.1. Skrzynski (1998) reported that 'Jonagold' apples from trees grafted on P.22 rootstock had considerably lower B concentration than those on P.2, P.14, P.22, P.60, M.9 and M.26.

4.1.5. Cultivar

According to Faust (1989), concentrations of B in leaf and fruit tissues of apple trees are influenced by cultivars. Wojcik et al. (1995) reported that apple cultivar had significant effect on B concentration in fruits; 'Gloster' apples contained considerably less B (8.6 mg kg^{-1} DM.) as compared to 'Jonagold' (13.4 mg kg^{-1} DM) and 'Idared' fruit (12.5 mg kg^{-1} DM). Also, Skrzynski (1998) found significant differences in fruit B concentration between apple cultivars. The highest B concentration in fruit flesh was noted in 'Kosztela' and 'Pilot' and the lowest in 'Szampion' fruit. However, Fallahi (1994) claims that apple cultivar has no effect on the rate of nutrient absorption by roots and that the differences in nutrient concentrations in leaves and apple fruit result from varying distribution of nutrients within the plant, being determined by their growth character (vigorous, forming lateral and spur shoots, position of shoots).

5. Transport and B Distribution within the Apple Tree

Boron taken up by cells of rhizodermis is transported across the cortex to the stele as $B(OH)_3$ (Raven, 1980). In most plant species, B is moved to the stele only through the apoplasm (cell walls and intercellular spaces). During radial transport to the stele, the part of B is bound by pectins of middle lamella and cell walls (Tanaka, 1967). Apoplasmic transport of B across the cortex is constrained by Casparian band in the walls of the endodermal cells, because this band has hydrophobic properties. Thus, in endodermis cells, B has to move into cytoplasm to be released into xylem. In apple trees, the radial transport of B to the stele may also take place in the symplasm through the plasmodesmata, as suggested by Hu and Brown (1994). This results from the fact that forming B- sorbitol and mannitol complexes are readily transported in cytoplasm (Brown and Hu, 1996).

The long-distance transport of B in the apple tree takes place in both the xylem and the phloem. Boron moves in xylem as $B(OH)_3$, as a result of the gradient in hydrostatic pressure (root pressure) and the gradient in the water potential within the plant (Bowen, 1972; Pate, 1975; Shelp et al., 1987). Generally, all factors reducing the transpiration rate of plants limit B movement in xylem. During transport in xylem vessels, the part of B binds with diols occurring in cell walls (Raven, 1980). Therefore, B accumulation in the upper parts of plants is not closely related to transpiration rate. It is particularly pronounced in plant species with a high status of pectins. In apple trees, B movement in phloem has a great importance (Brown and Hu, 1996; Picchioni et al., 1995). Boron is transported in the phloem mainly as borate complexes formed with sorbitol and mannitol (Brown and Shelp, 1997). Thus, in apple trees and also plant species having high contents of sorbitol and/ or mannitol, phloem is a predominant source of B for developing sinks (Shelp, 1987).

Boron distribution within a plant is generally consistent with the pattern of transpiration rate. In apple trees, B taken up by roots, is moved mainly to leaf and fruit tissues. Competition for B between leaves and fruits changes during the growing season. Van Goor and Lune (1980) reported that fruitlets, till four weeks after bloom, had a higher ability to accumulate B rather than leaf tissues. Later, B was accumulated preferentially to leaf tissues. Intensive B accumulation to fruitlets during the first four weeks of their growth results mainly from a high transpiration rate, which is due to an increased number of stomatas and lenticels on fruit surface unit. The strong accumulation of B into fruitlets can also be caused by intensive biosynthesis of hormones (mainly auxins) in this organ.

6. Diagnosis of B Nutritional Status

Chemical soil analysis indicates the potential availability of nutrients that roots may take up under conditions favorable for root growth and root activity. Extractions of soil B by hot water and 0.01 M $CaCl_2$ are now commonly used in most laboratories. Hot water and 0.01 M $CaCl_2$ solution extract B from soluble inorganic, organic and adsorbed inorganic pools (Gupta et al., 1985). Extraction of soil B by hot water is usually successful in predicting B uptake by plants. However, in plant species with low B requirement, hot water soluble B concentration often does not correspond with B uptake rate. In such cases, 0.05 M HCl, NH_4 acetate, mannitol and sorbitol are also used to determine the available

B in soil; although they are used only in a limited way. However, many soil tests are developed and recommended without demonstration that they indeed the predict plant response. We believe that soil tests should be also correlated against plant growth, yield or product quality.

In Poland, soil B is extracted with hot water. According to the Research Institute of Pomology and Floriculture in Skierniewice, symptoms of B deficiency in apple orchards can be observed when hot water-extractable B concentration drops below 0.3 mg kg^{-1} soil. The level of soil B from 0.6 to 1.5 mg kg^{-1} is believed to be optimal for high-yielding and good fruit quality. When the concentration of soil B is higher than 1.6 mg kg^{-1}, the risk of B toxicity is also correspondingly high. Soil samples from apple orchards are taken from a layer of 0-40 cm from surface of herbicide strips along tree rows. This is due to the fact that tree root system develops mainly under herbicide fallow. In apple orchards where fertigation is applied, soil solution B concentration is used as an indicator of soil B status. Soil solution in these orchards is taken from the depth of 0-40 cm by applying a vacuum, three weeks before beginning of bloom. According to the Research Institute of Pomology and Floriculture in Skierniewice, when the concentration of B in soil solution drops below 3 mM, B fertigation of apple trees is necessary. At B concentration of 20-40 µM in soil solution, B fertigation is not effective in increasing the fruit yield and improving the fruit quality. It is worth noting that determination of B in soil solution is a less precise index to predict B uptake rate rather than B extraction by hot water. This is due to the fact that soil solution B does not indicate about soil buffering capacity for B. Therefore, in apple orchards planted on light-textured soils with low buffering capacity for B, the determination of B in soil solution should be performed twice during a growing season (3 weeks before bloom and at petal fall).

The use of elemental analysis of plants to quantitatively assess the plant nutrient requirements was first proposed by Macy (1936). Plant analysis, in the strict sense reflects only the actual nutritional status of plants. Leaves, petioles, flowers and fruits are the mostly-used indicators to assess plant nutrition. In apple trees, leaf tissues are commonly used for determining B status. Leaf samples are collected 80-120 days after full bloom from the middle part of one-year-old shoots. However, leaf B concentration often does not correspond with cropping and fruit quality. Experiments carried out in Poland on nutrition and B fertilization of apple trees indicate that fruit tissues are more sensitive indices of tree B status than leaf analysis. Boron fertilization usually results in a relatively greater increase of B concentration in fruits rather than in leaf tissues. Differences in fruit B concentration usually persist throughout the growing season. Boron deficiency in apple fruit is observed at B concentration below 8 mg kg^{-1} DM. Optimal B level in apple flesh ranges from 12 to 25 mg kg^{-1} DM. Boron toxicity in apple fruit is closely influenced by specific cultivars. 'Elstar', 'McIntosh' and 'Jonathan' cultivars are very sensitive to high B concentration in apple flesh; negative effect of B on fruit quality is observed at 35 mg kg^{-1} DM. 'Delicious', 'Golden Delicious', 'Jonagold' and 'Szampion' cultivars are tolerant to B toxicity, since even at B concentrations in fruit tissues > 60 mg kg^{-1} DM, tree productivity and fruit quality are not decreased.

Boron concentrations in bud and blossom tissues can also be used for assessing the B status in apple trees. Preliminary experiments carried out in Poland showed that optimal B concentrations in buds, taken at the stage of bud burst, and in blossoms at stage of pink bud range from 50 to 80 and from 40 to 60 mg kg^{-1} DM, respectively. Lower B levels in these tissues result in poor fruit set and low fruit quality. It is worth noting that optimal B concentrations in plant tissues differ considerably, as influenced by environmental conditions. At high light intensity and long photoperiods, B requirements of plants are particularly high.

7. Boron Fertilization

The commercial guideline for soil application of B in Poland is a surface-broadcast of 1-5 kg ha.$^{-1}$, made once every three years. Soil B application is recommended when the soil B level is below 0.6 mg kg^{-1} or plants exhibit B deficiency symptoms. The lower levels of available B in soil or the more severe symptoms of plant B deficiency, the higher B rate should be applied. Boron rates higher than 5 kg ha^{-1}. increase the risk of B toxicity.

In Poland, borax (Na$_2$B$_4$O$_7$ 10H$_2$O) and boric acid are mostly applied B fertilizers in agriculture. Both borax and boric acid are easily dissolved in soil, which guarantees quick plant response. To avoid B toxicity, borates with low solubility (colemanite and borosilicate) are used; although the scale of application of these materials is inconsiderable as compared to total B fertilizer application. In apple orchards, B fertilizers are applied approximately 4-5 weeks before full bloom. Too early soil application increases the risk of B leaching. Therefore, soil application of B in autumn is not recommended.

In apple orchards on course-textured soils with available B of 0.6-0.8 mg kg^{-1} and without irrigation system, mixed fertilizers with B are recommended. However, B in mixed fertilizers should not exceed 0.5%. Application of mixed fertilizers containing small amounts of B improves the fruit quality (mainly fruit storability). This effect is especially pronounced in years with high temperatures and dry summers.

On soils with high pH, surface-broadcast application of B is ineffective in controlling plant B deficiency, since B is strongly adsorbed on clay materials and organic matter, which consequently reduces its availability to plants. Under these conditions, only foliar B application and B fertigation overcome plant B deficiency.

Despite the fact that in Poland fertigation in fruit orchards has been used since 1990, little is known about the efficiency of B fertigation on plant nutrition, yielding and fruit quality. Our observations indicate that fertigation is particularly effective in controlling B deficiency on soils with high pH. This is due to the fact that nutrient solution with pH 4.5-5.0 applied through irrigation system can quickly lower soil pH, consequently, increasing B availability to plants. On soils with B concentrations below 0.6 mg kg^{-1} or when B deficiency symptoms in plants occur, fertigation of B should be applied annually at the rate of 0.1-0.4 kg ha^{-1}. Sixty percent of total B rate is applied during three weeks before full bloom and the other part within two weeks after petal fall. However, in the case of non-uniform B application through the irrigation system, B phytotoxicity can easily occur. Therefore, a few times per season, B concentration with pH and EC (electrical conductivity) in the dripping points should be checked.

In Poland, foliar B sprays are commonly applied in apple orchards. Foliar B application is particularly beneficial because the uptake of exogenous B by leaf tissues is very high. Picchioni et al. (1995) reported that foliar uptake of labelled B was 88-96% within 24 h of application. Moreover, B export was closely related to the quantity of B absorbed and reached 50% within 6 h of B application. The uptake rate and B export are equal to or greater than that for urea. This is very likely due to similarities in physic-chemical properties of molecules of boric acid and urea (urea and boric acid are both non-charged and of similar molecular weight).

If the soil test level is 0.6 to 1.0 mg B kg^{-1} and B deficiency symptoms in apple trees do not occur, foliar sprays of B are recommended to be applied either in spring or autumn at rates of 0.4 kg and 1.3 kg B ha^{-1}, respectively. In spring, two B sprays are recommended; the first one at the stage of pink bud and the second immediately after petal fall. Autumn B spray is recommended after fruit harvest, approximately 3-4 weeks before leaf fall. In apple trees, exogenous B is readily transported from leaves to buds, shoots and branches and in the following year it is moved to developing sinks (bud and blossom tissues), which consequently increases fruit set and reduces fruit russeting. In order to improve apple blushing in Poland, summer B spray is recommended. This treatment is applied approximately 3-4 weeks before fruit harvest

(a)

(b)

Plate 1. Symptoms of boron deficiency. (a) Cracked 'Jonagold' apples; (b) Cracked 'Szampion' apples.

at a similar rate as in spring. Our observations indicate that the efficiency of this measure depends on the growing season. In years with high air temperatures during a few weeks before fruit harvest and with relatively low cropping, summer B spray does not affect fruit blushing. As crop load is high and air temperatures low at nights, summer B spray usually improves apple blushing. Summer B spray should be conducted at high solution volumes (1000-1500 L ha^{-1}), which improves fruit surface covered with B solution, particularly those located inside a tree canopy. Summer B spray should not be applied with Ca materials because uptake of exogenous B by fruit cells is strongly reduced.

8. Conclusions and Recommendations

Undoubtedly, plant B deficiency is the most important of all micronutrients. Positive responses to B application, which provide clear evidence of B deficiency, have been reported in over 80 countries and on 132 crops over the last 60 years (Shorrocks, 1997). In Poland, more than 70% of agricultural soils have B deficiency. It is estimated that about 50% of apple orchards in central Poland has symptoms of B deficiency. Boron deficiency in apple orchards occurs mostly on coarse-textured soils with low organic matter, and without an irrigation system. Therefore, under these soil conditions, B fertilization is a beneficial treatment that increases fruit yield and improves fruit quality. Boron deficiency in apple orchards can be readily prevented and corrected by broadcast, fertigation or foliar applications. However, necessity of B fertilization and choice of application mode of B should be related to B levels in soil and plant tissues (buds, blossoms, leaves and fruits). We assume that adequate B fertilization of apple orchards in Central Poland will increase the fruit yield by approximately 20-30% and also improve fruit quality considerably.

9. Acknowledgments

The author is grateful to Dr. Augustyn Mika for fruitful discussions and advice.

10. References

Atkinson, D. and S.A. Wilson, 1980. The growth and distribution of fruit tree roots: some consequences for nutrient uptake. p. 137-149. In: Atkinson D., J.E. Jackson, R.O. Sharples and W.M. Waller (eds), *The Mineral Nutrition of Fruit Trees*, Butterworths, London.

Bagheri, A., J.G Paull., A.J. Rathjen, S.M. Ali and D.B. Moody, 1992. Genetic variation in the response of pea (*Pissum sativum* L.) to high soil concentrations of boron. *Plant and Soil* 146: 261-269.

Barber, S.A. 1995. *Soil Nutrient Bioavailability*. John Wiley& Sons, INC, Toronto.

Berger, K.C. and E. Truog, 1945. Boron availability in relation to soil reaction and organic content. *Proceedings of the American Society for Soil Science* 10: 113-116.

Bingham, F.T., A. Elseewi and J.J. Oertli, 1970. Characteristics of boron absorption by excised barley roots. *Proceedings of the American Society for Soil Science* 34: 613-617.

Blatt, C.R. 1976. Phosphorus and boron interactions on growth of strawberries. *HortScience* 11: 597-599.

Bowen, J.E. 1972. Effects of environmental factors on water utilization and boron accumulation and translocation in sugarcane. *Plant and Cell Physiology* 13: 703-714.

Brooks, B.J. 1991. The adaptation of *Triticum turgidum L. var durum* (durum wheat) to South Australia. Honours Thesis, The University of Adelaide: 12-18.

Brown, P.H. and H. Hu, 1996. Phloem mobility of boron is species dependent: Evidence for phloem mobility in sorbitol-rich species. *Annals of Botany* 77: 497-505.

Brown, P.H. and B.J. Shelp, 1997. Boron mobility in plants. *Plant and Soil* 193: 85-101.

Cutcliffe, J.A. and U.C. Gupta, 1980. Effects of added nitrogen, phosphorus and potassium on leaf tissue boron concentration of three vegetable crops. *Canadian Journal of Plant Science* 60: 571-576.

Eck, P. and F.J. Campbell, 1962. Effect of high calcium application on boron tolerance of carnation, *Dianthus caryophyllus*. *Proceedings of the American Society for Horticultural Science* 81: 510-517.

El-Motaium, R., H. Hu and P.H. Brown, 1994. The relative tolerance of six *Prunus* rootstocks to boron and salinity. *Journal of the American Society for Horticultural Science* 119: 1169-1175.

Elrashidi, M.A. and G.A. O'Connor, 1982. Boron soption and desorption in soils. *Soil Science Society of America Journal* 46: 27-31.

Fallahi, E. 1994. Root physiology, development and mineral uptake. p. 19-30. In: Peterson A.B., R.G. Stevens (eds). *Tree Fruit Nutrition*. Good Fruit Grower, Yakima, Washington.

Fallahi, E., M.N. Westwood, M.H. Chaplin and D.G. Richardson, 1984a. Influence of apple rootstocks and K and N fertilizer on leaf mineral composition and yield in a high density orchard. *Journal of Plant Nutrition* 7: 1161-1177.

Fallahi, E., M.N. Westwood, D.G. Richardson and M.H. Chaplin, 1984b. Effects of rootstocks and K and N fertilizers on seasonal apple fruit mineral composition in a high density orchard. *Journal of Plant Nutrition* 7: 1179-1201.

Faust, M. 1989. *Physiology of Temperate Zone Fruit Trees*. John Wiley & Sons. New York. 335 p.

Fox, R.H. 1968. The effect of calcium and pH on boron uptake from high concentrations of boron by cotton and alfalfa. *Soil Science* 106: 435-439.

Francois, L.E. and R.A. Clark, 1979. Boron tolerance of twenty-five ornamental shrub species. *Journal of the American Society for Horticultural Science* 104: 319-322.

Goldberg, S. 1997. Reactions of boron with soils. *Plant and Soil* 193: 35-48.

Gupta, U.C. 1972. Effects of boron and lime on boron concentration and growth of forage legumes under greenhouse conditions. *Communications of Soil Science and Plant Analysis* 3: 355-365.

Gupta, U.C., Y.W. Jame, C.A. Campbell, A.J. Leyshon and W. Nicholaichuk, 1985. Boron toxicity and deficiency: a review. *Canadian Journal of Soil Science* 3: 381-409.

Gupta, U.C., J.A. MacLeod and J.D.E. Sterling, 1976. Effects of boron and nitrogen on grain yield and boron and nitrogen concentrations of barley and wheat. *Soil Science Society of America Journal* 40: 723-726.

Hu, H. and P.H. Brown, 1994. Localization of boron in the cell wall and its association with cell wall pectin. *Journal of Experimental Botany* 47: 227-232.

Jamjod, S. 1996. Genetics of boron tolerance in durum wheat. Ph.D. Thesis. The University of Adelaide, South Australia: 14-18.

Macy, P. 1936. The quantitative mineral requirements of plants. *Plant Physiology* 11: 749-765.

Marschner, H. 1995. *Mineral Nutrition of Higher Plants*. Academic Press, London. 889 p.

Musierowicz, A. 1960. Microelements in soils (in Polish). *Roczniki Gleboznawcze* 9: 8-15.

Olszewski, T., A. Mika and K. Szczepanski, 1994. Influence of orchard cultural practices on the mineral composition of apple leaves and fruit. IV. The influence of the time and method of young tree pruning and rootstock on leaf mineral composition. *Journal of Fruit and Ornamental Plant Research* 2: 9-20.

Pate, J.S. 1975. Exchange of solutes between phloem and xylem and circulation in the whole plant. p. 451-473. In: Zimmermann M.H. and J.A. Milburn (eds), *Encyclopedia of Plant Physiology, Vol. 1*. Transport in Plants. I. Phloem Transport. Springer-Verlag, New York.

Patel, N.K. 1967. Effect of various Ca-B and K-B ratios on the growth and chemical composition of aromatic strain of Bedi tobacco. *Indian Soil Science* 14: 241-251.

Peryea, F.J. 1994. Boron nutrition in deciduous tree fruit. p. 95-99. In: Peterson A.B. and R.G. Stevens (eds), *Tree Fruit Nutrition*. Good Fruit Grower, Yakima, Washington.

Picchioni, G.A., S. Miyamoto, 1991. Growth and boron uptake of five pecan cultivar seedlings. *HortScience* 26: 386-388.

Picchioni, G.A., S.A. Weinbaum, P.H. Brown, 1995. Retention and the kinetics of uptake and export of foliage-applied, laboreled boron by apple, pear, prune and sweet cherry leaves. *Journal of the American Society for Horticultural Science* 120: 28-35.

Raven, J.A. 1980. Short- and long-distance transport of boric acid in plants. *New Phytologist* 84: 231-249.

Schuman, G.E. 1969. Boron tolerance of tall wheat grass. *Journal of Agronomy* 61: 445-447.

Shelp, B.J. 1987. The composition of phloem exudate and xylem sap from broccoli supplied with NH_4^+, NO_3^- or NH_4NO_3. *Journal of Experimental Botany* 38: 1619-1639.

Shelp, B.J., V. Shattuck and T.A. Proctor, 1987. Boron nutrition and mobility and its relation to elemental composition of greenhouse grown root crops. II. Radish. *Communications of Soil Science and Plant Analysis* 18:203-219.

Shorrocks, V.M. 1997. The occurrence and correction of boron deficiency. *Plant and Soil* 193: 121-148.

Shu, Z.H., Y. Wendy and G.H. Oberly, 1991. Boron uptake by peach leaf slices. *Journal of Plant Nutrition* 14: 867-881.

Sinha, H., 1961. Effect of potassium-boron interactions upon growth and composition of soybeans and vegetable crops. Ph.D. dissertation, University of Missouri, Order No. 61-4098.

Skrzynski, J. 1998. Effect of selected vegetative apple rootstocks on concentrations of nutrients in 'Jonagold apple fruit' (in Polish). 1th Symposium of Mineral Nutrition of Fruit Crops: 41.

Smith, C.B., C.T. Morrow and G.M. Greene, 1987. Corking of 'Delicious' apples on four rootstocks as affected by calcium and boron supplied through trickle irrigation. *Journal of Plant Nutrition* 10: 1917-1924.

Sparr, M.C. 1970. Micronutrient needs—which, where, on what—in the United States. *Communications of Soil Science and Plant Analysis* 1: 241-262.

Tanaka, H. 1967. Boron adsorption by plant roots. *Pant and Soil* 27: 300-301.

Wildes, R.A. and T.F. Neales, 1971. The absorption of boron by disks of plant storage tissues. *Australian Journal of Biology Science* 24: 873-884.

Wojcik, P., A. Mika and D.Krzewinska, 1995. The effect of beyond root fertilization on yield and quality of apple fruit (in Polish). *Zeszyty Naukowe ISK* 2: 41-53.

Van Goor, B.J. and P. Lune, 1980. Redistribution of potassium, boron, iron, magnesium and calcium in apple trees determined by an indirect method. *Physiologia Plantarum* 48: 21-26.

Yau, S.K., M.M. Nachit, J. Ryan and J.Hamblin, 1995. Phenotypic variation in boron-toxicity tolerance at seedling stage in durum wheat (*Triticum durum*). *Euphytica* 83: 185-191.

4

Integrated Nutrient Management in Indian Soils for Sustainable Crop Production

C.L. Acharya and J.K. Saha

Indian Institute of Soil Science, Nabibagh, Berasia Road, Bhopal, M.P., India. Pin-462038, E-mail : jks@iiss.mp.nic.in

1. Introduction

Enhancement and maintenance of soil productivity is essential to sustain agriculture and meet the basic needs of the country's rising population. In the context of increasing food demand, some important issues that need to be addressed are:

  (i) Are the soil resources adequate to meet the food demands for the rapidly-increasing human population?

 (ii) What are the potentials and constraints of soil resources?

(iii) Is generic technology available that can be adopted to soil specific conditions for sustainable management of soil resources?

Such issues are very pertinent to developing countries such as India, where soil resources are poorly managed and the resource-poor farmers cannot afford the scientific inputs needed to increase food production. It must be remembered that the soil sources of the country are finite and non-renewable. Most of the potentially-cultivable land has already been brought under cultivation and majority of the remaining land is located in either inaccessible areas or in ecologically sensitive eco-regions. Hence, further increase in food production has to come from an intensification of cultivation and increase in the productivity of the existing land.

Agriculture is no longer an entity in itself. In present times, it has major environmental dimensions. Sustainable agriculture has been defined as a practice that involves the successful management of resources for agriculture to satisfy human needs, while maintaining or enhancing the quality of environment as well as conserving the natural resources. Greenland (1975) suggested five basic principles of soil management that are essential for sustainable crop production system.

- The chemical nutrients removed by crops must be replenished in the soil.
- Physical condition of the soil must be maintained.
- There must be no increase of soil acidity or toxic elements.
- Soil erosion must be controlled to be equal to or less than the rate of soil genesis.
- There must be no build up of weeds, pests and diseases.

2. Soil Resources of India

Indian soil broadly falls into five main groups, viz., red-yellow-lateritic soils (87.6 m ha.), black soils (73.2 m ha.), alluvium derived soils (74.3 m ha.), soils of desert region (30 m ha.) and soils of mountain (Himalayan and Shivalik) region (28.7 m ha.). In northern river plains, Inceptisol occur with Entisol. Due to the deposits from annual floods, these soils are very productive. In central India, most of the area is under Vertisol associated with Entisols and Alfisols. Due to the high content of swelling type (montmorillonite/vermiculite), these soils are difficult to manage during rainy season as well as post-rainy season. A major area in the southern peninsula, eastern and hill regions is predominated with Alfisols which occur in association with Inceptisol, Entisol and Ultisol. In the western part, most of the area is under Aridisol, which occurs, along with Entisol. These areas receive less than 400 mm rainfall and mainly comprise of deserts.

2.1. Degrading Soil Resources

The ever-increasing population pressure in India has brought intensive cultivation of land to the forefront, since there is little possibility of increasing the size of land area currently under cultivation. Intensive farming received an input from the introduction of high-yielding dwarf wheat from CIMMYT (Mexico) and semi-dwarf high-yielding rice varieties form IRRI (Philippines). In modern-day agriculture, fertilizer use is considered as indicator of agricultural production and most of the technologies have emphasized the widespread use of fertilizers as a source of nutrients. These intensive cropping systems remove about 554 to 932 kg of nutrients (N, P, K) per ha. annually. There is little doubt that fertilizer played a major role in bringing self-sufficiency in food production in the country. However, in this process of increasing productivity, farmers were lured to use fertilizers in a large scale. Consequently, the use of organic manures and green manures to supplement fertilizer N declined substantially. These and other associated problems of intensive cropping with cereal-cereal rotation raises a question of the sustainability of any system highly dependent on inorganic fertilizer input and the crops have started exhibiting signals of decline in yield.

Degradation of soil health has also been reported due to long-term imbalance in the use of fertilizer nutrients. Although the overall nutrient use ($N:P_2O_5:K_2O$) of 4:2:1 is considered ideal for Indian soils, the present use ratio of 6.8:2.8:1 is far too wide. This imbalance and low nutrient use has resulted in a wide gap between crop removal and fertilizer application.

With the objective of monitoring the depletion of soil fertility, All India Coordinated Research Project on Long-Term Fertilizer Experiments (LTFE) was launched in 1971 by the Indian Council of Agricultural Research, New Delhi in 11 centres located in 11 major agroclimatic regions. Results from these experiments at different places indicated lowering of soil pH due to continuous use of inorganic fertilizers for 15 years, particularly at Ranchi, Palampur and Jabalpur (Fig. 1). At Pantnagar (Hapludolls), there was an increase in soil pH mostly with control treatment, where it increased from an initial value

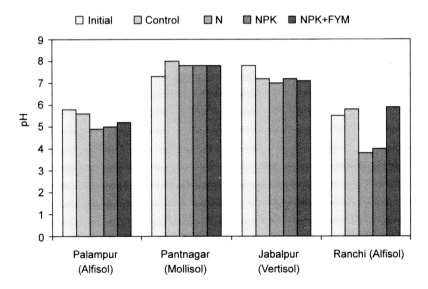

Fig. 1 Effect of long-term imbalance fertilizer use on soil pH (Nambiar et al., 1989; Sarkar et al., 1989; Acharya et al., 1988)

of 7.3 to 8.0. This was ascribed to a decrease in soil organic C due to cultivation, which was reduced from 1.48% to 0.76% with NP treatment after 15 years of cultivation. In another old experiment in acidic red loam soils of Ranchi, continuous application of N as ammonium sulphate for 28 years decreased the soil pH from 5.5 in 1956 to 3.9 in 1984; even the application of P and K along with N did not prevent this fall in pH (Sarkar et al., 1989). The increasing soil acidity due to continuous N and NP application reduced wheat yield to zero by 1975 and 1988 onwards, respectively. Application of lime along with NPK and farmyard manure (FYM) kept wheat yields at more than 5 Mg ha^{-1}. and more than 2.5 Mg ha^{-1}., respectively.

Soil organic matter is one of the most important fertility parameters, which determines the quality of soil. In Palampur, continuous cropping for 15 years without fertilizer or with only N fertilizer resulted in a significant decrease in the status of organic matter (Acharya *et al.*, 1988) (Fig. 2).

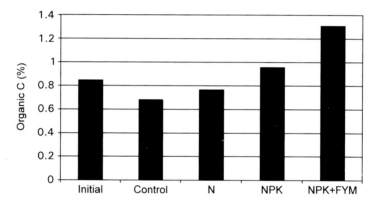

Fig. 2. Effect of 15 years of cropping on soil organic matter status (Acharya et al., 1988)

Long-term experiments in India have in general showed that available NPK declined in unmanured plots, while application of NPK fertilizer increased the available NPK in soil. When only N was applied, the P and K status in soils at all centres have decreased. Subramanian and Kumarswamy (1989) found that soils receiving N alone had the highest P fixation capacity while those receiving NPK with or without FYM had much less capacity of P fixation (Fig. 3). A decline in available S was recorded at several centres in plots where single superphosphate or FYM was not applied and significant reduction in rice yield due to S deficiency was noted at Barrackpore and Bhubaneswar.

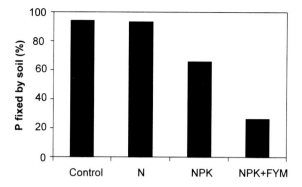

Fig. 3. Effect of nutrient management practices on P fixation by the soil (Subramanian and Kumarswamy, 1989)

Similarly, yields of corn and wheat at Palampur and those of soybean and wheat at Jabalpur showed a decline during continuous cropping without manure due to deficiency of the micronutrients. A decline in available Zn as a result of continuous cropping in unfertilized as well as fertilized plots was noted at several centres, particularly Pantnagar, Ludhianna and Jabalpur resulting in a decrease in crop yields. In Ranchi, continuous application of N or NP for 27 years resulted in significant reduction in available Zn status in alfisol (Fig. 4).

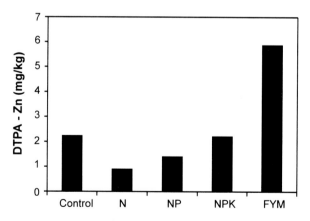

Fig. 4. Available Zn in soils after 27 years of cropping with maize–wheat system on alfisol at Ranchi (Lal and Mathur, 1989)

Not only available nutrient status, but also physical parameters of soil deteriorated due to long-term imbalance inorganic fertilizer use. In Kanke (Alfisols), 27 years' cropping with inorganic fertilizers resulted in a significant increase in the bulk density of soils (Table 1). In another experiments on red, namely, loam soil of Mandya, application of N alone for 10 years gave low sugarcane yield and resulted in low organic and poorer physical properties (Table 2).

Table 1. Effect of fertilizer treatments on soil properties of an alfisol (Kanke)

Treatment	Bulk density (g cm^{-3})		Water-holding capacity (%)	
	1970	1984	1970	1984
Check	1.44	1.54	31.8	33.3
N	1.46	1.54	38.3	39.3
NP	1.40	1.51	37.4	36.2
NPK	1.36	1.46	33.9	40.5
NPK + lime	1.48	1.50	31.6	42.4
FYM	1.43	1.32	33.5	47.2
Initial (1956)	1.45		31.5	

Source: Sarkar et al. (1989)

Table 2. Effect of manure and fertilizer application for 10 years on the yield of sugarcane and soil (red loam) properties

Treatment	Av: cane yield (Mg ha^{-1}.)	Bulk density (g cm^{-3})	Structural index	Maximum water-holding capacity (%)	Organic C (%)
Initial soil properties	-	1.84	30.2	30.4	0.46
Check	28	1.98	24.3	28.7	0.28
N as Am. Sulfate	38	1.93	27.1	29.2	0.41
N as urea	42	1.92	28.2	30.1	0.43
P	45	1.78	29.2	33.5	0.40
NPK	114	1.75	36.1	36.1	0.68
FYM at 25 Mg/ha	56	1.72	38.7	37.8	0.79
NPK+FYM at 25 Mg/ha	120	1.86	47.1	38.6	0.81

Source: Rabindra et al. (1985); Rabindra and Gowda (1986)

3. Integrated Nutrient Management (INM) in Indian Agriculture

INM is a concept that aims at managing and supplying nutrient to plants through various sources such as nutrient reserves in soil as well as inorganic fertilizers and organic sources to an optimum level for sustaining the desired crop productivity. This optimizes the benefit from all possible sources of plant nutrients in an integrated manner. The discussion in the early pages clearly indicates that intensive agriculture with low and imbalance supply of plant nutrient has resulted in a deterioration of native soil fertility, posing a serious threat to long-term sustainability of different production systems. The interactive advantage of combining the use of all possible sources of nutrients and their scientific management in INM has proved superior to the use of each of its components for optimum growth, yield and quality of different crops and cropping systems as well as maintaining or improving soil quality parameters. Thus, the basic concept of integrated nutrient management (INM) system is the maintenance and possible improvement of soil fertility for sustainable crop productivity on a long-term basis and also to reduce the fertilizer input cost. Researches on the role of different components of INM in increasing the productivity as well as soil fertility are discussed below:

3.1. Role of Organic Manure in INM

Organic manures have been time-tested materials to improve the fertility and productivity of soils. In the last 15-20 years, these have been incorporated into INM for intensive cropping sequences in contrast to the subsistence level of production in the past. The potential nutrient ava ilability from major organic sources is estimated at 10.5 to 16.2 m tonnes, out of which only 3.9 to 5.7 m tonnes is available for agricultural use (Table 3). Organic manures are not just sources of nutrients; they have profound, sometimes even the dominant effect on the physical properties of soil. The beneficial effects of such properties on crop yields have rarely been given due economic importance.

Table 3. Potential and possible nutrient (NPK) availability from major organic resources

Resources	Nutrients (mt)	
	Potential	For agriculture use
Crop residues	5.6–8.7	1.7–2.6
Animal dung	3.4–5.7	1.0–1.7
Night soil	1.5–1.8	1.2–1.4
Total	10.5–16.2	3.9–5.7

Source: NAAS (1996)

Data from long-term experiments have revealed that additional yields of different crops could be realized over and above soil test-based optimum NPK rates if only 10-15 tonnes of FYM was annually supplemented along with NPK doses (Table 4). The average yield with 100% NPK+FYM was even more than the average yield obtained with 150% NPK. Besides increasing the yield of crops, continuous application of FYM has been found to improve the soil quality parameters in different agro-ecoregions. These experiments in different agro-ecoregion involving a number of cropping system and soil types have also showed a remarkable improvement in soil organic matter (SOC) as a result of the continuous application of NPK or NPK plus FYM (Table 5).

Table 4. Average grain yield of crops (Mg ha^{-1}) over the years in long-term experiments on yield stability and productivity

Soil type	Control	NPK 100%	NPK 150%	NPK 100% + FYM
Inceptisol	1.13	3.13	3.50	3.64
Vertisol	1.05	3.25	3.35	3.50
Mollisol	2.50	4.65	4.80	5.40
Alfisol	0.55	2.73	3.12	3.37

Source: Swarup (1998)

Table 5. Effect of continuous cropping and manuring on organic C status (g kg^{-1}) of soils under LTFE

Location	Initial	Control	N	NP	NPK	NPK+FYM
Ludhiana	2.1	3.6	3.6	3.7	3.9	4.2
Jabalpur	5.7	5.7	6.8	7.5	7.5	11.2
Bangalore	5.5	4.4	4.5	4.9	5.6	7.5
Ranchi	4.5	3.0	3.0	3.1	3.5	4.0
Bhubaneswar	2.7	3.8	4.6	5.6	5.6	8.4
Palampur	8.5	6.8	7.7	8.6	9.6	13.1

Source: Swarup and Wanjari (2001); Acharya et al., (1988)

Besides increasing crop yields, organic manure application resulted in the enhancement of soil fertility. In all the centres of LTFE, regular application of FYM resulted in the highest increase in soil organic matter status (Fig 5). In Ranchi, continuous application of FYM for 24 years improved soil pH, and resulted in higher available soil moisture content and lower bulk density of soil as compared to the treatments supplying nutrients only through inorganic fertilizers (Table 1 and Fig. 1). Base saturation was the lowest with the continuous ammonium sulphate application and the highest with lime+NPK or FYM. Available Zn and Cu content in soils were highest in plots receiving FYM (Lal and Mathur, 1989) (Fig. 4). Improvement in the physical and chemical properties of soils as well as yields of sugarcane were noticed with integrated nutrient management involving FYM in 10 years' study in Udic Haplustalf at Mandya (Table 2). In a 45-year study (from 1932 to 1978, only with FYM) on a vertisol (Pune) seed cotton and sorghum yields, SOM status, available N, P, K and maximum water holding

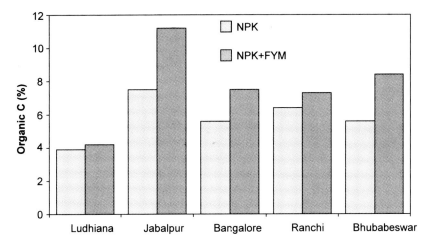

Fig. 5. Effect of continuous manure application on changes in soil organic matter status (after 27 years) in different locations

capacity were significantly higher in FYM (@ 6.2 Mg ha^{-1}) applied plots as compared to unmanured plots (Table 6). Significant improvement in soil physical properties, viz., mean weight diameter, available water content and infiltration capacity of the soil was observed in alfisol of mountainous region due to application of lime and organic matter for 13 years under maize-wheat cropping system (Table 7). About 200 trials conducted on farmers' fields on rice-wheat system under All India Coordinated Agronomic Research Project have shown that the combined use of 12 t FYM ha^{-1} and 60 kg N ha^{-1} produced rice grain yield equivalent to that obtained with 120 kg N ha^{-1} (Table 8).

Table 6. Physico-chemical properties of a vertisol and mean yield of cotton and sorghum grown in rotation after 45 years under rain-fed condition (Vertisol, Pune)

	No manure	*FYM @ 6.2 Mg ha^{-1}*	*LSD$_{0.05}$*
Seed cotton (Mg ha^{-1}.)	0.15	0.31	*0.03*
Sorghum grain (Mg ha^{-1}.)	0.84	1.08	*0.22*
Organic C (%)	0.56	1.14	*0.04*
Total N (%)	0.05	0.06	*0.01*
Available N (ppm)	75.0	92.3	*2.26*
Available P (ppm)	11.1	14.5	*0.56*
Available K (ppm)	290	333	*47*
Maxm. Water-holding capacity (%)	57.3	64.3	*1.3*
Moisture retention at 0.33 bar (%)	42.8	44.8	*2.6*

Source: Khiani and More (1984)

Table 7. Effect of fertilizer treatments on soil properties of an alfisol (Palampur)

Treatment	Bulk density (g cm^{-3})	MWD (mm)	Infiltration rate (cm hr^{-1})	Available water (g 100g^{-1})
Check	1.19	1.78	8.5	9.8
N	1.14	3.17	9.0	9.3
NP	1.24	2.80	10.9	11.3
NPK	1.19	3.04	10.6	11.0
NPK + lime	1.14	3.24	11.2	11.7
NPK + FYM	1.03	3.93	18.9	12.5
CD$_{0.05}$	0.18	0.88	1.03	

Source: Acharya et al. (1988)

Table 8. Effect of the combined use of organic and inorganic fertilizers on crop yields (Mg/ha.) in rice-wheat rotation

Fertilizer treatment		Bhagalpur, alluvial, mean of 87 trials		Manipur, red & yellow, mean of 96 trials	
Rice	Wheat	Rice	Wheat	Rice	Wheat
F$_0$N$_0$	F$_0$N$_0$	2.18	1.57	3.33	0.57
F$_0$N$_{120}$	F$_0$N$_{60}$	4.21	2.75	4.41	1.04
F$_{12}$N$_{60}$	F$_0$N$_{60}$	4.14	2.95	5.44	1.33

Source: Kulkarni et al. (1978)

The kinetic relationship between the addition of organic carbon (through unharvested crop biomass and externally-applied farmyard manure) and storage in a Vertisol (Typic Haplustert) was studied in a field experiment of 7 years' duration under soybean-wheat cropping situation at Bhopal (Kundu et al., 2001). The average annual contribution of C input from soybean was 21.65% and from wheat was 32.32% of the harvestable above ground biomass. Net increases in the contents of the soil organic C at 0-15 and 15-30 cm depths were observed in all treatments. The annual rate of soil organic C increase ranged from 85 to 739 kg C ha^{-1} at 0-15 cm and 54 to 149 kg C ha^{-1} at 15-30 cm soil depth. The observed annual rate of change in soil organic C at 0-30 cm was positively correlated with the gross annual C input to the 0-30 cm soil horizon (Fig. 6). About 18% of the annual gross C input was incorporated in the soil organic matter. The study concluded that growing soybean-wheat in the rotation on Vertisol of central India is not only a SOC restorative process but also helps to sequester C into the soil.

3.2. Role of Green Manure in INM

Green manuring in rice fields in India has received special attention for the contribution that it can make towards the N needs of rice, which has

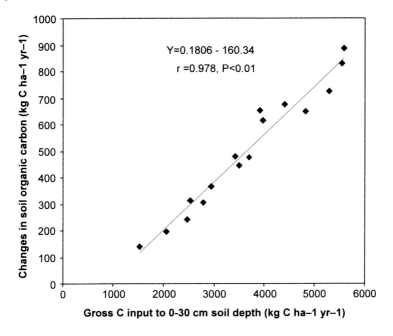

Fig. 6. Changes in SOC content as affected by estimated gross C input to 0-30 cm depth of soil layer

become the key input since the introduction of high-yielding heavy N demanding varieties and also due to the fact that the efficiency of fertilizer N in rice is low. Results from a large number of field experiments showed that the grain yield increase for rice is equivalent to 25-60 kg N ha^{-1} or more, depending on the biomass incorporated and N content of green manuring crop (Goswami et al., 1988). A significant increase in SOM status due to continuous green manuring practice has also been reported by various workers (Table 9). There was

Table 9. Effect of green manure on soil organic carbon under different land use systems

Land use	*Treatment*	*Organic C (%)*
Red soil, rice-rice	Control	0.43
(10 years)	50% from inorganic + 50% through green manure (*Sesbania aculeata*)	0.90
Sodic soil, rice-wheat	Fallow-rice-wheat	0.23
(7 years)	Green manure (*Sesbania aculeata*) rice-wheat	0.38

Source: Hegde (1996); Swarup (1998)

significant increase in organic C (from 0.58% to 0.67%) and total N content (from 0.021% to 0.054%) due to the incorporation of green manure (Sahu and Nayak, 1971).

A study conducted on 19 farmers' fields in Ropar and Patiala districts of Punjab for 2 years with green manuring (cowpea or *Sesbania aculeata*) before maize crop indicated that green manuring nearly substituted 60 kg N ha^{-1} for the succeeding maize crop. There was a slight build up of available N, P and K content in green manured plots as compared to the plots receiving fertilizer N (60-120 kg ha^{-1}) (Table 10).

Table 10. Effect of green manuring and fertilizer nitrogen on the changes in organic C content and available nutrients after maize harvest during 1983 (2 years' study on Maize crops)

Treatment	Organic C (%)	Available N	Available P	Available K	Yield (q/ha.)
N0	0.55	154	17.9	178	16.8
N60	0.55	155	17.6	178	22.5
N90	0.56	157	17.4	177	29.4
N120	0.56	158	16.9	177	30.5
Cowpea	0.57	169	19.9	190	21.5
Cowpea + N60	0.58	167	19.7	189	27.7
Sesbania	0.57	168	19.7	189	20.2
Sesbania + N60	0.57	167	19.7	189	28.2
$CD_{0.05}$	0.007	47	1.22	4.0	3.06

Source: Bhandari et al. (1989)

3.3. Role of Legume in INM

Soil nitrogen is the major plant nutrient usually lacking in vertisol in semi-arid tropics. Nodulating legumes benefit the succeeding non-legume crops in terms of increased yields in comparison with the yield of non-legumes after non-legumes. Such increased yields are attributed to enrichment of soil N due to N_2 fixation and the N-conserving effect of legumes. The response to N application in mustard declined after 40 kg N ha^{-1} if preceding crops were legume; while if grown after pearl millet (a non-legume), there was good response upto 80 kg N ha^{-1} (Kumar and Singh, 1988). A long-term experiment was initiated in 1983 on vertisol at ICRISAT to examine the effectiveness of different types and proportions of grain legume intercropping system rotations, wherein a grain legume intercropping system (cowpea/pigeonpea) has consistently benefited the subsequent sorghum crop to the level equivalent to 40 kg N ha^{-1}. Pigeonpea had substantial effect on increasing surface soil N content by 140 mg kg^{-1}; while soil N in a continuous sorghum+chickpea system increased by only 25 mg N kg^{-1} in 8 years (Table 11). In contrast, a continuous non-legume system (sorghum+safflower) resulted in a loss of 25 mg N kg^{-1} of soil, indicating the long-term benefits to both soil and non-legumes from the inclusion of grain legume.

Table 11. Total N content (ug g^{-1}) of the 0-15 cm depth of soil under different cropping Systems (After 8 years)

Rotation	1983	1991	Change
Legumes every year			
S+PP S+PP	546	686	140
S+PP S-CP	566	635	69
S-CP S-CP	536	561	25
Legumes in alternate years			
S+PP S-SF	543	640	97
S+PP S-SF	559	610	51
S-CP S-SF	540	548	8
No legumes			
S-SF S-SF	537	513	– 24
Standard error a±		19.9	18.0

S = sorghum; PP = pigeonpea; CP = chickpea; SF = safflower; – = sequential; + = intercrop.
Source: Laryea et al. (1995)

3.4. Role of other Organic Residues

Effect of organic cycling of subabul (*Leucaena leucocephala*) in winter sorghum was studied on Typic Chromustert at Solapur (Narkhede and Ghugare, 1987). Subabul produced a dry matter of 1.97 Mg ha^{-1}, which contained 3.86% N. Thus, subabul loppings added 76 kg N ha^{-1}; the loppings were applied on the surface. The addition of subabul loppings increased the available N in the surface soil layer by 44.5% over control (Table 12). This also increased grain and straw yield, N and P uptake and moisture use efficiency by sorghum and left more moisture in the soil at sorghum harvest. Crop residues leave substantial residual effect on succeeding crops in a cropping system. Organic N is slowly mineralized and about 30% N, 60-70% P and 75% of K is likely to be available to the first crop and the rest to subsequent crop (Gaur, 1992). A significant residual effect of rice straw was observed in a third crop of groundnut when applied to the first crop of groundnut in a groundnut-maize-groundnut sequence (Table 13). A study carried out on changes in soil fertility parameters due to crop residue incorporation through cowpea or pearl millet in both red sandy loam and red loamy sand soil (Venkateswarlu, 1987). The crop residues equivalent to 20 kg N ha^{-1} were applied in cowpea and pearl millet rotation. All the soil physical parameters and available N and P improved due to crop residue incorporation after 5 years (Table 14). Available K was not affected in sandy loam soil but was somewhat lowered in loamy sand soil.

Table 12. Effect of organic recycling on sorghum and organic carbon and moisture content in soil

Characteristics	Control	Subabul loppings added	$CD_{0.05}$
Grain yield (Mg ha^{-1})	1.2	1.4	0.13
Stover yield (Mg ha^{-1})	3.3	3.6	0.30
N uptake (kg ha^{-1})	3.4	4.0	0.67
P uptake (kg ha^{-1})	1.4	1.7	0.29
Moisture use efficiency (kg grain mm^{-1} ha^{-1})	5.4	6.7	-
Soil organic carbon (%)	0.42	0.47	-

Source: Narkhede and Ghugare (1987)

Table 13. Residual effect of rice straw applied to first crop of groundnut on pod yield and oil content in third crop of groundnut in groundnut-maize-groundnut sequence

Rice straw (t/ha^{-1})	Pod yield (q/ha^{-1})	Oil content (%)
0	16.2	43.4
2	17.8	44.3
4	18.6	45.2
6	20.6	46.0
$CD_{0.05}$	1.0	0.9

Source: Reddy et al., (1990)

Table 14. Physical and chemical properties as affected by crop residue incorporation

Soil parameters	Red sandy loam		Red loamy sand	
	Initial	After 5 yrs of incorporation	Initial	After 5 yrs of incorporation
MWD (mm)	2.63	3.04	3.21	4.19
Hyd. Conductivity (cm ha^{-1})	4.19	10.50	2.34	8.00
Bulk density (g cm^{-3})	1.81	1.69	1.73	1.65
Moisture content at field capacity (g/100 g)	15.7	19.0	12.4	12.9
Av. N (kg ha^{-1})	284	307	235	286
Av. P$_2$O$_5$ (kg ha^{-1})	5.5	14.0	6.6	14.0
Av. K$_2$O (kg ha^{-1})	256	254	269	234

Source: Venkateswarlu (1987)

Results of a long-term experiment on the integrated use of fertilizer, manure and crop residues in light-textured soil of Haryana growing pearl millet–wheat indicated that organic C content in soil increased and C:N ratio of soil organic matter decreased in soil amended with manure, crop residue and green manure as compared to fertilizer alone in ten years. A maximum increase in organic C content was observed due to wheat straw, followed by FYM and green manure (Table 15). Soil biological activity parameters were also increased due the application of these organic materials. Meelu et al. (1994) reported that the incorporation of sesbania green manure along with wheat and rice straw not only counteracted the adverse effect of crop residues on rice but also improved the organic matter content and physical properties of the soil (Table 16). Similarly, sorghum stubble co-incorporated with subabul loppings (50:50) at the rate 5 t ha^{-1} not only helped in increasing the crop yields, but also soil nutrient status as well as organic C content (Table 17). In a study undertaken in the hills of northwest India, Acharya et al. (1998) observed considerable increases in earthworm population in the surface layer due to both incorporation of

Table 15. Effect of integrated nutrient management on chemical and biological parameters of soil

Treatment	Org. C (%)	C:N ratio	Biomass C (mg kg^{-1})	Dehydrogenase (μg NH$_4$-N g^{-1} soil hr^{-1})
N$_0$P$_0$	0.45	11.3	147	38
N$_{60}$P$_{30}$	0.51	11.1	195	37
N$_{90}$P$_{45}$	0.58	10.9	213	39
N$_{120}$P$_{60}$	058	9.4	225	35
N$_{90}$P$_{45}$ + N$_{30}$ as FYM	0.63	9.1	273	39
N$_{90}$P$_{45}$ + N$_{30}$ as wheat straw	0.65	8.7	423	67
N$_{90}$P$_{45}$ + N$_{30}$ as green manure	0.62	9.7	317	34
LSD$_{0.05}$	*0.01*	*0.9*	*15*	*4*

Source: Singh (2000)

Table 16. Effect of green manuring and wheat straw management in rice on grain yield and soil properties in rice-wheat rotation during 1988-1993

Treatment to rice	Grain yield (Mg ha^{-1}.)		Soil organic C (%)	Bulk density (g cm^{-3})	Waterstable aggregates >1 mm (%)	Mean weight diameter (mm)
	Rice	Wheat				
Fert. N	6.3	4.3	0.38	1.55	19.8	1.42
GM	6.3	4.3	0.42	1.43	26.6	1.58
Straw + fertil.	6.0	4.4	0.48	1.40	26.7	1.56
Straw + GM	6.5	4.3	0.50	1.36	28.2	1.68

Source: Meelu et al. (1994)

Table 17. Effect of organics and inorganics on rabi sorghum and chickpea yields and soil properties

Treatment	Sorghum yield, kg ha⁻¹ (3 yrs' average)	Chick yield kg ha⁻¹ (residual effect)	Org. C (g kg⁻¹)	Available nutrient content (kg ha⁻¹)		
				N	P	K
Sorghum stubble @ 5 t ha⁻¹	841	672	5.4	200	10.5	533
Sorghum stubble + subabul loppings (50:50) @ 5 t ha⁻¹	1098	788	6.0	215	11.2	566
Sorghum stubble @ 5 t ha⁻¹ + 20 kg N ha⁻¹	1060	670	5.3	208	10.8	570
RDF (50-25-0)	1101	637	4.8	197	11.0	428
Control	825	452	4.1	158	7.11	456

Source: Bellaki and Badanur (2000)

crop residues as well as mulching with minimum tillage, as compared to farmers' practice of conventional tillage with no crop residue incorporation (Table 18).

Table 18. Earthworm population (number m^{-2}) in surface (0-15 cm) layer at sowing and flowering stage of wheat

Treatment	Number of earthworms	
(Mulch-tillage)	Sowing	Flowering
Incorporation	20	8
Minimum tillage	21	10
Farmers' practice	4	1
$LSD_{0.05}$	1.7	1.4

Source: Acharya et al., (1998)

3.5. Role of Biofertilizers

Biofertilizers, are an important component of INM, may help in increasing crop productivity by way of increased N fixation, enhanced availability of nutrients through solubilization or increased absorption, stimulation of plant growth through hormonal action or antibiosis, or by decomposition of organic residues. Among the biofertilizers, *Rhizobium*, in association with legume (symbiosis), seems to have the maximum potential in INM. Its ability to contribute N is not only realized in measurable terms, but it is far more consistent than non-symbiotic *Azotobacter* or *Azospirillum*. The contribution of legume-*Rhizobium* symbiosis to the production system varies, depending on a number of physical, environmental, nutritional and biological factors. Though extensive information has been generated on the inoculation benefits from rhizobial strains on a number of pulses, groundnut and soybean crops, data from on-firm trials is meager. Multilocational trials conducted on different pulse crops on farmers' fields gave variable results, ranging from 4 to 24% (Table 19). Indian soils are generally low to medium in available P content, though its total content is sufficient. Phosphate solubilizing bacteria (PSB), therefore, hold great promise as they solubilize the unavailable forms of P into soluble forms by release of a variety organic acids. Out of 37 field trials conducted during 1960s on a variety of crops in different locations, 10 trials showed a significant increase in yields of wheat, rice, maize, chickpea, pigeonpea, soybean, groundnut and berseem (Sundar Rao, 1968). In another experiment, inoculation of PSB resulted in a significant increase in the grain yield of chickpea, P uptake and protein content (Table 20).

Table 19. Response of pulses to rhizobial inoculation at the farmers' fields

Crop	No. of Locations	No. of trials	Increase over uninoculated control	
			Kg ha^{-1}. (range)	Percentage (average)
Chickpea	3	10	120-520	10-24
Lentil	2	8	40-110	15-17
Pigeonpea	3	16	20-80	8-14
Moong bean	5	21	30-180	4-17
Urad bean	3	14	50-80	12-17

Source: Khurana and Dudeja (1997)

Table 20. Response of chickpea to phosphate solubilizing bacteria

Treatments	Seed yield (kg ha^{-1}.)	P uptake (kg ha^{-1}.)		Protein content (%)
		Grain	Stalk	
No P	1555	5.42	6.52	19.86
PSB	1790	6.65	8.20	21.00
30 kg P$_2$O$_5$ ha^{-1}	2082	7.62	9.53	22.49
30 kg P$_2$O$_5$ ha^{-1} + PSB	2205	8.77	10.51	22.74
60 kg P$_2$O$_5$ ha^{-1}	2280	8.70	10.53	22.67
60 kg P$_2$O$_5$ ha^{-1} + PSB	2377	9.33	10.62	22.42
CD$_{0.05}$	104	0.60	0.68	0.69

Source: Ghonsikar and Shinde (1997)

4. Conclusions

Results from research on enhancing the crop productivity and soil fertility through integrated nutrient management has generated the following conclusions:

1. Soil acidity is an important soil constraint to crop productivity in many Indians soils. In these soils, continuous application of nitrogenous fertilizer alone leads to further aggravation of soil acidity. Introducing organic manure or lime in the nutrient management programme can prevent this decline in soil pH.

2. Resource-poor farmers generally apply only nitrogenous fertilizer (mostly through urea) as urea is a cheaper fertilizer. Continuous N fertilizer application declines available P and K status as well as physical properties of the soil, resulting in the decline in soil productivity.

3. Organic matter status in cultivated Indian soils is generally low. Cultivation of initially high organic matter containing soil (such as forest soil) reduces organic matter content. Fertilizer application prevents this decline in soil organic matter due to root biomass

addition. Continuous application of organic manure improves soil organic matter content and associated physical and chemical properties of the soil.

4. In soils containing marginal amount of Zn, continuous intensive cropping can lead to a decrease in productivity due to the appearance of their deficiencies in soils and plants. Organic manure application prevents such constraints.

5. Green manuring, particularly with legume crop, not only meets a part of the N requirement by crops but also builds up soil organic matter and N content.

6. Crop residue incorporation in soil—though it may depress yield of crops—can significantly enhance crop productivity in the long run by improving soil properties. However, the depressing effect of crop residues can be removed by co-incorporating green manure like *Leucaena leucocephala* or *Sesbania aculeata*.

7. Biofertilizers, though a promising component of integrated nutrient management, have an inconsistent effect on crop productivity, mostly due to ineffective competition with native organisms.

5. References

Acharya, C.L., Bishoni, S. and Yaduvanshi, H.S. 1988. Effect of long-term application of fertilizers, and organic and inorganic amendments under continuous cropping on soil physical and chemical properties in as Alfisol. *Indian J. Agric. Sci.* 58: 509-516.

Acharya, C.L., Kapur, O.C. and Dixit, S.P. 1998. Moisture conservation for rainfed wheat production with alternative mulches and conservation tillage in the hills of north-west India. *Soil and Tillage Research* 46: 153-163.

Bellakki, M.A. and Badanur, V.P. 2000. Residual effects of crop residues in conjunction with organic, inorganic and cellulolytic organisms on chickpea grown on vertisol. *J. Indian Soc. Soil sci.* 48: 393-395.

Bhandari, A.L., Sharma, K.N., Kapur, M.L., and Rana, D.S. 1989. Supplementation of N through green manuring for maize growing. *J. Indian Soc. Soil Sci.* 37: 483-486.

Gaur, A.C. 1992. Bulky organic manures and crop residues. p.36-51. In:. H.L.S. Tandon (ed.), *Fertilizers, Organic Manures, Recyclable Wastes and Biofertilizers*, FDCO, New Delhi.

Ghonsikar, C.P. and Shinde, V.S. 1997. Nutrient management practice in pulses and pulse based cropping system. p.91-124. In: C.P. Ghonsikar and V.S. Shinde (eds), Nutrient Management Practice in Crops and Cropping Systems, Scientific Publishers, India.

Goswami, N.N., Prasad, R., Sarkar, M.C., and Singh, S. 1988. Studies on the effect of green manuring in nitrogen economy in a rice-wheat rotation using ^{15}N technique. *J. Agric. Sci.* (Camb.) 111: 413-417.

Greenland, D.J. 1975. The magnitude and importance of the problem. p.3-7. In: D.J. Greenland and R. Lal (eds), *Soil Conservation and Management in the Humid Tropics*, John Wiley & Sons, New York, USA.

Hegde, D.M. 1996. Integrated nutrient supply on crop productivity and soil fertility in rice (*Oryza sativa*) rice system. *Indian J. Agron.* 41:1-8.

Khiani, K.N. and More, D.A. 1984. Long-term effects of tillage operations and farmyard manure application on soil properties and crop yield in a vertisol. *J. Indian Soc. Soil Sci.* 32: 392-393.

Khurana, A.L. and Dudeja, S.S. 1997. Biological nitrogen fixation technology for pulses production in India. Indian Institute of Pulse Research, Kanpur, India.

Kulkarni, K.R. Hukeri, S.B. & Sharma, O.P. 1978. Proc. India/FAO/Norway Seminar on Development of Complimentary use of Mineral Fertilizers & Organic materials. Ministry of Agriculture & Cooperation, New Delhi.

Kumar, A. and Singh, Y. 1988. Response of oilseeds to P in northern India. Paper presented at FAI(NR)-PPCL seminar on 'Importance of P fertilizer with special reference to oilseeds and pulses in Northern India' Lucknow. 21-22 June.

Kundu, S., Singh Muneshwar, Saha, J.K., Biswas, A.K., Tripathi, A.K. and Acharya, C.L. 2001. Relationship between C addition and storage in a Vertisols under soybean-wheat cropping system in sub-tropical central India. *J. Plant Nutr. Soil Sci.* 164: 483-486.

Lal, S. and Mathur, B.S. 1989. Effect of long-term application of manure and fertilizers on the DTPA-extractable micronutrients in acid soil. *J. Indian Soc. Soil Sci.* 37: 588-590.

Laryea, K.B. Anders, M.M. and Pathak, P. 1995. Long-term experiments on alfisols and vertilisols in the semiarid tropics. p.267-292. In: R. Lal and B.A. Stewart (eds), *Soil Management-Experimental Basis for Sustainability and Environmental Quality, Advances in Soil Science*, CRC, Boca Raton, USA.

Meelu, O.P., Singh, Y., and Singh, B. 1994. Green manuring for soil productivity improvement. *World Soil Resources Reports* No. 76, FAO, Rome.

NAAS 1996. *Agricultural Scientists' perceptions on plant nutrient needs, supply, efficiency and policy issues;* 2000-2025, National Academy of Agricultural Sciences, New Delhi.

Nambiar, K.K.M., Soni, P.N., Vats, M.R., Sehgal, D.K. and Mehta, D.K. 1989. Ann. Rept. 1985-86/1986-87, All India Coordinated Research Project on Long-Term Fertilizer Experiments (ICAR)., New Delhi, India.

Narkhede, P.L. and Ghugare, R.V. 1987. Organic recycling in drylands II. Effect of organic cycling of subabul on yield, nutrient uptake and moisture utilization by winter sorghum. *J. Indian Soc. Soil Sci.* 35: 417-420.

Rabindra, B. and Gowda, H. 1986. Long range effect of fertilizer, lime and manures on soil fertility and sugarcane yield on a red sandy loam soil (Udic Haplustalf). *J. Indian Soc. Soil Sci.* 34:200-202.

Rabindra, B., Narayanaswamy, G.V., Janardhan Gowda, N.A., and Shivanagappa 1985. Long range effect of manures and fertilizers on soil physical properties and yield of sugarcane. *J. Indian Soc. Soil Sci.* 33:704-706.

Reddy, M.S.S., Reddy, S.R., Reddy, M.G. and Reddy, M.G.R.K. 1990. Residual effect of organic matter and fertilizer applied to groundnut (*Arachis hypogaea*) and maize (*Zea mays*) on succeeding groundnut in groundnut-maize-groundnut sequence. *Indian J. Agron.* 36: 298-300.

Sahu, B.N. and Nayak, B.C. 1971. Soil fertility investigation in Bhubaneswar long-term fertility trial. *Proc. International Symposium in Soil Fertility Evaluation*, Indian Society of Soil Science, New Delhi.

Sarkar, A.K., Mathur, B.S., Lal, S., and Singh, K.P. 1989. Long-term effects of manure and fertilizers on important cropping systems in sub-humid, red and laterite soils. *Fert. News* 34: 71-80.

Singh, M. 2000. Progress Report. All India Coordinated Research Project on Microbial Decomposition and Recycling of Farm and City Wastes. Indian Institute of Soil Science, Bhopal.

Subramanian, K.S. and Kumaraswamy, K. 1989. Effect of continuous cropping and fertilization on the phosphate fixation capacity of soil. *J. Indian Soc. Soil Sci.* 37: 682-686.

Sundar Rao, W.V.B. 1968. Phosphorus solubilization by microorganisms. p.210-229. In: *Proceedings of All India Symposium on Agricultural Microbiology,* University of Agricultural Sciences, Bangalore, India.

Swarup, A. 1998. Emerging soil fertility management issues for sustainable crop productivity in irrigated systems. p.54-68. In: A. Swarup, D. Damodar Reddy and R.N. Prasad (eds) Proc. National Workshop a *Long-term Soil Fertility Management through Integrated Plant National Supply Systems.* Indian Institute of Soil Science, Bhopal, India.

Swarup, A. and Wanjari, R.H. 2001. Lessons from long-term fertility experiments. All India Coordinated Research Project on Long-Term Fertilizer Experiments to Study Changes in Soil Quality, Crop Productivity and Sustainability, Indian Institute of Soil Science, Bhopal, India.

Venkateswarlu, J. 1987. Soil fertility management of red soils. p.115-121. In: *Alfisols in the Semi-Arid Tropics-Constraints Workshop,* International Crops Research Institute for the Semi-Arid Tropics, Patancheru, AP., India.

5

Role of Phosphorus in Carbon Uptake, Fruit Yield and Quality in Strawberry

Parmjit Singh[1], Zora Singh[2] and M.H. Behboudian[3]
[1]*Orange Agricultural Institute, New South Wales Agriculture,*
Orange NSW 2800, Australia
[2]*Horticulture/Viticulture, Muresk Institute of Agriculture, Division of*
Resources and Environment, Curtin University of Technology, GPO Box
U1987, Perth, Western Australia 6845, Australia
[3]*Institute of Natural Resources, Horticultural Science, Massey University,*
Private Bag 11 222, Palmerston North, New Zealand
E-mail : Z.Singh@curtain.edu.au

1. Introduction

The major issues for the world farming communities are declining productivity and profitability from traditional farming enterprises and increasing costs of production. Sustainable management of natural and resources is, therefore, of utmost importance for increasing and sustaining productivity and reducing the production costs. Updating our knowledge of plant mineral nutrition, including that of phosphorus (P) physiology, will help in adopting cultural practices that are cost effective and safe for the environment. P maintains a special status in this respect because its over-application will result in environmental pollution. This chapter is, therefore, devoted to the P physiology of the

---

Abbreviations: ATP: Adenosine triphosphate; Ca: Calcium; CO_2: Carbon dioxide; *Fm*: Maximum fluorescence of dark-adapted leaves; *Fm'*: Maximum fluorescence of light-adapted leaves; *Fv/Fm*: Optimum quantum yield of PS II; DF/Fm: Actual quantum yield of PS II; K: Potassium; m: meter; Mg: Magnesium; N: Nitrogen; P: Phosphorus; Pi: Inorganic phosphorus; *qP*: Photochemical quenching; *qN*: Non-photochemical quenching; s: second.

strawberry plant, which is one of the most popular horticultural crops. Besides presenting the physiological responses to P, we will also explore certain aspects of fruit quality that are affected by the level of this element in the plant.

2. P-A Key Element for Plant Growth

Phosphorus (P) is an important nutrient element required for plant growth. Its deficiency leads to reduced plant growth, reduced yield and poor quality fruit. On the other hand, excess application of P to fruit crops, including strawberry, can deteriorate produce quality, increase the cost of production and cause environmental damage, but without economic benefit. P deficiency in strawberry starts as stunted growth, dark green foliage with bluish-purple colourations. The bluish colour begins in small veins, gradually spreads to whole veins and ultimately colours the entire leaf. Root development appears normal initially but is not abundant. Top growth is retarded to a greater extent than root growth, hence the root/shoot ratio of P deficient plants increases as compared to those plants supplied with adequate amounts of P (Ulrich et al., 1992; Singh, 1998). Flowers and fruits from phosphorus-deficient plants tend to be smaller than normal, and fruit occasionally develops albinism in susceptible varieties. Deficiency of phosphorus can be corrected either with the application of P at planting or as a side dressing at the onset of the deficiency (Ulrich et al., 1992; Hancock, 1999). The application of P to strawberry can be adjusted according to the leaf analysis in order to maintain the plant in the sufficiency range (Table 1). Singh (1998) reported the threshold P requirement for carbon uptake and biomass accumulation in strawberry and reported that the threshold P value for above ground plant growth of strawberry is 0.15% on leaf dry weight basis, while the threshold P value for root growth is 0.10%.

Table 1. Sufficiency ranges of P for foliar nutrient levels in strawberry

Sufficient P (% in leaf DM)	Reference
0.25 – 0.40	Pritts and Handley, 1998
0.3 – 0.5	Vock, 1996
0.15 – 1.3	Ulrich et al., 1992

Eccessive use of P in strawberry production is not only an economic loss but excess P runs into waterways as well as leaching into the ground water, thus contaminating the drinking water. This will have adverse effects on the health of humans as well as animals (Wignarajah, 1995).

P may become a limiting factor for plant production in the coming years due to declining P resources worldwide and it may also become uneconomical to mine it for agricultural use. P demand will increase to fulfill the increasing food requirements of the fast-growing world population. Therefore, controlled P fertilization of crops is necessary. With this aim in mind, soil and leaf analyses are practised to get information about the P status and plant growth. P availability is also influenced by climatic and other factors such as soil texture, organic matter and pH, indicating that P fertilizer recommendation should be based on climatic and soil characteristics (Lacertosa et al., 1999).

A positive carbon balance is required for optimum plant growth and development. Carbon uptake is increased with a corresponding increase in photosynthetic rate (Baxter et al., 1994), which is the result of an increase in leaf area and net assimilation over the same period of time (Baxter et al., 1995). Therefore, a positive carbon balance will be achieved by higher photosynthetic rates rather than respiration rates and a large well-illuminated leaf area. Net photosynthesis as well as respiration rate of leaves and leaf area development depend to a large extent on the availability as well as uptake of essential nutrients, apart from carbon (Plesnicar et al., 1994; Rao et al., 1993; Singh et al., 1997). Therefore, the nutrient status through leaf analyses is a valuable indicator of plant efficiency to the yield. On the other hand, plants have some mechanisms to adapt to a certain extent to nutrient deficiencies (Ciereszko et al., 1996; Rao et al., 1993).

P plays a key role in the energy transfer necessary for the functioning of metabolic processes in plants (Marschner, 1995). It is also part of nucleotides and reproduction. P limitation lowers the inorganic phosphate level in the tissue (Ciereszko et al., 1996; Rao et al., 1993; Singh et al., 1997); strongly affects carbon uptake and partitioning and, in doing so, determines the growth and yield (Fredeen et al., 1989; Radin and Eidenbock, 1986; Rao et al., 1986). Photosynthesis and carbon uptake are differently affected under short- and long-term periods of P limitation. Response to P limitation may be somewhat variable among the species, as levels of P that reduce net photosynthesis in sucrose accumulating species may show little influence on starch-accumulating species (Foyer and Spencer, 1986; Walker and Sivak, 1985). Under P deficiency, net photosynthesis may (Fredeen et al., 1989; Plesnicar et al., 1994; Singh and Lenz, 1997) or may not be reduced (Ciereszko et al., 1996; Foyer and Spencer, 1986). The quantum efficiency of spinach leaves decreased (Brooks, 1986), but that of soybean leaves was not affected by P deficiency (Fredeen et al., 1989).

3. P Supply, Plant Metabolism and Plant Composition

Metabolic processes such as cell division, cell expansion, net photosynthesis, respiration, dry matter accumulation and partitioning between different plant organs as well as nutrient uptake require endogenous energy (Pearcy and Björkman, 1983). As P is needed for growth and metabolism of plants, its deficiency leads to a hinderance in energy generation resulting in a decline in leaf area development, net photosynthesis, dry matter accumulation and nutrient uptake. P deficiency can impair the functioning of metabolic processes in plants through its effect on energy transfer (Marschner, 1995). The increased accumulation of starch in strawberry leaves under low P results in a decline in nutrient concentration of the leaves, which to a certain extent, may be due to a diluting effect (Chen and Lenz, 1997). Figure 1 presents a simplified model showing the reactions of photosynthetic carbon metabolism in which P has a regulatory role.

The P requirement for optimal growth of strawberry is in the range of 0.3 – 0.5% on dry weight basis during the active growth stage (Bergmann, 1993; Vock, 1996). Pritts and Handley (1998) reported the optimal range of P for strawberry growth of 0.25–0.40%, whilst Singh (1998) found that a minimum threshold value of 0.15% in leaf dry matter is optimum for growth and dry matter accumulation of strawberry (Table 2). Plants exhibit deficiency symptoms and reduced growth when P level in the leaves is less than the minimum threshold

Table 2. Effect of adequate and low P supply on N, P, K, Ca, and Mg concentration (% in dry matter) in the fully-expanded leaf blades of strawberry over a 28-day period.

Treatment	Days after treatment establishment			
	7	14	21	28
N concentration				
Control	3.38 a	2.67 a	2.71 a	2.40 a
Low P	2.48 b	2.05 b	2.00 b	1.99 b
P concentration				
Control	0.67 a	0.55 a	0.51 a	0.56 a
Low P	0.19 b	0.15 b	0.09 b	0.05 b
K concentration				
Control	2.58 a	2.28 a	2.30 a	2.22 a
Low P	2.30 b	2.06 a	2.12 a	1.97 a
Ca concentration				
Control	1.17 a	1.09 a	1.23 a	1.42 a
Low P	1.15 a	1.04 a	1.05 b	1.16 b
Mg concentration				
Control	0.55 a	0.53 a	0.51 a	0.55 a
Low P	0.52 a	0.43 b	0.42 b	0.44 b

The values represent means of three replicates from day 0 to 21 and of five replicates for day 28 after treatment. For the column averages, different letters denote significant differences ($P=0.05$).

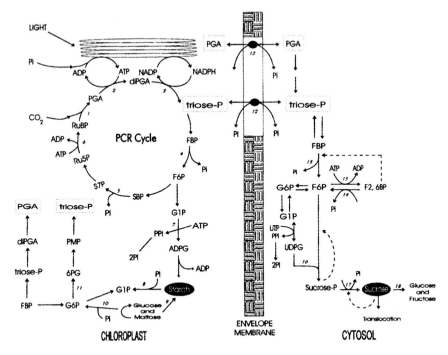

Fig. 1. A simplified model depicting the reactions of photosynthetic carbon metabolism in which Pi has a regulating function or in which energy-rich phosphates and the corresponding phosphate esters are involved (Singh et al., 1997). Due to these functions, strict compartmentation and regulation of the Pi level in the metabolic pool are essential for photosynthesis in leaf cells. Fixed carbon inputs and reducing equivalents converge in the PCR cycle. Two major branch points of the PCR cycle lead to the production of starch in the chloroplast and the export of triose-P to the cytosol via the Pi translocator, located on the inner envelope of the chloroplast membrane. Synthesis of sucrose in the cytosol is linked to the release of Pi, which is returned to the stroma via the Pi translocator in exchange for triose-P. The dashed arrows indicate possible feedback mechanisms. The reactions are catalyzed by enzymes numbered as follows: 1. Rubisco; 2. PGA kinase; 3. NADP-G3P dehydrogenase; 4. FBPase; 5. SBPase; 6. Ru5P kinase; 7. ADPG PPase; 8. Phosphorylase; 9. 0-amylase; 10. hexokinase; 11. NADP-G6P dehydrogenase; 12. Pi translocator; 13. FBPase; 14. F2,6BPase; 15. F6P-2-kinase; 16. SPS; 17. SPPase; and 18. invertase. (Adapted from Handbook of photosynthesis. M. Pessarakli (eds). Marcel Dekker, Inc. New York.).

level, and toxicity symptoms when P concentration exceeds 1% in leaf dry matter. The high accumulation of P in the soil induces a micronutrient deficiency, particularly that of zinc, iron and aluminium by affecting their transport to the plant (Bergmann, 1993). Excessive P application to the soils favouring excessive leaching causes accumulation of P into the soil and reservoirs of drinking water in the form of phosphates.

What is the effect of P supply on leaf area development in strawberry? The leaf area per plant is reduced under low P. Decline in leaf dry weight under low P can be accredited to a reduction in leaf number as well as leaf cell expansion (Fredeen et al., 1989; Singh, 1997). Reduction in leaf area occurs before the photosynthetic rate per unit leaf area is affected (Fredeen et al., 1989; McDonald et al., 1986; Singh et al., 1997). High net photosynthesis at this stage could be due to high stomatal conductance. ATP is involved in stomatal movement, therefore, low stomatal conductance under low P may probably be due to an inhibition of ATP synthesis (Zelitch, 1969). More allocation of the photosynthate to non-P carbon compounds than to sugar phosphates also helps to maintain high rates of photosynthesis by Pi, freeing up the mechanism under low P availability (Rao and Terry, 1995). Figure 2 shows the effect of leaf P concentration on the leaf number, leaf area per plant, leaf dry weight per plant and leaf area ratio over a 28-day period.

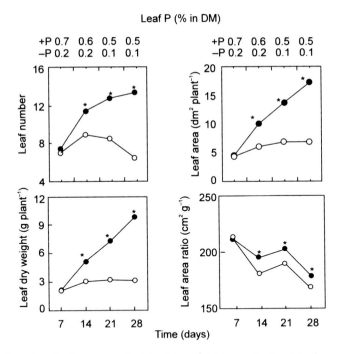

Fig. 2. Leaf number, leaf area per plant, leaf dry weight per plant and leaf area ratio of strawberry over a 28-day period under adequate and low P supply (Singh et al.,1997). Each value is a mean of 4 replicates up to day 21 after treatment and of 8 replicates at the end of the experiment. Closed symbols represent plants supplied with adequate P and open symbols that of low P plants. The asterisks at a growth stage indicate significant differences between the two treatments using Tukey HSD-multiple range test at $P<0.05$.

4. P and Leaf Gas Exchange Characteristics

The gas exchange of strawberry leaves is significantly impaired under low P due to lack of phosphate ions and increased carbohydrate concentration, particularly starch in the photosynthetic system (Singh, 1998). Increased starch accumulation in P deficient leaves inhibits the export of triose-P from the chloroplast to the cytosol and thus reduces the rate of photosynthesis (Furbank et al., 1987; Heineke et al., 1989). As the internal CO_2 concentration is not reduced but is increased under P deficiency in strawberry (Singh, 1998), it can be proposed that low stomatal conductance was not the sole cause of reduced photosynthetic rate but the photosynthetic system is damaged. Brooks (1986) found that the partial pressure of CO_2 in the intercellular spaces did not change significantly with changes in P nutrition.

Even though net photosynthetic rate per unit leaf area was not quickly affected under low P, the CO_2 compensation point rose as the total CO_2 assimilating leaf area was found to be somewhat reduced in strawberry (Singh, 1998). An increase in CO_2 compensation point under low P is similar to the effect of low P in *Populus maximowiczii* (Houman et al., 1990) and in *Glycine max* (Lauer et al., 1989). The reason for an increase in CO_2 compensation point under low P in strawberry is not yet clear. However, it may be related to either the decreased leaf area under low P (Singh et al., 1997) or due to higher rates of respiration (Lauer et al., 1989).

McDonald et al. (1986) found that low P reduced net photosynthesis, transpiration rate and stomatal conductance and enhanced internal CO_2 concentration, which may be due to the lack of phosphate ions in the photosynthetic system or an accumulation of large amounts of non-structural carbohydrates. Singh (1998) and Chen and Lenz (1997) found high rates of net photosynthesis in leaves of plants that received sufficient P under control (9-11 mmol CO_2 m^{-2} s^{-1}), which was due to optimal availability and distribution of nutrient elements. Tolbert and Zelitch (1983) attributed the reduced net photosynthetic rate under P deficiency either to starch accumulation or to a hinderance in energy transfer in CO_2 assimilation in P deficient leaves. Figure 3 shows the effect of leaf P concentration on gas exchange properties of leaf blades over a 28-day period.

5. P and Chlorophyll Fluorescence

A positive relationship exists between chlorophyll fluorescence, leaf chlorophyll content and leaf P status. Both chlorophyll fluorescence and leaf chlorophyll content decline when leaf P concentration is very low in strawberry leaves (Singh, 1998), showing that only severe P

Leaf P (% in DM)

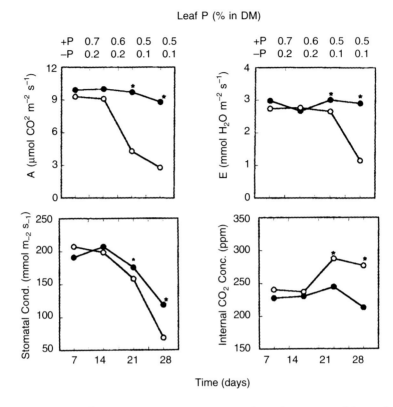

Fig. 3. Net photosynthesis, transpiration rate, stomatal conductance, and internal CO_2 concentration of fully expanded leaf blades of strawberry over a 28 day period (Singh et al.,1997). The values are means of 8 replicates up to day 21 after treatment and of 16 replicates at the end of the experiment. Closed symbols represent plants supplied with adequate P and open symbols that of low P plants. The stars indicate significant differences between the two treatments using Tukey HSD-multiple range test at $P<0.05$.

deficiency has a negative impact on chlorophyll fluorescence and leaf Chl content in strawberry. The optimal quantum yield of fully-expanded leaves should be between 0.8-0.83, as found in dark adapted, non-photoinhibited leaves of a wide variety of species. The Fv/Fm ratio of 0.816 to 0.854 showed the potential maximum quantum efficiency of PS II of strawberry leaves (Singh, 1998). Low P resulted in reduced optimal as well as actual quantum efficiency of PS II of strawberry leaves, as indicated by significantly reduced values of Fv/Fm and $DF/Fm´$. A decline in Fv/Fm under low P could be due to an increase in Fo (Kamaluddin and Grace, 1992), which shows the damage of PS II reaction centers (von Willert et al., 1995). In strawberry leaves, the reduction in Fv/Fm is associated with an increase in qN (Buwalda and

Noga, 1994). Changes in qP were associated with the changes in both net photosynthesis and photochemical yield reflecting changes in the oxidation potential of PS II reaction centres (Ottander et al., 1993). The above findings suggested that low P caused a high proportion of PS II reaction centres to close or, perhaps, even become non-functional. Plants under low P were then unable to achieve full photosynthetic potential. Figure 4 shows the effects of leaf P concentration on chlorophyll fluorescence of strawberry leaf blades over a 28-day period.

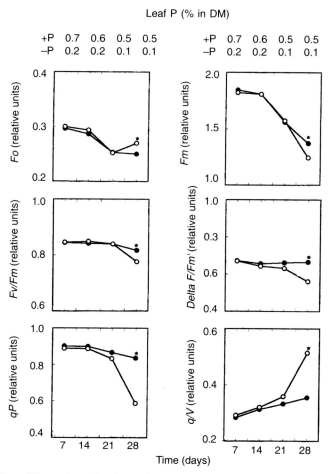

Fig. 4. Effect of P supply on Fo, Fm, *FvlFm*, qP, and qN of fully expanded leaf blades of strawberry over a 28 day period (Singh et al., 1997). Each value is a mean of 8 replicates up to day 21 after treatment and of 16 replicates at the end of the experiment. Closed symbols represent plants supplied with adequate P and open symbols that of low P plants. The asterisks at a growth stage indicate significant differences between the two treatments using Tukey HSD-multiple range test at P<0.05

P is an essential component of nucleic acids and phospholipids (Marschner, 1995). ATP, which is partly created during the light reactions of photosynthesis, serves as a source of energy and phosphate groups for metabolic reactions. Restricted ATP synthesis under low P increases the pH gradient and, therefore, enhances non-photochemical quenching and thylakoid energization. Due to a lower rate of photosynthesis in phosphate-deficient leaves, a greater proportion of the absorbed light, being in excess of that required to support CO_2 assimilation, is dissipated by radiationless thermal de-exitation, causing an enhanced non-photochemical quenching (Plesnicar et al., 1994).

6. Dry Matter Accumulation and Distribution

P deficiency causes a decline in dry matter accumulation in the leaves, petioles, fruits, and stem as well as total biomass accumulation in the plant. This may be due to less number of leaves (Lynch et al., 1991; Singh, 1998) and reduced illuminated leaf area as well as due to reduced photosynthetic potential for maximum yield (Bazzaz, 1990). The root growth in strawberry is not so much affected as the shoot growth by P deficiency (Singh, 1998). A greater partitioning of dry matter to the roots under low P have been observed for most other species (Cakmak et al., 1994; Plesnicar et al., 1994), which may probably be attributed to higher transport and utilization of photoassimilates in the roots (Fredeen et al., 1989).

Low P reduces shoot and root dry matter but increases the proportion of dry matter partitioned into the roots in strawberry as in many other plants (Chen and Lenz, 1997). The differences in fruit and root dry weight resulting in different root:shoot dry weight ratios in response to low P are related either to the efficiency of plants to export photosynthates to the roots (Cakmak et al., 1994; Ciereszko et al., 1996) or to the increased nitrogen concentration in the roots under low P (Singh, 1998).

Strawberry fruit act as very large sinks for assimilates (Nishizhawa and Hori, 1988), accumulating 40-50% of total dry weight (Lenz and Bünemann, 1967; Forney and Breen, 1985; Olsen et al., 1985). Under low P, more assimilates are exported towards the fruit, thereby resulting in relatively greater allocation of dry matter to the fruit. The harvest index for dry weight could increase from 0.33 under normally fed plants to 0.48 for plants supplied with low P (Singh, 1998). This shows that decreased vegetative growth under low P might have increased the translocation of assimilates to the reproductive organs (Nishizhawa, 1994).

The inoculation of mycorrhizal fungi in strawberries is reported to be beneficial to increase P content and dry weight accumulation, particularly at low P levels and during the reproductive phase, although this practice is rarely followed commercially (Holevas, 1966; Hughes et al., 1978; Dunne and Fitter, 1989; Khanizadeh et al., 1995; Hancock, 1999).

7. Carbon and Starch Accumulation and Distribution

Carbon concentration in the leaves, stem and roots can increase under low P, which may probably be due to high starch concentration (Eissenstat et al., 1993) that is accumulated (Chen and Lenz, 1997). The inefficiency of low P plants to degrade starch during the night exhibits certain metabolic effects of its deficiency. The high carbon concentration, however, cannot increase the total carbon uptake as the uptake efficiency is decreased but the utilization efficiency as well as carbon production efficiency is increased.

A higher starch accumulation in strawberry leaves (Knecht and Oleary, 1983; Singh, 1998) and/or due to diluted N concentration (Ceulemans and Mousseau, 1994) could cause a decline in leaf P concentration. Brouwer (1962) found that the requirements of organs nearest to the source of scarce essential elements are satisfied at the expense of those far away from the source. The uptake of total N, P, K, Ca and Mg per plant can be lower under low P, which may be due to reduced nutrient uptake efficiency, although the nutrient use efficiency was found to be higher in P deficient plants (Singh, 1998). Chen and Lenz (1997) found similar results in strawberry under ambient atmospheric CO_2 concentration and P deficiency, which is either due to reduced transpiration rate (Singh et al., 1997), or increased accumulation of carbohydrates in the leaves (Lauer et al., 1989), or even reduced total biomass accumulation (Fredeen et al., 1989).

Decrease in the concentrations of carbohydrates in the roots during growth could be due to increased export of carbohydrates into fruits and developing leaves (Nishizhawa and Hori, 1986, 1988). Relatively higher distribution of total non-structural carbohydrates in the stem, fruits and roots exhibited more accumulation in these organs under low P. A decrease in sucrose concentration in the roots may be due to higher accumulation in the fruits as sucrose is the major translocated carbohydrate in strawberry plants which increases greatly during fruit development (Forney and Breen, 1985). Starch, sucrose and glucose concentrations in the roots can decrease during growth, may be due to more assimilates exported to the fruit as the concentrations of glucose, fructose, and sucrose are increased in the fruit. High concentration of starch as well as sucrose in the roots and, hence, lower sucrose/starch ratio under low P indicates extensive translocation of photoassimilates from the shoot to the roots.

Increase in leaf starch concentration under P deficiency (Cakmak et al., 1994; Radin and Eidenbock, 1986; Singh, 1998; Sritharan, 1991) is suggested to be a reflection for P-induced limitation of sink activities in the shoot, such as lower leaf expansion (Radin and Eidenbock, 1984). This may be a result of more photosynthates being produced than could be utilized in growth of P starved plants, leading to higher concentrations of starch in the leaves or increased starch synthesis in the chloroplasts under light conditions and reduced starch degradation or higher triose-P or triose-P/inorganic P ratios in chloroplasts at the expense of sucrose synthesis (Heldt et al., 1977). High concentration of starch and sugars in the roots of P deficient plants suggest that root growth is not reduced in P-starved plants by impaired assimilate transport to the roots. The increase in sucrose import to the roots is the result of increased sink strength (Farrar, 1992).

8. P and Fruit Quality

Fruit quality depends on appearance (size, shape, colour, gloss and freedom from defects and decay), texture (firmness, crispness, and juiciness), flavour (sweetness, sourness, aroma and off-flavours), nutritive value (vitamins, minerals, dietary fibre and phytonutrients), and on chemical composition (Haffner et al., 1997; Kader, 2001). These characters depend to a great extent on the nutrient supply. Szymoniak (1925) reported that P increases the size and firmness of strawberry fruit and reduces the proportion of culls. Singh (1998) observed reduced strawberry fruit yield when leaf P concentration was 0.15% on dry weight basis, which may be due to reduced water content (Awang et al., 1993). Prange and DeEll (1997) reviewed the effects of various preharvest factors including nutrition, which affect postharvest quality of berry crops. Fruit maturity at harvest had a significant effect on fruit quality and its shelf life (Mokkila et al., 1997). They found that fully ripe strawberries had slightly higher vitamin C and sugar contents and lower organic acid content than ¾-ripe strawberries. The shelf life of fully ripe strawberries was less compared to ¾-ripe strawberries. The content of anthocyanins responsible for fruit colour increases with fruit maturity before harvest and with further ripening during storage (Robbins and Fellman, 1993). The latter authors reported that juice pH, titratable acidity and the content of soluble solids contribute to fruit flavour. Titratable acidity decreases and total soluble solids increases with maturity and during storage. Velemis et al. (1997) found that the dry matter (DM) content of the fruit at harvest significantly affected the soluble solids content (SSC) of the fruit after storage, and that kiwifruits with the highest DM content (18-19%) at harvest had the highest SSC (16%), and fruit with the lowest DM (14-15%) had the lowest SSC (14%) after 6 months of cold storage. Therefore, the fertilizer programme should match the nutritional requirement to ensure a uniform and timely maturity of fruits to get potential market benefits.

9. Conclusions

From the information presented above, it can be concluded that P is an important nutrient element for plant growth, yield and quality of strawberries. The suboptimal use of P may reduce strawberry yield and also have an adverse impact on fruit quality. On the other hand, excess usage of P may not be of any benefit to the yield and quality of strawberry, but may have a negative economic and environmental effect. Therefore, horticulturists, land managers and others involved in strawberry production should make careful decisions about P application to strawberries. This could include making soil analysis before planting suckers and making leaf tissue analysis in order to meet the later crop P requirements. Some alternate fertility improving options such as green manuring, use of farmyard manure and use of mycorrhiza for increasing P availability to plants should be included in the management decisions.

There is also a growing pressure on farmers to demonstrate environmentally-friendly production methods and to do more than merely market their produce as clean and green. Some international markets such as Birdseye, Sainsburys, Tesco's and Safeway stores in the UK, now "strongly suggest' their supplying farmers to demonstrate environmental stewardship of farm resources, particularly fertilizers and chemicals. Optimum P use will help farmers capture market benefits arising from improved environmental management and, reducing their operational costs. It will also assist farmers to cope with and in some cases, also take advantage of changing trends in the environmental awareness of consumers, regulators, financiers and the general public.

10. References

Awang, Y.B., J.G. Atherton and A.J.Taylor. 1993. Salinity effects on strawberry plants grown in rockwool. 2. Fruit quality. *J. Hort. Sci.* 68: 791-795.

Baxter, R., S.A. Bell, T.H. Sparks, T.W. Ashenden and J.F. Farrar. 1995. Effect of elevated carbon dioxide on three grass species from montane pasture. III. Source leaf metabolism and whole carbon partitioning. *J. Expt. Bot.* 46: 917-929.

Baxter, R., M. Gantley, T.W. Ashenden and J.F. Farrar. 1994. Effect of elevated carbon dioxide on three grass species from montane pasture. II. Nutrient allocation and efficiency of nutrient use. *J. Expt. Bot.* 45: 1267-1278.

Bazzaz, F.A. 1990. The response of natural ecosystem to the rising global CO_2 levels. *Ann. Rev. Ecol. Syst.* 21: 167-196.

128   *Parmjit Singh et. al.*

Bergmann, W. 1993. *Ernährungsstörungen bei Kulturpflanzen.* G. Fischer Verlag, Stuttgart.

Brooks, A. 1986. Effects of phosphorus nutrition on ribulose-1,5-bisphosphate carboxylase activation, photosynthetic quantum yield and amounts of some Calvin cycle metabolites in spinach leaves. *Aust. J. Plant Physiol.* 13: 221-237.

Brouwer, R. 1962. Distribution of dry matter in plant. *Neth. J. Agric. Sci.* 10: 361-376.

Buwalda, J.G. and G. Noga. 1994. Intra-plant differences in leaf chlorophyll fluorescence parameters in perennial fruiting plants. *New Zealand J. Crop Hort. Sci.* 22: 373-380.

Cakmak, I., H. Christine and H. Marschner. 1994. Partitioning of shoot and root dry matter and carbohydrates in bean plants suffering from phosphorus, potassium and magnesium deficiency. *J. Expt. Bot.* 45 (278): 1245-1250.

Ceulemans, R. and M. Mousseau. 1994. Effects of elevated atmospheric CO_2 on woody plants. *New Phytol.* 127: 425-446.

Chen, K. and F. Lenz. 1997. Responses of strawberry to doubled CO_2 concentration and phosphorus deficiency- I. Distribution of dry matter, macronutrients and carbohydrates. *Gartenbauwissenschaft* 62: 30-37.

Ciereszko, I., M. Gniazdowska, M. Mikulska and A.M. Rychter. 1996. Assimilate translocate in bean plants (*Phaseolus vulgaris* L.) during phosphate deficiency. *J. Plant Physiol.* 149: 343-348.

Dunne, M.J. and A.H. Fitter. 1989. The phosphorus budget of a filed-grown strawberry (*Fragaria* X *ananassa* cv. Hapil): evidence for mycorrhizal contribution. *Annl. Applied Biol.* 114: 185-193.

Eissenstat, D.M., J.H. Graham, J.P. Syvertsen and D.L. Drouillard. 1993. Carbon economy of sour orange in relation to mycorrhizal colonization and phosphorus status. *Annals Bot.* 71: 1-10.

Farrar, J.F. 1992. The whole plant: Carbon partitioning during development. In: *Carbon Partitioning Within and Between Organisms*, C.J. Pollock, J.F. Farrar and A.J. Gordon (eds.). Environment plant biology series, BIOS Scientific Publishers Ltd., pp. 163-179.

Forney, C.F. and P.J. Breen. 1985. Dry matter partitioning and assimilation in fruiting and deblossomed strawberry. *J. Amer. Soc. Hort. Sci.* 110: 181-185.

Foyer, C. and C. Spencer. 1986. The relationship between phosphate status and photosynthesis in leaves. Effects on intracellular phosphate distribution, photosynthesis and assimilate partitioning. *Planta* 167: 369-375.

Fredeen, A.L., I.M. Rao, and N. Terry. 1989. Influence of phosphorus nutrition on growth and carbon partitioning in *Glycine max*. *Plant Physiol.* 89: 225-230.

Furbank, R.T., C.H. Foyer and D.A. Walker. 1987. Regulation of photosynthesis in isolated spinach chloroplasts during orthophosphate limitation. I894: 552-561.

Haffner, K., S. Vestrheim and M. Fjelltun. 1997. Quality criteria of greenhouse or field-grown strawberry cv.Korona and Elsanta. *Erwerbsobstbau* 39 (3): 65-71.

Hancock, J.F. 1999. *Strawberries.* CAB International, Willingford, UK, pp. 90-110.

Heineke, D., M. Stitt and H.W. Heldt. 1989. Effects of inorganic phosphate on the light dependent thylakoid energisation of intact spinach chloroplasts. *Plant Physiol.* 91: 221-226.

Heldt, H.W., C.J. Chon, D. Maronde, A. Herold, Z.S. Stankovic, D.A. Walker, A. Kraminer, M.R. Kirk and U. Heber. 1977. Role of orthophosphate and other factors in the regulation of starch formation in leaves and isolated chloroplasts. *Plant Physiol.* 59: 1146-1155.

Holevas, C.D. 1966. The effect of vesicular-arbuscular mycorrhiza on the uptake of soil phosphorus by strawberry (*Fragaria sp.* Var Cambridge Favorite). *J. Hort. Sci.* 41: 57-64.

Houman, F., D.L. Godbold, W. Shasheng, and A. Hüttermann. 1990. Gas exchange in *Populus maximowiczii* in relation to potassium and phosphorus nutrition. *J. Plant Physiol.* 135: 675-679.

Hughes, M., L.W. Martin and P.J. Breen. 1978. Mycorrhizal influence on the nutrition of strawberries. *J. Amer. Soc. Hort. Sci.* 103: 179-181.

Kader, A. A. 2001. Quality assurance of harvested horticultural perishables. *Acta Hortic.* 553: 51-55.

Kamaluddin, M. and J. Grace. 1992. Photoinhibition and light acclimation in seedlings of *Bischofia javanica*, a tropical forest tree from Asia. *Annal Bot.* 69: 47-52.

Khanizadeh, S., C. Hamel, D. Buszard and D.L. Smith. 1995. Effect of three vesicular-arbuscular mycorrhizae species and phosphorus on reproductive and vegetative growth of three strawberry cultivars. *J. Plant Nutrition.* 18: 1073-79.

Knecht, G. and J.W. Oleary. 1983. The influence of carbon dioxide on the growth, pigment, protein, carbohydrate and mineral status of lettuce. *J. Plant Nutrition.* 6: 301-312.

Lacertosa, G., V. Lateana, N. Montemurro, D. Palazzo and S. Vandia. 1999. Soil fertility and plant nutritional status of strawberry in the Basilicata region, Southern Italy, D. Anac and P. Martin-Prevel (eds). *Improved Crop Quality by Nutrient Management.* Kluwer Academic Publishers, Dordrecht, Netherlands pp. 159-162.

Lauer, M.J., G.S. Pallardy, G.D. Blevins and D.D. Randall. 1989. Whole leaf characteristics of phosphate deficient soybeans (*Glysine max* L.). *Plant Physiol.* 91: 848-854.

Lenz, F. and G. Bünemann. 1967. Bezichungen zwischen dem vegetativen und reproductiven Wachstum in Erdbeeren (Var. Senga Sengana). *Gartenbauwissenschaft* 32: 227-236.

Lynch, J.A., A. Lauchli and E. Epstein. 1991. Vegetative growth of the common bean in response to phosphorus nutrition. *Crop Sci.* 31: 380-387.

Marschner, H. 1995. *Mineral Nutrition of Higher Plants.* Academic Press, London.

McDonald, A.J.S., A. Ericsson and T. Lohammar. 1986. Dependence of starch storage on nutrient availability and photon flux density in small birch (*Betula pendula* Roth.). *Plant Cell Environ.* 9: 433-438.

Mokkila, M., K. Rendell, J. Sariola, M. Hägg, U. Häkkinen and v.d.H. Scheer. 1997. Improvement of the postharvest quality of strawberries. *Acta Horticulturae* 2 (39): 553-557.

Nishizhawa, T. 1994. Comparison of carbohydrate partitioning patterns between fruiting and deflorated June-bearing strawberry plants. *J. Japan Soc. Hort. Sci.* 62 (4): 795-800.

Nishizhawa, T. and Hori, Y. 1986. Translocation of ^{14}C assimilates from leaves of strawberry plants in vegetative stage as affected by leaf age and leaf position. *J. Japan Soc. Hort. Sci.* 54: 467-476.

Nishizhawa, T. and Hori, Y. 1988. Translocation and distribution of ^{14}C-photo assimilates in strawberry plants varying in developmental stage of the inflorescence. *J. Japan Soc. Hort. Sci.* 57: 433-439.

Olsen, J.L., L.W. Martin and P.J. Pelofske. 1985. Functional growth analysis of field grown strawberry. *J. Amer. Soc. Hort. Sci.* 110 (1): 89-93.

Ottander, C., T. Hundal, B. Andersson, N.P.A. Hunner and G. Öquist. 1993. Photosystem II reaction centres stay intact during low temperature photoinhibition. *Photosyn. Res.* 35: 191-200.

Pearcy, R.W. and O. Björkman. 1983. Physiological effects. In: *CO_2 and Plants,* E. R. Lemon (eds). Westview Press Inc. Boulder, Colorado, USA, pp. 65-105.

Plesnicar, M., R. Kastori, N. Petrovic and D. Pancovic. 1994. Photosynthesis and chlorophyll fluorescence in sunflower (*Helianthus annuus* L.) leaves as affected by phosphorus nutrition. *J. Exp. Bot.* 45: 919-924.

Prange, R.K. and J.R. DeEll. 1997. Preharvest factors affecting postharvest quality of berry crops. *HortSci.* 32:824-830.

Pritts, M.P. and D. Handley. 1998. Strawberry Production Guide for the Northeast, Midwest and Eastern Canada. NRAES-88, Northeast Regional Agricultural Engineering Service, Ithaca, New York.

Radin, J.W. and M.P. Eidenbock. 1984. Hydraulic conductance as a factor limiting leaf expansion of P deficient cotton plants. *Plant Physiol.* 75: 372-377.

Radin, J.W. and M.P. Eidenbock. 1986. Carbon accumulation during photosynthesis in leaves of nitrogen- and phosphorus-stressed cotton. *Plant Physiol.* 82: 869-871.

Rao, I. 1997. The role of phosphorus in photosynthesis. In: *Handbook of Photosynthesis,* M Pessarakli (eds). Marcel Dekker, Inc. Newyork, Basel, Hongkong, pp. 173-194.

Rao, I.M., J. Abadia and N. Terry. 1986. Leaf phosphate status and photosynthesis in vivo: Changes in light scattering and chlorophyll fluorescence during photosynthetic induction in sugar beet leaves. *Plant Sci.* 44: 133-138.

Rao, I.M., A.L. Fredeen and N. Terry. 1993. Influence of phosphorus limitation on photosynthesis and carbon allocation and partitioning in sugar beet and soybean grown with a short photoperiod. *Plant Physiol. Biochem.* 31: 223-231.

Rao, I.M. and N. Terry. 1995. Changes with time following increased supply of phosphate to low-phosphate plants. *Plant Physiol.* 107: 1313-1321.

Robbins, J.A. and J.K. Fellman.1993. Postharvest physiology, storage and handling of red raspberry. *Postharvest News and Information* 44:53N-59N.

Singh, P. 1997. Influence of phosphorus and potassium supply on chlorophyll content, gas exchange, chlorophyll fluorescence, leaf area development, dry matter accumulation and nutrient uptake in kohlrabi (*Brassica oleracea* var. gongylodes L.). M. Agr. Thesis, University of Bonn, p. 100.

Singh, P., N. Keutgen and F. Lenz. 1997. Phosphorus deficiency in kohlrabi (*Brassica oleracea* var. gongylodes L.) - Influence on leaf gas exchange and chlorophyll fluorescence, N.El Bassam, R.K. Behl and B. Prochnow (eds). *Proceedings of the International Conference on Sustainable Agriculture for Food, Energy and Industry*, James and James (Science Publishers) Ltd., London, Vol. 1, pp. 433-438.

Singh, P. and F. Lenz. 1997. Leaf gas exchange and chlorophyll fluorescence of rape (*Brassica napus* var. napus L.) in relation to potassium nutrition. *Proceedings of the 11th World Fertilizer Congress* (Fertilization for sustainable plant production and soil fertility).

Singh, P. 1998. Carbon uptake and biomass production as affected by leaf area development and photosynthetic rate in kohlrabi and strawberry under adequate and low phosphorus supply. PhD Thesis, University of Bonn, p. 110.

Sritharan, R. 1991. Growth and nitrate accumulation in kohlrabi (*Brassica oleracea* var. gongylodes L.) as influenced by light, water, nutrient supply and CO_2 concentration. PhD Thesis, University of Bonn, Bonn, p. 149.

Szymoniak, B. 1925. Fertiliser test for strawberries. *Louisiana Expt Sta. Rpt.*; 52-53.

Tolbert, N.E. and I. Zelitch. 1983. Carbon metabolism. In: *CO_2 and Plants,* E.R. Lemon (eds). Westview Press Inc., Boulder, Colorado, USA, pp. 21-64.

Ulrich, A., M.A.E. Mostafa and W.W. Allen. 1992. Strawberry Deficiency Symptoms: A Visual and Plant Analysis Guide to Fertilization. Bulletin 1917, Division of Agriculture and Natural Resources, University of California, Oakland, California, pp. 42-45.

Velemis, D., M. Vasilakakis and M. Manolakis. 1997. Effect of dry matter content of the kiwifruit at harvest on storage performance and quality. *Acta Hortic* 444:637-42.

Vock, N. 1996. *Growing Strawberries in Queensland.* Department of Primary Industries, Queensland.

von Willert, J. Matyssek and Herppich. 1995. *Experimentelle Pflanzenökologie. Grundlagen und Anwendungen*, Georg Thimie Verlag Stuttgart, New York.

Walker, D.A. and M.N. Sivak. 1985. Can phosphate limit photosynthetic carbon assimilation in vivo? *Physiol. Veg.* 23: 829-841.

Wignarajah, K. 1995. Mineral nutrition of plants. In: *Handbook of Plant and Crop Physiology*, M. Pessarakli (eds.). Marcel Dekker, New York, pp. 193-221.

Zelitch, I. 1969. Stomatal control. *Ann. Rev. Plant Physiol.* 20: 329-350.

6

Nutrition of Tropical Horticulture Crops and Quality Products

B.C. Ghosh and S. Palit

Agricultural and Food Engineering Department
Indian Institute of Technology
Kharagpur, West Bengal 721302, India

1. Introduction

All fruit crops have strict requirements for a balanced fertilization management, without which their growth and development and the quality of the economic plant parts will be affected. A fruit is attractive to the consumer if it possesses aesthetic qualities of flavor, color and texture. Nutrient management in maintaining the desired growth and quality of a fruit crop has a great significance. Generally, fruit contains a high percentage of fresh weight as water and exhibits relatively high metabolic activity when compared to other groups of seed-producing crops. Such metabolic activity continues after harvest and at postharvest period, thus rendering most fruits highly perishable. Therefore, short self-life and perishability of the fruits pose a greatest problem to transportation and marketing and, ultimately, in maintaining quality as per consumers' choice. In a majority of fruit crops, the quality has largely maintained through a judicious management of nutrients along with cultural management at harvest and postharvest processes. The aim of this introductory chapter is to provide information on the behavior of tropical fruit crops on their biochemical changes at pre-harvest and postharvest stages in relation to nutrient management.

Fruits can be classified as climacteric or non-climacteric on the basis of their respiration pattern during ripening, wherein climacteric fruit displays a rapid increase in respiratory activities during ripening. However, a magnitude of the peak can vary enormously between the fruits. Fruits with the highest respiratory rate such as banana and mango tend to ripen most rapidly and, hence, are extremely perishable.

In contrast, non-climacteric fruits like pineapple, lemon, grape, etc., simply exhibit a gradual decline in their respiration during ripening.

1.1. Biochemical Processes in Fruits

1.1.1. Respiratory Substrate

Two major respiratory substrates found in fruits are sugars and organic acids, which form a major contribution to the overall flavor of the fruit. The most common sugars are fructose, glucose and sucrose and prevalent organic acids constitute malate and citrate.

1.1.2. Flavor Changes

Two senses, taste and smell, account for the perception of flavor. Thus, the flavor of the fruit depends on the complex interaction of sugars, organic acids, phenolics, tannins and more specialized flavor compounds, including a wide range of volatiles. Individual fruit differ in their relative contents of sugars and acids (Ulrich, 1970; Whiting, 1970). Both sugar and organic acids originate from photosynthetic assimilates of plants and green fruits and accumulate mostly during development and ripening stages. Some fruits like banana accumulate a bulk of their carbohydrate as starch prior to onset of ripening, while others continue to accumulate sugar from the plant during ripening. In some other fruit crops like grapes, carbohydrate accounts for a large part of their flavor. Fruits which accumulate their assimilate prior to ripening can be harvested at the mature green stage and still attain an acceptable flavor on ripening. This is important when considering the need for early harvest of fruit and marketing to optimize the self-life. Fruits like grape, which depend on the plant for assimilation during ripening fail to develop their full flavor if harvested at the green stage. Commercially, grapes become unacceptable if not fully ripe. For some fruit crops, the commercial harvest can be made at the green stage since improved flavor can often be obtained by allowing the fruit to initiate ripening prior to picking. In general, levels of acids decline during ripening (Ulrich, 1970; Whiting, 1970) and level of sugars within fruit tend to increase. Increase in sugar level either increases sugar importation from the plant or to the mobilization of starch reserves within the fruit, depending on the type of fruit. In some cases, a small percentage of sugar accumulation in the fruit may occur when gluconeogenesis occurs during ripening. In many fruits breakdown of starch to glucose, fructose and sucrose is a characteristic of the ripening effect. In fruit, the major starch degrading enzymes, α and β-amylase and starch phosphorylase have been identified. The breakdown of sucrose is probably mediated by the action of invertase, which is widespread in fruit and often increases in activity during ripening.

However, the activity is not always associated with an increase in the metabolism of sucrose.

1.1.3. Phenolics and Flavor Volatiles in Fruit

Although sugar and acids constitute the major components in fruit flavor, other specialized important compounds are phenolics and flavor volatiles. Several phenolic compounds have been isolated from different fruits. The phenolics and flavor volatiles, though usually present at relatively low levels, are important since they are thought to provide the characteristic flavor and aroma of different fruits. The flavor volatile profile of any fruit is usually very complex and the nature of the volatile compounds is very diverse. Some important chemical groups present are alcohols, aldehydes and esters (Nursten, 1970). The biosynthesis is further complicated by the fact that while some volatiles are synthesized in the intact fruit, others are produced only when the fruit tissue is macerated.

1.1.4. Color Changes

Not all fruits change color during ripening. For example, some varieties of banana and mango remain green. In general, color change is associated with ripening and can be brought about by the degradation of chlorophyll. Such a process occurs in degreening of lemons. However, in most fruits, the loss of chlorophyll is accompanied by the biosynthesis of one or more pigments, usually either anthocyanin or carotenoid. However, there is no correlation between chlorophyll loss and pigment synthesis during ripening. The chlorophyll degradation is involved in enzyme and chemical reactions. The chlorophyll becomes soluble and oxidizes chemically to the colorless purine and chlorine products. Anthocyanins are a diverse range of pigments which give rise to colors from red to blue. As, for example, orange fruits are known to contain 30 different anthocyanin kinds of pigments. Anthocyanins are derived from flavonoid compounds. Carotenoid pigments are localized within the chloroplast, which is called choromoplast. In fruit crop, the beta-carotene and lycopene is synthesized during ripening.

1.1.5. Texture Changes

Softening of a fruit during ripening is a major quality attribute that often indicates its shelf-life. Fruit softening could be due to loss of turgor, degradation of starch, or breakdown of the fruit cell wall. Loss of turgor is associated with postharvest dehydration of the fruit and has commercial importance during storage. In fruits like banana, the degradation of starch results in a pronounced textural change. Ingeneral, the texture of fruit change during the ripening process of most fruits is thought to be the result of cell wall degradation. Carbohydrate polymers make up 90-95% of the structural component of the wall; the remaining 5–10% being largely hydroxyproline–glycoprotein (HPRG). The carbohydrate polymers can be grouped together as cellulose, hemicellulose or pectins.

Changes in cell wall consist of dissolution of the pectin-rich middle lamella region of the cell wall. At a biochemical level, major changes can be observed in the pectic polymer of the wall. During ripening, there is a loss of neutral sugars and some loss of arabinose (Tucker and Grierson, 1987). A cell, in general, seems to have relatively high proportions of galacturonic acid, galactose and arabinose, suggesting that they are relatively rich in pectin.

1.1.6. Control of Ripening

Ripening is primarily a catabolic process comprising a breakdown in the cellular organization and processes (Blackman and Parija, 1928). It was found that in fruits like banana, protein and nucleic acid synthesis continued during ripening (Richmond and Biale, 1967). It has also been shown that cell wall carbohydrate polymer also continued to be synthesized during ripening (Mitcham et al., 1989).

Ripening is probably under the control of plant growth regulators (Bruinsma, 1983). Five major growth regulators that affect the ripening in one way or other if applied exogenously are auxin and gibberellins. Cytokinin generally act to retard ripening, while ethylene and abscisic acid act to enhance the ripening process. Ethylene, found to be associated with ripening, is being considered as a ripening hormone. Climacteric fruits are characterized by the production of a high level of ethylene during ripening. Non-climacteric fruits do not exhibit high level of production of ethylene but simply a decline in production from the mature green to ripe stage of development. A damage fruit responds with an increase in ethylene synthesis. Besides responding to endogenous ethylene, both climacteric and non-climacteric fruits will also respond to exogenous ethylene.

Green climacteric fruit also respond to ethylene with an increase in respiration and eventually, by autocatalytic ethylene production, resulting in a shortening of the shelf-life of the fruit. Preserved fruit under controlled atmospheric storage condition fail to produce the normal level of ethylene, fail to change color or soften but do exhibit the normal changes in sugars and acids seen during ripening (Goodenough et al., 1982). It is, thus, possible that only some ripening events such as color and texture changes are under the control of ethylene, while others including flavor development may be independent of ethylene. It is illustrated that ripening can be considered as a coordinated rather than as a linked set of biochemical pathways.

1.2. Role of Nutrient Elements on Biochemical Processes

Essential mineral elements play an important role in the growth and development of plant and of the deficiency symptoms that might accrue from a shortage of these elements. These elements are obtained from organic and inorganic sources of fertilizer materials which may supply one or more essential elements to make important foods, enzymes, hormones and vitamins. The complex fertilizer materials exist in the soil system in a form to supply the elements to the plants, which absorbs and uses them. For example, nitrogen is part of the molecule of all protein as well as a part of the molecule of both chlorophyll a and b. In soil system, nitrogen exist in many types of compounds and crop plants absorb and use nitrogen in the simple form of nitrate ion and ammonium ion. These ions are absorbed and used by plants in making nitrogenous compounds. Therefore, the raw organic and inorganic source of materials are all chemical compounds or parts of chemical compounds which contain one or more essential elements and are absorbed and utilized by the plants for growth and development. Essential elements include carbon, oxygen, hydrogen, nitrogen, phosphorus, potassium, sulphur, calcium and magnesium, which are grouped as macro-elements because of their greater requirement by the plant. Other essential elements like iron, manganese, boron, zinc, copper and molybdenum are grouped as micro-elements because of their low requirement by the crop.

Carbon, Oxygen, Hydrogen: The carbon, oxygen and hydrogen contained in plants are obtained from carbondioxide and water, converted into simple carbohydrates by photosynthesis and ultimately elaborated into amino acids, proteins and protoplasm. These three elements form about 92% of the dry weight of plants. Apart from having a major structural role, the elements-carbon, hydrogen and oxygen-play a key role in providing energy required for the growth and metabolism of plants. The bulk of the energy required for this process is derived from the oxidative breakdown of carbohydrates, fats and proteins during cellular respiration. The intricate process by which the plant, making use of solar energy synthesizes sugar and starch from atmospheric carbondioxide and soil water is a unique process which is characteristic of green plants.

Other mineral elements are derived from the soil minerals. The form at which the nutrient elements is abosorbed and their structural composition in plant tissue is presented in Table 1. The plant content of these elements is affected by several factors and their composition in crops varies considerably. The physiological role of the mineral elements is illustrated below:

Table 1. Form of nutrient element absorbed by plant and composition as structural element in plant tissue

Source	Nutrient element	Form of nutrient element composition	Composition as structural element in plant tissue
Air, water	Carbon	Co_3^{2-}, HCo_3^-	45 percent
	Oxygen		43 percent
	Hydrogen	H^+, OH^-	6 percent
Soil	Nitrogen	NH_4^+, NO_3^-	1-3 percent
	Phosphorus	PO_4^{3-}, HOP_4^{2-}, $H_2PO_4^-$	0.05-1.0 percent
	Potassium	K^+	0.3-6.0 percent
	Calcium	Ca^{2+}	0.1-4.0 percent
	Magnesium	Mg^{2+}	0.05-1.0 percent
	Sulphur	SO_3^-, SO_4^{2-}	0.05-1.5 percent
	Iron	Fe^{2+}, Fe^{3+}	10-1000 ppm
	Manganese	Mn^{2+}, Mn^{3+}	5-500 ppm
	Zinc	Zn^{2+}	5-100 ppm
	Copper	Cu^+, Cu^{2+}	2-50 ppm
	Boron	BO_3^{3-}, $HB_4O_7^-$	2-75 ppm
	Molybdenum	$HMoO_4^-$	0.01-10 ppm
	Chlorine	Cl^-	100-250 ppm

Nitrogen: The major role of nitrogen in the plant is its presence in the structure of the protein molecule. Besides, it is found in important molecules as purines, pyrimidines, porphyrins and co-enzymes. They help in protein synthesis and are also found in chlorophyll pigments and the cytochromes, which are essential in photosynthesis and respiration. The supply of nitrogen is related to carbohydrate utilization.

Phosphorus: Phosphorus is a constituent of nucleic acids, phospholipids, the co-enzymes NAD and NADP and, most important, a constituent of ATP and other high-energy compounds. Phospholipids along with protein are significant constituents of cell membrane. The co-enzymes NAD and NADP are important in oxidation and reduction reactions and are responsible for functioning essential plant processes like photosynthesis, respiration, nitrogen metabolism, carbohydrate metabolism and fatty acid synthesis.

Potassium: The best-known function of the potassium is its role in water relation in plant and stomatal opening and closing. Potassium is an essential activator for different enzymes involved in the synthesis of certain peptide bonds and carbohydrate metabolism.

Calcium: The well-known role played by calcium in the plant is as a constituent of a cell wall in the form of calcium pectate. The middle lamella is composed primarily of calcium and magnesium pectate. Calcium is also important in the formation of cell membrane and cell structure. Calcium also acts as an activator for a number of enzymes.

Magnesium: Magnesium is a constituent of chlorophyll molecule without which photosynthesis could not occur. Two essential roles played by magnesium in the plant is found in the processes of photosynthesis and carbohydrate metabolism. The magnesium acts as an activator for enzymes involved in carbohydrate metabolism and in the synthesis of the nucleic acid (DNA and RNA) from nucleotide polyphosphates.

Sulphur: The most obvious function of sulphur is its participation in protein structure in the form of sulphur-bearing amino acids, cystine, cysteine and methionine. Sulphur is involved in the metabolic activities of sulphur-bearing vitamins, biotin, thiamine and co-enzyme A. Sulphur is also important in Fe-S proteins in photosynthesis, nitrogen metabolism and ferredoxin synthesis.

Iron: Iron plays a number of important roles in the overall metabolism of the plant. It is incorporated directly into the cytochromes, into compounds necessary to the electron transport system in mitochondria, and into ferredoxin, which is indispensable to the light reaction of photosynthesis. Iron is also required in the synthesis of enzymes involved in chlorophyll synthesis. This element has been identified as a compound of various flavoproteins active in biological oxidations.

Manganese: Manganese is an essential element both for respiration and nitrogen metabolism. It is also known to activate a wide variety of enzymes concerned in cellular oxidation—reduction, hydrolysis and group transfer. Manganese has been suggested to function in the development and multiplication of chloroplast.

Copper: Copper is a metal activator of several enzymes participating in the cellular oxidation-reduction reactions. It acts as a component of phenolases, laccase and ascorbic acid oxidase.

Zinc: Zinc is necessary for the formation of tryptophan, the precursor of *indole-3- acetic acid* (IAA) and is a component of two enzyme systems: (1) the glyco-glycine di peptidases necessary for the formation of proteins; and (2) the dehydrogenases necessary for the glycolysis of sugar in terminal phase of respiration. Thus, zinc is necessary for cell elongation, for the formation of proteins and for the oxidation phase of respiration.

Boron: Boron is known to be required for proper development and differentiation of tissues. Boron functions in cell maturation by regulating the formation and lignification of the cell wall. Boron also affects tissue hydration, thus, disturbing the water economy of plants. Boron is associated with the reproductive phase in plants and its deficiency is often found to be associated with sterility and malformation of reproductive organs. As for metabolic role boron is involved in carbohydrate metabolism, particularly in the translocation of photosynthates—the sugar.

Molybdenum: Molybdenum is a constituent part of the enzyme nitrate reductase concerned with reduction of nitrate to nitrite in gaseous nitrogen fixation and nitrate assimilation. Nitrogen fixation is considered to be an important chemical reaction next to photosynthesis and respiration. Deficiency of molybdenum has been shown to decrease the concentration of sugars, particularly reducing sugars, suggesting an involvement of molybdenum in carbohydrate metabolism. This element also plays an important role in pectin metabolism.

Chlorine: Chlorine has been shown to be involved in the oxygen evolution in primary photosynthetic reactions. It has also been claimed to be involved in cyclic photophosphorylations, but not much information is available about its precise involvement in the process. Chlorine in excessive quantity has a detrimental effect by thickening and rolling of leaves and poor storage quality of the product.

1.3 Fertilizer use in Crop Production

Horticultural fruits are composed of certain chemical elements, often referred to as plant nutrients. An inadequate supply of one or more of these essential nutritional elements limits the yield and quality of a crop. Soils are seldom able to supply all the nutrients that a horticultural crop requires. Moreover, different horticultural crops have been found to differ markedly in their nutritional requirements apart from the fertility status of soil. Certain other factors such as the moisture content of the soil have been found to have a profound influence in mobility of nutrients, nutrient uptake and thereby response of a crop to fertilizer. The objective of manuring and fertilizer application is to improve the nutritional status of the soil by increasing the store of nutrients present and thus to raise the yield and quality of the product to a higher level. The fertilizer response is related to crop and soil characteristics, fertilizer management and its economic use.

Tropical Fruits

2. Mango

2.1. Introduction

The Mango (*Magnifera indica.* L) is an important tropical fruit, indigenous to South Asia, specially at the base of Malayan Archipelago from where it has spread to almost all parts of the world. It belongs to the family of Anacardiaceae. Mango is a medium to large evergreen tree with an open or dense symmetrical canopy; long taproot and dense fibrous root. The plant height ranges from 9 m to 31 m. The inflorescence is much branched, bearing both male and hermaphrodite flowers. The fruit is a fleshy drupe with variable shape (nearly round, oval or ovoid-oblong), size (0.6 to 2.3 kg) and comes in various colors (Sukonthasing et al., 1991). The fruits are single seeded, which may be either mono or poly-embryonic.

Mango is produced principally in the developing countries of the tropics. In Asia, mango occupies about 1 million hectares area with an annual production of 18.54 million metric tonnes, accounting for 79% of the total production. India is a leading mango-producing country, covering 51% of world production and exports this fruits largely in the processed form as juice, slices or pickles. In contrast, the Philippines and Mexico mainly export the fresh fruit. Mango grows well in humid as well as in dry climate within a temperature ranging from 24° to 30°C (Whiley et al., 1989). Cold temperature limits crop production. Microclimate factors also influence mango production (Chang et al., 1996). Mango can do well in a wide range of rainfall ranging from 25 cm to 250 cm. It is suggested that the total rainfall and its distribution during April and May in Taiwan are key factors affecting mango production. With an increase in light intensity from 4,000 to 30,000 lux,

Table 2. Mango production in the world (Majumder et al., 2001)

Countries		Production ('000 MT)
Africa		1904
N. C. America		2040
	Mexico	1461
South America		891
	Brazil	456
	Venezuela	147
Asia		18536
	Bangladesh	187
	Cyprus	2142
	India	12000
	Indonesia	605
	Pakistan	914
	Philippines	950
	Thailand	1350
	UAE	173
Australia		35
	World	23455

there is an increase in the leaf surface area (Parisot, 1988). The chlorophyll and N content in leaf increase with increase in percentage shading.

Mango can grow well in deep and well-drained soil and in a wide range of soil types from alluvial to lateritic. The suitable soil reaction ranges between 5.5 and 7 (Whiley, 1984), but it can be grown commercially upto 8.5 with proper nutritional management (Crane and Campbell, 1991). However, it is sensitive to saline condition.

2.2. Composition and Uses

The composition of mango fruit differs with type of cultivars and stage of maturity. Unripe green mangoes are reported to have 90% moisture, 0.7% protein, 0.1% fat and 8.8% carbohydrate. Sugars in mango comprise sucrose, glucose, fructose and maltose (Anonymous, 1994). Besides, it also contains xylose (Sankar, 1963), arabinose (Wali and Hassan, 1965), sedoheptulose and mannoheptulose (Ogata et al., 1972). After fruit setting, soluble protein content was found to decrease upto 44 days and thereafter increase till 96 days (Tandon and Kalra, 1983). There are twelve amino acids identified in mango fruit including the essential ones such as alanine, aspertic acid, lysine, leucine, cystine, valine, arginine, phenylalanine, and methionine (Elahi and Khan, 1973), which accounted for 10–40 mg/100 gm edible pulp. During the early growth stage (1-4 weeks after fruit setting), the number of amino acids are less and it increases to a maximum during fruit development stage (10–13 weeks after fruit setting). Some amino acids like aspartic acid, glutamic acid, phenylalanine, alanine and histidine were found throughout the development of fruit. Proline and glycine are observed during the later part of fruit growth. Leucine and threonine are observed at the initial growth stage, while lipid content in peel and pulp vary from 0.75 to 1.70% and 0.80 to 1.36%, respectively (Pathak and Sarada, 1974). Total kernel lipid in cultivar Alphonso amounted to 11.6% of the dry kernel (Hemavathy et al., 1987). Tannin is also present in a small amount in the flesh (0.16%) and in skin (0.105%), which is responsible for the astringency in the mango fruit. Mango peel is also a good source of pectin (Narasimha Char et al., 1989). From the nutritional point of view, mango is a rich source of vitamin A (Singh, 1960). The carotenoid pigments, β-carotene (provitamin-A) increase with ripening. Mango also contains a fair amount of vitamin C, which generally varies with cultivars and age of the fruit (Sanyal et al., 1991). Palaniswamy et al. (1974) reported a higher amount of ascorbic acid content in young fruit and the concentration decreases with the age of fruit. The ascorbic acid concentration also decreases with duration of storage at room temperature (Gofur et al., 1994). In green mango, the presence of folic acid (vitamin B) content upto 3.6 µg / 100 gm and vitamin B_1 (thiamine) and vitamin B_2 (riboflavin) ranging between 35–63 µg/100 gm and 37–73 µg/100 gm fresh weight, respectively, has been reported by Ghosh (1960).

The aroma and flavor of mango fruit vary widely with cultivars (Wilson et al., 1990). MacLeod et al., (1988) revealed that monoterpene hydrocarbons constituted nearly 49% (w/w) of the total volatile aroma constituents in mango variety Kensington. The important fruit volatiles are monoterpene hydrocarbons (91.3%) and esters (7.6%), which are predominant along with a two more abundant component, á-pinene (67.2%) and α–phellandrene (11.0%). Wong Siew (1994) reported two major chemical classes containing ester (71.3%) and alcohols (23.2%), besides the major constituents of ethyl 3-methyl–butanoate (40%).

The seed kernel contains fat (7.5 to 8.8%), protein (6.1 to 6.8%), crude fibre (1.3 to 2.4%) and ash (2.2 to 2.8%). The main fatty acids of the kernel fat are palmitic (6.9 to 7.3%), stearic (44.3 to 44.4%), oleic (38.9 to 42.1%) and linoleic (4.5 to 7.4%) acids (Augustin and Ling, 1987). Mango seeds contain a higher Iodine and saffonification value and high unsaffonifiable matter than roots and flowers (Khan et al., 1994).

Mango is used as fresh table fruits as well as processed products. Young and unripe mango fruits are mostly used for culinary purpose as well as preparing pickles. Ripe fruits are utilized for preparation of squash, jam, mango pulp, flex, mango candy, and also as canned fruit and juice (Gowda and Ramanjaneya, 1995). The various plant parts are put to several other uses, viz., tender leaves, ash of burnt leaves as a household remedy for curing burns and scales, fumes from burning leaves for relief from throat problem; fresh and dried flowers have curative properties for treating diarrhoea and chronic dysentery and also produce mangiferine and tannin (16 to 20%) from bark (Bhatt and Shah, 1985). The wood can be used for furniture, packing boxes, matchboxes, boats (Anonymous, 1962). From mango peels and stones, good quality vinegar containing 4.5–5% acetic acid could be obtained (Ethiraj and Suresh, 1992). The mango stone can also be an important source of edible fat (Narasimha Char and Azeemoddin, 1989).

2.3. Nutrient-Deficiency Symptoms of Mango

Deficiency of one or more nutrient elements may cause an abnormal appearance of the growing plant. If a plant is lacking in a particular element, more or less characteristic symptoms may appear. Deficiency of nitrogen, in the mango crop cultivar Langra has shown a reduction in fruit size with premature drop (Sen et al., 1947). Phosphorus deficiency has shown symptoms of tip burn and marginal necrosis in leaf, while potassium deficiency symptom caused die back, which proved fatal to the plant (Singh, 1957). Calcium deficiency caused shortening of internodes, while a dark yellow line was observed between necrotic and living tissues in Mg-deficient leaves. Sulphur deficiency resulted in mosaic-like white taints over the entire surface of older leaves. Mallik and Singh (1959) observed that Zn deficiency markedly reduced the leaf size and inhibited plant growth appreciably. At times, the Zn deficiency leads to severe malformation in mature mango trees, which had become unproductive. The deficiency of B and Cu also showed similar symptoms. In Israel, the occurrence of iron deficiency is common, resulting in the yellowish green color in the young leaf blade at early growth stage of crop, and at later stage, the new leaves ceased to grow with gradual die back (Kadman and Gazit, 1984).

2.4. Physiological Disorder

Soft nose: In acid sandy soil, the occurrence of soft nose—a physio-logical disorder of—mango occurs when the tree receives either low or excess doses of N fertilizer. On the other hand, in calcareous rock soil, the incidence was marginal, irrespective of the level of N fertilizers. The calcium content of leaves on the rock soil was 2–3 times more than that of sandy soil (Young and Minar, 1961). There is a negative correlation between calcium level in the leaves and soft nose incidence in mango fruit. Young et al. (1965) stated that if the level of calcium is maintained at 2.5% or slightly higher, the incidence of soft nose would be reduced appreciably.

Mal formation: The application of N, P and K fertilizer has a bearing on the incidence of malformation of the inflorescence in mango (Jagirdar and Sheikh, 1970). Incidence of malformation was observed in mango trees received P alone or P and K. Prasad et al. (1965) found that application of N, P and K in the ratio of 9: 3: 3 reduced the incidence of floral malformation. Application of chelated iron @ 1 lb per tree has corrected the incidence of malformation and resulted in increased flowering (Minessy, 1971).

Black tip: Mango necrosis or black tip is a common physiological disorder in mango grown in India. It originates as yellow spots on the fruit and later on, turns into brown and finally black. Such a physiological disorder is also variety specific. Singha (1961) reported that cultivar Dashehari is a susceptible cultivar to black tip disorder. Spraying of borax was found to restrict the disorder at the yellow tip stage.

2.5. Nutrient Management on Growth, Yield and Fruit Quality

Macro-nutrients: The amount of N, P and K absorbed per 1000 kg of fruits have estimated to be 0.98 – 1.5 kg N, 0.25 – 0.57 kg P_2O_5 and 1.37

– 2.19 kg K_2O (Zhaong et al., 1994). The beneficial effect of nitrogen in mango in increasing the vegetative growth and yield is reported by several workers. In Florida, Young and Miner (1960) reported that a higher dose of nitrogen application to Kent mango trees growing on sandy soil increased fruit yield by three times. Spraying urea of 4% resulted in an increase in the length of terminal shoots, number of leaves, leaf area and leaf nitrogen content of varieties Langra and Chausa (Singh et al., 1973). The individual fruit weight was increased with spraying of urea, as observed by Tiwari and Rajput (1975). Singh (1975) recorded an improvement in fruit size, acidity, ascorbic acid content, sugar and TSS contents in fruit of variety Chausa by spraying urea at 2% concentration at 4 months' interval. Urea spraying in mid-November reduced the incidence of flower bud malformation, increased the number of perfect flowers, improved pollen viability and increased fruit set. Following this practice, the fruit yield was increased by 8 times as compared to control (Shawky et al., 1978). Besides increase in growth and yield of mango, application of urea by spraying also reduced the fruit drop and improved the physico-chemical composition of fruits of varieties Langra, Dashehari and Totapuri (Rajput and Tiwari, 1979). After pruning, foliar spray of urea at 3% concentration on 60-year-old Fazli tree resulted in a marked improvement in fruit yield (Mallick et al., 1985). A survey on mango nutrition revealed that orchards with a higher N status tended to produce green fruits and those with lower N status usually resulted in the production of yellow fruits (Smith 1989; McKenzie 1994). Vega and Molina (1999) recorded fruit yield of 2197 kg/ha. in the first year with application of 80 kg N/ha. while in the second and third year 60 kg N produced yields of 3087 kg and 3965 kg/ha. respectively. According to Kanwar et al. (1987), N dose of 100 g per tree per year was sufficient for good tree growth. The fruit quality of mango in terms of pulp percentage was increased in cultivar Bombay Alphanso when the crop received nitrogenous fertilizer.

Application of phosphorus improved the yield of mango, when applied through orthophosphoric acid at 0.5% concentration alone or in combination with 2% urea during September, November and March. Reddy (1984) reveals that recovery of the applied P by the fruit was highest (12.7%) after application of orthophosphoric acid and urea when applied in March, while it was lowest when similar treatment was imposed during September.

Application of K to mango cultivar Hindi-Be-Sinnara resulted in increase in fruit number per tree, fruits per panicle and TSS content (Abd-El-Al et al., 1994). Mallik and De (1951) observed that the beneficial effect of N was realized on adding P and K along with N. Combined application of all the three major nutrient elements improved fruit size, acidity, ascorbic acid, sugar and TSS content of fruit (Singh, 1975). However, Malhi and Nijjar (1985) noted that mango showed maximum response to N followed by P and K.

The fertilizer dose varies with age of plant. One-year-old mango plant requires 75 gm of N, 110 gm of P_2O_5 and 55 gm of K_2O per plant (Singh, 1967). This fertilizer dose needs to be increased by the same amount each year. However, during the non-bearing stage, the plants should be supplied with 73 gm of N, 18 gm of P_2O_5 and 68 gm of K_2O per tree per year. A full grown fruit-bearing tree requires 726 gm of N, 182 gm of P_2O_5 and 670 gm of K_2O per plant in a year (Roy et al., 1951). The largest increase in trunk circumference was obtained with 1 kg N, 0.4 kg P_2O_5 and 1 kg K_2O per tree, while the maximum fruit yield was obtained with 0.5 kg N, 0.4 kg P_2O_5 and 1.5 kg K_2O per tree (Suriyapananont, 1992). The best fertilizer combination for mango, variety Dashehari was found to be 1.5 kg N, 0.75 kg P_2O_5 and 1.0 kg K_2O per plant per year (Sharma et al., 1993). Higher dose of N (1.5 kg) and low dose of K_2O (0.5 kg) per plant promoted vegetative growth, whereas, a higher dose of both N (1.5 kg) and K_2O (1.5 kg) per plant promoted fruiting in mango variety Fazli (Banik, 1997a). The fertilizer should be applied at vegetative flush and fruit bud differentiation stage (Singh, 1960).

Micronutrients: Deficiency symptoms of Fe, Mn, Zn, B and Cu have been observed in the mango cultivars Haden and Zill in Florida (Smith and Scudder, 1951). Zinc deficiency was first reported from Florida (Lynch and Ruehlu, 1940) and, subsequently, from Israel (Oppenhgimer and Gazit, 1961). Zinc deficiency is not reported to affect the yield adversely but spraying zinc sulphate at 0.8% concentration showed significant increase in the length of terminal shoots, leaf area, fresh and dry weight of fruit of Chausa mango (Rajput et al., 1976). In case of boron deficiency in soil, boron should be applied at least three months before the onset of flowering (Coetzer et al., 1992; Janse Van Vuuren et al. 1992). At flower initiation stage, spraying of boron @ 3 gm/litre improved yield per tree in mango variety Tomy Atkins. Fruit quality could be improved by combine application of boron (0.4%) and urea (1%) to young mango plants (Banik et al., 1997b). The combine spraying of zinc and boron significantly increased the TSS, sugar and ascorbic acid content of the fruits (Rath et al., 1980).

3. Banana

3.1. Introduction

Banana (*Musa sp*) is one of the important commercial fruit crops in the tropics. It belongs to the family Musaceae. More than 600 types of musa germplasms comprising wild forms and cultivated species have been reported. The edible banana is believed to have originated in the hot, tropical regions of Southeast Asia. Spreading from India to Papua New Guinea to Bangladesh, Philippines, Indonesia and Srilanka, it is also grown in Columbia, Brazil, Uganda, Zaire and Kenya. India, Uganda, Ecuador, Brazil and Columbia account for 44% of world production. Banana is grown in an area of 10 million/ha. with an annual production of 88 million tonnes. In Asia, banana is mostly produced in India, Philippines, Thailand and Indonesia. At the global level, the banana production has increased by 99% from 46 million tonnes in 1968 to 88 million tonnes in 1998. Growing world exports of banana are due to increasing exports from Latin America and the Caribbean Island. The export from Africa and Asia is less than 15%. The leading importers of banana are the United States, Germany, Japan and other European countries.

Banana is grown in all type of soils provided with high water holding capacity and adequate organic matter content. However, alluvial soils with high clay content are best for banana cultivation (Durmanov, 1974). The climatic requirement of banana is specific to temperature, solar radiation, rainfall distribution and relative humidity. A temperature range of 25 – 35°C is ideal for vegetative growth of banana. The crop cycle becomes longer in cool subtropic than the warmer areas. The growth of banana crop ceases below 14°C (Robinson and Anderson, 1991). Temperature has a profound influence on nutrient absorption by roots (Turner and Lahav, 1985). It has been observed that a rise in day/night temperature from 17/10°C to 33/26°C increases the absorptions of nutrients like N, P, K, Ca, Mg, Na, Cl, B, Mn, Zn and Cu. Increase in day and night temperature beyond 37/30°C cause reduction in the uptake of all the elements except Na. Temperature also influence largely on partitioning of nutrients, particularly Ca, Na, Mn, Zn, Fe and B (Veerannah et al., 1974). Growth of banana plants is severely damaged at temperature below 3°C, resulting in yellowing of leaves and degradation of chlorophyll. Root growth stops when soil temperature falls below 11.5°C and maximum growth has been observed at 23.5°C (Robinson and Alberts, 1989). Low temperature also prevents the emergence of inflorescence through the pseudostem stalk, which causes 'choke throat' in banana. This phenomenon is common in high altitude banana growing areas. Fruit growth has been reported to be optimum at 28 – 30°C (Turner and Lahav, 1983). Temperature below 13°C and above 40°C during fruit growth and development causes peel discoloration. Reddish brown streaks occur just below the epidermis of fruits. Such fruits ripen much more slowly than the normal fruit. Fruits at advance maturity stage, if exposed to extreme temperature of below 13°C and above 40°C, get severely affected with under peel discoloration. Atmospheric temperature above 40°C and a leaf temperature of about 47°C cause drying of leaf tissues, resulting in blackening of leaf (Taylor and Sexton, 1972). High temperature of 40°C and above induces premature ripening of green fruits and also caused mixed ripening. The shelf-life of banana fruit reduces at high atmospheric temperature (40 – 45°C). The pulp of such fruits collapse into a liquid mushy consistency and fruits become non-edible, which is turned as yellow pulp. Sunburn in fruits is a common phenomenon in the tropics.

In subtropical climate where frost occurs occasionally, banana can be grown with risk of low productivity. The rate of emergence of new leaf increases as photoperiod increases from 10 to 14 hrs (Allen et al., 1988). Both low and excessive light is harmful to the banana plant, causing bleaching of leaves. In subtropical condition, shading of banana plantation delays shooting and reduces yield. High density planting reduces photosynthetic efficiency when shaded leaf receives 50% intercepted light and ultimately, the yield of individual banana plant is reduced. Under high water deficit, banana leaf turns yellowish green and dies when the leaf water potential falls to – 0.26 Mpa (Kallarackal et al., 1990). Prolong water stress condition in banana plantation resulted in stunted growth, choked bunches and short fingers (Holder and Gumbs, 1983). The positive response on yield was recorded with supplemental irrigation (Asoegwu and Obeiefuna, 1987).

3.2. Composition and Uses

Banana is the most nourishing fruit, and contains many essential nutrients including minerals and vitamins. It is a rich source of energy and can provide 368 kilo joules/100 gm fruit. The pulp of banana contains higher amount of water and is also rich in K. Composition of banana is presented below.

Banana has multiple uses from fresh consumption to culinary purposes. The flower bud and also the central core of the pseudostem are made into tasty dishes. Various processed products like banana powder, flour, chips, jam, jelly and wine can be prepared from it.

3.3. Nutrient-Deficiency Symptoms of Banana

Nitrogen deficiency in banana leads to chlorosis of leaves. Murray (1959) recorded the development of inward chlorosis from the mid rib followed by necrosis of the tissue. Both in Dwarf Cavendish and Robusta banana, the leaves became pale green in color due to nitrogen deficiency (Murray, 1959; Martin-Prevel and Charpentiar, 1963).

Table 3. Composition of banana

Composition	Banana (Stover and Simmonds, 1987)	Cooking banana (Singh and Uma, 1996)
% pulp fresh weight		
Water	75.7	72.9
Carbohydrate	22.2	25.1
Protein	1.1	1.4
Fat	0.2	0.6
Ash	0.8	0.71
mg/100 g pulp fresh weight		
Phosphorus	27	21.3
Potassium	460	-
Calcium	7	23
Magnesium	36	-
Sulphur	34	-
Iron	-	0.7
Thiamine	0.04	0.02
Riboflavin	0.07	0.01
Niacin	-	0.73
Pantothenic acid	0.26	-
Pyridoxine	0.51	-
Ascorbic acid	10	30.1
Vitamin A (IU)	Medium	-

Phosphorus deficiency showed a necrotic appearance in the lower leaves and restricted plant growth (Freiberg, 1955). In P deficient soil, the plant showed stunted growth and severe leaf scorching, vascular discoloration in the center of rhizome and improper root development. Simmonds (1959) observed the development of inward chlorosis from the mid rid and withering of leaves due to phosphorus deficiency. Older leaves, however, showed a dark green color with a bluish bronze-like tinge (Martin-Prevel and Charpentiar, 1963).

Potassium deficiency leads to brown to purple brown flecks along the edge and upper surface of petiole. The leaves curve with the tip pointed towards the base and the plant growth is stunted (Freiburg and Steward, 1960). They have also observed that in K-deficient leaves, the protein synthesis was arrested and breakdown of protein was enhanced with glutamine and asparagines. Martin-Prevel and Charpentiar (1963) observed that potassium deficiency produced malformed bunches, smaller in size with less number of hands and fingers. Twyford (1965) noted yellowing of leaves and dying of shoots and finally reduction in bunch weight and inferior fruit quality due to deficiency of potassium.

Hasselo (1961) reported that the occurrence of premature yellowing in 10-month-old Lacatan banana was associated with low K content in leaf.

The critical level of Ca for its deficiency was found to be less than 0.25% in the first leaf. Calcium deficiency symptoms are characterized by narrow bands of marginal chlorosis of lower leaves, which gradually turn into necrotic hollows in the entire margin and the leaves become smaller with reduction in weight (Freiberg and Steward, 1960).

In case of magnesium deficiency, the color of the leaf turns yellowish, showing brown spots on the leaf margin (Murray, 1959). However, Freiberg and Steward (1960), observed interveinal chlorosis, extending towards the youngest leaves and restricting the increase in plant height. Chalker and Turner (1966) observed leaf yellowing, retarded growth and reduction in yield in Mg deficient plant. Moreira and Hiroce (1978) reported fruits of magnesium deficient plant do not ripen well and are tasteless. The deficiency of magnesium can be diagnosed when the level of Mg remains at 0.22% or less.

Sulphur deficiency in banana delays the greenness of newly-emerged leaves and thickening of secondary veins and leaf. Sometimes, the leaf size gets reduced with a rosette appearance.

The characteristic symptoms of iron-deficient plants are formation of abnormal leaves, yellow brown coloration, appearance of streaks on older leaves, short and malformed fruits. Wardlaw (1940) called this malady as plant failure, which generally occurs in alkaline soil at pH 7.5 – 9.2. The chlorotic leaves contains 3.4 ppm iron as compared to 8.9 ppm in normal leaves.

The zinc deficient plant showed a bunchy foliage, leaf deformity and increase in length and breadth ratio of leaf. The leaves became entirely white or show whiteness at the base. In case of Zn deficiency, the fruit development was slow and the fingers were twisted, shorter, thinner and lighter green in color.

Manganese deficiency symptom is characterized by the appearance of marginal interveinal chlorosis of the young leaves and with advancement in age, the whole leaf turns brown. The fruits of the affected plants develop a dark brown or black color (Jordine, 1962).

In case of copper deficiency, the plants showed an overall drooping appearance with shortened intervals between leaf petioles emerging from the pseudostem. The leaves drooped and showed an umbrella-like appearance. The size of the leaf was reduced and decreased in length and breadth ratio. Norton (1963) reported a stunted growth of banana plants due to boron deficiency. The newly-emerging leaves showed malformation with incomplete laminar development. Under prolong deficiency, the root and flower formation became inhibited, resulting in complete mortality of the plant (Norton, 1965).

3.4. Physiological Disorder

Excess of potash as compared to nitrogen content in soil results in the abnormal development of immature banana fruits, called yellow pulp disease. Application of fertilizer at high N, P ratio develops a degrain disease, which shows a drooping of ripe banana fruit on the bunch due to the rotting of the pedicels. Martin-Prevel and Montagut (1966) reported the problem of finger drop of banana, which was mostly due to an adverse climatic condition, favoring excessive photosynthetic activity in the post-flowering period.

3.5. Nutrient Management on Growth, Yield and Quality

Macronutrients: Nitrogen is most important limiting factor for growth and development of banana. Sharma (1984) recorded maximum pseudostem girth, bunch weight, number of fruits per bunch by application of 187.5 g N per plant to the soil in June/July and 187.5g N per plant applied by spraying in 12 splits starting from October/ November. This practice induced early flowering. Trials conducted at Indian Institute of Horticulture Research, Banglore with cultivar Robusta showed maximum bunch weight (13.46 kg) with an estimated yield of 41.5 t/ha. when 150 g N was applied per plant as compared with 9.78 kg bunch weight and 30.18 t/ha. yield in control. Nitrogen application @ 300 g per plant significantly increased plant height, pseudostem girth, pseudostem weight and leaf weight. However, a higher N rate (400 g N/plant), depressed all these parameters (Anjorin and Obigbesan, 1983). For Hill banana, Mustaffa (1983), however, found 150 g N/plant/year to be an optimum dose. In New South Wales, recommended fertilizer dose was 200 lb N, 22 to 44 lb P and 74 lb K/ acres/year (Anon., 1969). The number of hands and fingers were increased as a result of application of N, while K increased the weight, volume and density of fruit. The fruit quality was appreciably improved when K was combined with N (Singh et al., 1974). Echeverri-Lopez and Garica-Reyes (1976), in their experiment with banana cultivar Dominico, estimated 500 kg/ha. of complete fertilizer (12-6-22 ratio of N, P, K) as the most economic and efficient treatment which was applied two months after planting. Chu (1968) suggested the optimum N, P, K ratio as 1:1:3 and the optimum fertilizer dose of N and K were 160 gm and 480 gm/plant. Dressing with 200 lb N, 50–100 lb P and 450 lb K per acre per year in 3–4 split applications were recommended by Leigh (1969). The maximum yield of main crop and ratoon crop of banana cultivar Giant Governor was recorded by an annual application of 240 gm N, 90 gm P_2O_5 and 480 gm K_2O per plant (Chattopadhyay and Bose 1986). In a fertilizer trial with different doses of potassium on Giant Cavendish banana in Cost Rica, Garita and Jaramillo (1984) noted the highest yield of banana of 66.4 tonne/ha./yr with 750 kg K_2O/ha./ yr. Riascos et al., (1996) studied the response of banana to different potassium sources and found that potassium nitrate was the best source by increasing the bunch weight and fruit size.

Application of phosphorus did not influence the growth of banana. However, Ramaswamy (1976) reported that application of phosphorus @ 60 gm/plant improved pseudostem height, hand number/bunch and bunch weight as well as accelerated the flower initiation. The fruit quality, however, was not affected. Application of potassium has a pronounced effect on fruit acidity and TSS (Jambulingam et al., 1975). Higher rate of K application @ 360 gm per plant significantly increased the pseudostem height, induced early flowering and maturity with good graded bunches of Robusta banana. Koen (1976) recorded an optimum yield of high quality fruits with an annual application of 370 gm potassium ammonium nitrate and 450 gm potassium chloride per plant. In Cuba, application of 750 gm K per plant produced the highest yield with an average bunch weight of 38.7 kg and 164 bananas per bunch. Obiefuna (1984) noted that K applied @ 300 gm per plant at 19/ 20th leaf stage increased the bunch weight (73.9%), number of marketable fingers (33.7%) and finger weight (44.2%) per plant as compared to control.

Nitrogen should be applied in splits at shorter interval, while phosphorus at planting and potash in two split doses, i.e. at planting and at flower initiation stage. Summerville (1944) has met the following observations:

1. Minerals should be abundantly available at the time of planting and at the time of initiation of ratoons.
2. Potassium applied in the first two months influences the production of number of hands.
3. The uptake of phosphorus is more rapid when the plants are 60 – 90 days old.
4. Third and fifth months are ideal for fertilization and split application more than three times are most beneficial .

Micronutrients: Experimental evidences shows that micro-nutrient elements such as Cu, Zn, Mo, B and Mn are necessary for the healthy growth of banana. Srivastava (1964) reported that Zn and Cu are indispensable for banana. Application of Cu and Zn at 4 ppm each, improved the root and vegetative growth of banana. Langenegger and Reynolds (1986) found that application of Zinc sulphate in the soil corrected the zinc deficiency.

4. Citrus

4.1. Introduction

Citrus fruits include oranges (*Citrus sinensis*), lemons (*Citrus limon*) and limes (*Citrus aurantifolo*) and comprises the world's leading fruit crop. It is adaptable to a wide range of soil, topography and climatic conditions and grown in a number of countries (Reitz, 1984). The major citrus producing countries are Spain, the USA, Israel, Morocco and South Africa. In Israel, 80% of agricultural export value lies in citrus. In Japan, citrus occupies the first place among the fruit crops. Citrus production at the global level is increasing faster than consumption.

Citrus can be grown well in a wide variety of soil and climatic conditions. The ideal pH level for citrus ranges between 5.5 to 7.5. It is sensitive to high pH and presence of excessive salt in the root zone and cannot thrive well in saline alkaline soil. The poor performance of any citrus orchard is mainly due to poor physical condition of the soil and imbalance of nutrition (Dhingra and Kanwar, 1963). Kakde (1956) observed that the performance of orange orchards directly depends on the physical and chemical properties of soil.

Climatic factors like temperature, moisture and light intensity are of principal importance to citrus, of which temperature plays most crucial role. Most of the citrus are least resistant to cold injury, while the mandarins can tolerate low temperature. Hot wind velocity during flowering and fruit setting are highly detrimental for good bearing and causes fruit drop and sunburn of fruit. Huang et al. (1993) reported flower fall at high temperature greater than 28°C. Humidity has a bearing on the physical character of the fruit, while temperature imparts fruit quality. Low humidity usually favors better color development of fruits, while under high humid condition, fruits are more juicy with thin rind (Cooper et al., 1963). Citrus tree is sensitive to light. Trees exposed to high light intensity exhibit pale foliage color, while those under shade develop deep green foliage. Partial shading results in higher fruit quality and complete shade bears poor quality fruits. Under high mean annual temperature, citrus fruits mature early, the fruit size becomes big and the acidity development in fruit juice remains lower. Fluctuation between day and night temperatures intensifies color development and accentuates sugar accumulation and acid formation. Best quality citrus fruit grows in semi arid and subtropical regions with less than 20 inches of rain and under irrigated condition. In California, the citrus orchards are mainly found at a lower altitude at elevation below 700 m and that of in Spain at 250 m. In Southeast Asia, the trees flourish at altitude upto 200 m. Drought condition caused 20% reduction in leaf expansion and reduced photosynthesis and plant growth (Mataa et al., 1998). Floral shoots and number of flowers per branch exhibited a strong relationship to drought stress (Pire and Rojas, 1999).

4.2. Composition and Uses

Different species of citrus fruits have different chemical compositions. In the sweet group (*Citrus sinensis*) and in the mandarin group (*Citrus reticulate*), the principal constituents of the edible portions are sugars (glucose and sucrose) and acids (primarily citric acid and little of malic acid). The rind of citrus is rich in pectin and certain essential oils and also contains certain glucosides (hesperidin in oranges and lemons). Citrus fruits contain a considerable amount of ascorbic acid and vitamin C. The TSS in the fruit juice in most of the sweet group of citrus varies from 8 – 12%, while the titratable acidity ranges from 0.5 – 1.5%. For marketable fruits, the TSS and acid ratio of 8:1 is considered to be optimum. In lemons and limes, the titratable acidity of fruit juice usually ranges from 5-6%. The vitamin C content in the fruit juice of different citrus species varies from 25 – 85 mg/100 ml of juice.

Besides the demand of orange as fresh fruit, orange marmalade is also an important product of sour and sweet oranges. The juice from all types of citrus fruits are bottled and canned in a large scale. The production of canned juice and pulp and the frozen concentrated juice has been increasing throughout the world. The flower, leaf and rind of citrus contains oils of good fragrance and possesses good commercial value. Lemon and orange oils are important citrus oils used as a flavoring agent. Other commercial products are citric acid and pectin prepared from cull.

4.3. Nutrient-Deficiency Symptoms of Citrus

Nitrogen-deficient plants show under size of young leaves, fragile and are pale in color. Shedding of old leaves is found in such trees and their foliage becomes sparse. Chapman et al. (1945) reported that nitrogen deficiency resulted in smoother skins and greater juice content. Reig-Felin et al., (1963) noted that N deficiency resulted in a very thin rind, which developed pits after 40 days of storage. Jones and Smith (1964) opined that adequate amount of nitrogen are required for flowering and fruit setting of citrus.

Phosphorus is an important element that contributes to growth and development, flowering, fruiting and fruit quality of citrus trees. Phosphorus deficiency caused a reduction in foliage density and deteriorated the physico-chemical characters of fruit, stunted growth, premature abscission of older leaves, low yield and late maturity of fruits. Phosphorus deficiency affects the quality of fruits with thick peel and hollow centers with exceptionally high in acid content.

The deficiency of potassium causes chlorosis of mature leaves, gumming of twigs, premature wilt of leaves and fruit shedding and reduction in fruit size. In lemons, K deficiency caused a yellowing or bronzing of parts of the leaf. Chapman and Brown (1943) noticed that in case of acute K deficiency, leaves became twisted and new lateral shoots were weak and spindly.

Calcium deficiency in citrus is not a common occurrence in field condition, its deficiency results in low juice content and higher TSS in fruit. In some instances, premature drop of foliage and dieback of twigs followed by weak growth of lateral buds, reduced fruit growth and fruits with thicker rind in orange (Yokomizo, 1974).

Magnesium deficiency is observed in acidic and highly-leachable soil. Martin and Page (1965) reported that soil containing 5-10% exchangeable Mg is considered deficient for citrus. In acute deficiency, the tree leaves become yellow and yellow bands are observed on either side of midrib.

Zinc deficiency occurs widespread in citrus-growing regions of the world. It is popularly known as frenching in Florida. In South Africa and the USA, Zn deficiency is a common occurrence. The mottling of leaves is a typical symptom of Zn deficiency (Haas, 1936). Bain (1941) described partial chlorosis between the veins and lateral veins remain green. Degree of chlorosis increases with severity of deficiency. The chlorosis pattern is quite distinct by developing white yellow between main lateral veins. Leaves become small, pointed and narrow in acute cases. The tree shows a bushy appearance with rapid dieback and reduction in size in plant growth. Fawusi and Ormrod (1975) reported chlorosis and stunted growth in citrus, resulting due to Zn deficiency.

Mn deficiency is indicated by the development of green or yellowish green areas between major veins. Unlike Zn deficiency, the leaf size remains normal and the veins and small adjacent areas of the veins remain green. In severe Mn deficiency, defoliation, loss of vigor and low yield occurs. However, Mn deficiency commonly occurs in combination with Fe or Zn deficiency. In soil, with acidic in nature, is prone to Mn deficiency.

Copper deficiency symptoms in citrus are known as 'red rust', 'die back', 'multiple bud' or 'peach leaf'. Copper deficiency occurs widely in Florida (Jones and Smith, 1964), California (Bradford et al., 1962), South Africa (Beyers and Jonbert, 1952), Sicily (Majorana, 1960). Bain (1941) described the development of unusually dark green in Cu-deficient plants, die back of twigs and reddish luster in mature leaves of the shoots. Under severe deficiency, reddish brown gummy excrescences is covered on the affected twigs. Orange shows the deficiency symptoms with scabby lesions on the rind, while lemons show a lack of juiciness. Bar-Akiva and Lavon (1967) noticed a high ratio of chlorophyll a:b in Cu deficient Eureka lemon leaves.

In case of B-deficiency, the plants become less vigorous with sparse vegetative growth and abundant blooms. Fruits usually become hard and gum deposits (brownish in color) are observed on the fleshy coat of the fruit. The central core or pith of the fruits may be gum soaked and the seeds may abort and turn dark. Fruits of all ages may show deficiency symptoms with brown discoloration in the white portion of the rind and unusually thick albedo on young fruits, small size of fruit, with thick rind and high sugar content (Sato et al., 1962).

Iron deficiency, which is called as iron chlorosis shows symptoms of reduction in leaf size, dieback of twigs and reduction in growth and yield. Wallihan and Garber (1966) observed dieback of twigs in iron deficient crop.

In molybdenum deficiency, symptom become conspicuous in four-month-aged plant where light green spots appear (Jones and Smith, 1964). The Mo-deficient leaves usually drop and the tree shows thin look. With premature defoliation, Mo-deficient plants are adversely affected but the quality of the fruit remains unaltered.

4.4. Nutrient Management on Growth, Yield and Quality

Citrus is a nutrient sensitive crop and inadequate plant nutrition causes serious disorder in citrus fruit. Chapman et al., (1945) reported that about 18 tonnes of citrus fruit removed 21 kg nitrogen, 5 kg phosphorus, 41 kg potassium, 19 kg calcium, 3.6 kg magnesium, 2.3 kg Sulphur, 40 gm boron, 9 gm copper, 50 gm iron, 13 gm manganese and 13 gm zinc. Thus, all the essential elements play an important role in proper growth and development of citrus including fruit quality.

Macronutrients: Nitrogen requirement of citrus crop depends on factors such as soil type, fertility status of soil and the type of variety grown. In Florida, N application @ 150 to 200 lb/acre was found to be an optimum dose for maximizing the production of high quality orange fruit (Smith, 1959). The optimum N dose for mature Navel oranges in Queensland was 900 g/tree provided it was supplied in winter (Mungomery et al., 1980). Collado Fernandez (2000) suggested that the application of N as ammonium nitrate could be used commercially for production of small size fruits, which are desirable for marketing with thinner rind. In Philippines, Felizardo et al., (1961) recommended application of 2.5-3 kg of ammonium sulphate to 5-year-old citrus trees.

Nokrashy et al., (1977) noted that in 10 years of Balady orange trees, the yield increased with N doses but in 12-year-old trees the yield was maximized at 600 g N/tree and then declined. In Egypt, Shawky et al., (1979) recommended application of 1200 g N/ tree in four splits to Navel oranges which produced yield 48% more than traditional N dose @ 600 gm/tree applied in three splits. In China, suitable fertilizer dose for mandarin was reported to be 675 – 750 kg N, 375-450 kg P_2O_5 and 375-450 kg K_2O/ha. (Chen et al., 1998). Ouyang (1998) investigated on time of application of fertilizer on yield of 12-year-old Satsuma mandarin cultivar Weizhang trees. Nitrogen fertilization influenced the peel thickness in citrus fruits. Ali (1952) found that ammonium sulphate alone increased peel thickness in five cultivars of sweet orange, but higher juice content and TSS were obtained when ammonium sulphate and FYM were applied together. Bouma (1959) recorded that peel thickness of Washington Navel oranges reached a peak at 11 weeks after blossoming but thereafter, it was progressively thinner until the 23rd week, the rate of thinning, however, increased with the decrease in N supply. Dry weight of peel was increased rapidly with high dose of N nutrient. Dasberg et al. (1983) found that N dose did not affect fruit weight of Shamouti oranges but a higher dose developed thicker peels and delayed development of orange color. The organic N produced lower soluble solids and acids in oranges than when an integrated organic and inorganic source of N was applied. Further, an ammonium sulphate showed higher soluble solids than application of nitrate of soda. The content of juice and vitamin-C and external appearance of orange were not affected much by N sources. Analysis of Egyptian Balady mandarin fruits after harvest and after storage for two weeks reveals that the fruit weight was highest and the peel percentage was lowest in crop receiving combine organic and inorganic N. The use of only organic N delayed the development of rind color and of optimum flavor. Application of only inorganic N reduced total acidity of the fruits. Total Soluble Solid increased in fruits when the crop received combined organic and inorganic sources of N. However, juice color and vitamin-C content were not affected by N sources. Fruit maturity could be faster by 15 days under organic or combined treatment of organic and inorganic N. Split application of N has little effect on internal fruit quality of citrus but has appreciable effect on rind color. Reuther and Smith (1954) observed loss of greenness in Hamlin oranges in Florida when single dose of N was applied during autumn as compared to split application. Studies in California showed that regreening of Valencia oranges was minimized when N was applied only during late winter or early spring. N applied during summer greatly accentuated green color on the rind (Jones and Embleton, 1959).

Several studies have been made on the effect of N dose on fruit quality of different kinds of citrus. Reitz and Koo (1960) found that a high dose of N application improved the green color of fruit and increased acidity of the fruit and delayed maturity. Soluble solid and vitamin-C content of four cultivars of mandarins were found to be negatively correlated with N doses, while juice content, acidity and TSS/acid ratio were not affected (Singh and Agrawal, 1960). Jones et al. (1963) reported that leaf N levels above 2.6% reduced the fruit quality but this was unavoidable when S content exceeded 0.185% or B content exceeded 250 ppm in Valencia orange sampled during September-October. De Fossard and Lenz (1967) reported that high doses of N application reduced the storage quality of fruits at a temperature range of 20-25°C. In general, fruits from trees receiving low N doses improved storage quality. In Australia, Stanard (1973) suggested application of N at low dose (0.8 lb N/tree) in order to obtain a high yield and good quality fruits of Washington Navel oranges. He pointed out deterioration in fruit quality with application of high dose of N. Further, high N dose resulted in thicker rind or peel. Lenz (1967) found that growth of rind in Valencia Late oranges were encouraging at the beginning and during the ripening period at high level of N, while the reverse was true for edible part of the fruits. The sugar content of juice was low and the acid content was high in N treated crop, while, sugar: acid ratio was favorable in untreated crop. Bhattacharya et al. (1973) recorded the highest fruit weight and more number of fruits in Kagzi lime (*C. aurantifolia*) receiving an annual application of 454 g N/plant as Calcium Ammonium Nitrate. However, the juice quality in terms of TSS, vitamin-C content and acidity percentage was significantly higher in controlled fruits. The number of seeds per fruit was increased with application of N.

Excess application of N sometimes causes toxicity to flowering and fruiting of citrus trees. The toxicity may develop due to presence of toxic principles in the fertilizers. Urea N generally contains a small quantity of biuret, which may cause 'yellow tip' type of toxicity, which is similar to B toxicity. Jones and Embleton (1954) showed limits of 0.25% biuret for spray grade urea and 2 to 5% for soil applied urea. Jones and Steinacker (1955) suggested that urea containing more that 0.25% biuret should not be used as a citrus foliage spray. Maximum benefit from urea spray was obtained when lemons were sprayed with urea in a cool or dull whether.

Significant response to P application has been reported from various citrus-growing areas of the world. In contrast, partial or no response is also quite common (Hernandez, 1983; Orozco- Romero and Sepulveda-Torres, 1983). Phosphorous application showed improvement in foliage color, fruit setting, reduces fruit size and better fruit quality in lemon fruit. On the other hand, Iwasaki and Owada (1960) found no marked influence of P on yield of Satsuma mandarin and lemon. Several studies showed that P fertilization affects fruit quality, depending upon type of citrus fruit and level of P used. Increasing fruit size by P application was observed by Sato et al. (1958) in Satsuma mandarin, Rosselet et al. (1962) in Valencia orange and Suzuki et al. (1977) in Satsuma mandarin. Phosphorous fertilization reduced rind thickness (Moss, 1972). Application of P also increased the percentage of juice in orange (Moss, 1972).

Potassium deficiency can be corrected through application of fertilizer like KCl, K_2SO_4, etc. Reece and Koo (1975) noticed that K requirement for various types of citrus fruits were not the same. They found out the dose of K @ 160 kg/ha./year in case of Hamlin orange. In Florida, the commercial rate of K for citrus varies from 100–200 lb K/acre/year (Smith, 1966). Chu (1963) obtained a yield increase of 13.6 and 43.5% in *Citrus poonensis* with application of 200 and 300 g K_2O/plant, respectively, over the control. Similarly, the yield of Imperial mandarins was increased by 12.3 and 22% by application of 0.5 and 2 kg K/tree as K_2SO_4, respectively (Chapman, 1982). Both soil and foliar application of K was found effective in enhancing growth and increase in yield. Jones et al. (1973) found that the yield of young lemon trees was increased with either soil application of K_2SO_4 or foliar application of KNO_3 in an orchard where leaf K content was low. Ganje et al. (1966) observed that foliar application of KNO_3 could alleviate the visual symptom of K deficiency and increased K content appreciably in leaf of Navel oranges. In mandarin also, application of KNO_3 was found to be an ideal source of K fertilizer for foliar application. The best time for foliar spray of K was found to be after spring flush but well before winter dormancy. The beneficial effect of K fertilizer on growth and yield of citrus were demonstrated by several researchers. Further, there is a direct relationship between supply of K and fruit size of citrus, which has been demonstrated by several workers (Weir, 1967; Reese and Koo, 1975). The result revealed that the fruit size increased with increase in dose of K. Early maturing cultivars are less sensitive to K than late maturing ones. Valencia orange seems to be more responsive to K level than other cultivars and fruit size may increase even with K content in leaf remains above 2%. High levels of K retard degreening and hasten regreening (Embleton et al., 1966). Sakamoto and Okuchi (1963) noticed peel color development in Satsuma orange fruits on trees receiving 300 g K_2O/tree/year. Similarly, rind thickness and the rag of the pulp were increased with rising levels of K (Bar-Akiva, 1975). Jones et al. (1973) found that the peel thickness of lemon decreased with increase in K level. The incidence of peel disorder (creasing) is related to high N and low K doses (Reese and Koo, 1975). Creasing in oranges particularly in Valencia orange seems to be associated with low K status of the tree (Bester, 1975). Sites and Deszyck (1952) noticed reduction in the incidence of creasing in Valencia and Hamlin oranges as the rate of K supply was increased. Embleton et al., (1966) in lemons and Weir et al., (1978) in Valencia orange found that K increased juice percentage, while no significant effect of K on juice content was observed by Calvert (1969) in Hamlin and Temple orange and Bar-Akiva and Gotfried (1972) in Valencia orange. The TSS is also affected by K nutrition. The soluble solids tended to decrease with increase in K levels. Reese and Koo (1974) demonstrated a decreasing trend in TSS in K fertilized trees. The evidences showed that K increased acid concentration in the juice of citrus. Potassium also increased ascorbic acid content of the juice as demonstrated by Embleton et al. (1966) in Eureka lemon, Reitz and Koo (1960) in Valencia orange. Chu (1963) reported that the flavor of the fruit in K-treated citrus (*Citrus poonensis*) was sour at harvest but improved with storage. Kesterson et al. (1977) reported that K fertilization reduced peel oil content in citrus.

In acid red soil (pH 4.5-4.7), both slate lime and lime stone application in addition to NPK increased yield and improved fruit quality in mandarin. Anderson (1971) found that soil pH and calcium affected markedly the yield of Valencia orange trees. Haas and Brusca (1954) noticed that low calcium resulted in smaller fruit size with highest percentage of peel in Valencia. On the other hand, an increase in Ca was accompanied by an increase in the fresh weight in the fruit pulp.

In citrus, Mg deficiency symptoms were observed when the concentration of Mg in leaf was 0.2% or less. Koto and Takeshita (1957) found that the critical Mg content in Satsuma orange leaves was 0.3%. Martin and Page (1965) reported that leaf magnesium values of Mg deficient sweet orange trees varied from 0.15–0.2%. Datuadze (1976) showed that when the leaf Mg content declined to 0.18–0.06% the Mg deficiency appears in trees. Plessis and Smart (1982) observed that leaf Mg content was increased sharply by spray of $Mg(NO_3)_2$ twice. Application of MgO @ 1.0 kg/tree showed marked and long lasting effect of Mg content in leaf. The detrimental effect of Mg deficiency is associated with loss of photosynthetic efficiency of leaf and defoliation. Application of Mg to Mg-deficient plant corrected the chlorosis and restore health and vigor of plant. Kuznetsov and Treshchov (1979) reported that higher rates of $MgSO_4$ (200 g/tree) increased fruit dry matter and reduced pulp ratio in young mandarin.

Micronutrients: Zinc deficiency in citrus can be corrected by application of Zn directly to the soil or by foliar spray to the crop. Mottle leaf in mandarin due to Zn deficiency was cured by 1-3 sprayings of 10 lb $ZnSO_4$ and 5 lb lime in 100 gallons of water applied during February-March, May-June, and September-October. The effect of Zn application lasted for three to four years (Chowdhury, 1954). Russo and Raciti (1955) suggested spraying citrus trees with 600 g $ZnSO_4$ plus 300 g Na_2CO_3 plus adhesive per 100-litre water for correction of Zn deficiency. Kanwar and Dhingra (1962) found that foliar spray of 0.6% $ZnSO_4$ reduced chlorosis and raise the Zn content of leaves. Zn is indispensable for growth, flowering and fruiting of citrus. Dwarf and mottle leaves recover fast after application of Zn. Samoladas (1964) reported that applied Zn stimulated photosynthetic activity, increased yields and controlled irregular bearing of citrus. In Georgia, soil and foliar application of Zn to citrus improved yield in young mandarin trees by 13.7% over the control (Talakvadze, 1973). Chanturiya (1975) reported 25.8% increase in yield in young mandarin receiving Zn @ 3 gm/tree. Labanauskas et al. (1963) noticed that foliar spray of Zn on moderately Zn deficient trees increased the amount of ascorbic acid per unit volume of the juice but decreased the percentage of juice by weight. Bacha (1977) observed increase in TSS content of Succary orange following application of Zn, while Talakvadze (1973) found that Zn application improved the storability of Satsuma fruits.

Several workers reported successful control of Cu deficiency symptoms in citrus by spraying with $CuSO_4$. Ishihara et al. (1974) tried several Cu containing fungicides to correct deficiency on Satsuma mandarins. Bordeaux mixture spray at 0.4-0.5% was found to be most effective and Cu deficiency symptoms disappeared within two years. More vigorous growth and increase in fruit production of Satsuma mandarin were recorded following treatments containing Cu fungicides. Sugar content and sugar: acid ratio of the fruit were also increased (Ishihara, 1974). Bacha (1977) showed that the vitamin-C content of Baladi orange fruit increased following spray with $CuSO_4$ at 250 ppm.

Manganese deficiency can be corrected through foliar spray of 0.4-0.6% $Mn(SO_4)_2$ in spraying during summer when new growth occurs. Bar-Akiva (1965) found that Mn-oxide at 0.5% improved yield by 93% when applied for correction of Mn deficiency in Valencia and Shamouti oranges. Labanauskas et al. (1963) reported an increase in the yield and TSS of fruits in oranges with foliar application of Mn to moderately Mn deficiency orange trees. Koto et al. (1963) also reported a positive correlation between Mn content of leaves and sugar content and sugar: acid ratio of the fruits.

Deficiency of Boron can be corrected by foliar application of 0.1% Borax or boric acid. Sato et al. (1962) found that both soil and foliar application of B @ 37.5-75.0 g/tree corrected the B deficiency symptom in *C. natsudaidai*. Leonard (1952) stated that B application should not exceed 200 ppm in order to avoid a toxic effect on the citrus trees. Application of B reduced the acidity of the fruit. Beneficial effect of applied B on mandarin was observed on increasing the yield by 15-20%. Boron treatment increased the sugar content in fruits and reduced ascorbic acid content. Ishihara et al. (1965) showed that B application increased fruit and seed weight, improved seed development and lowered the sugar: acid ratio of fruit.

The deficiency symptoms of Molybdenum can be corrected through foliar application of sodium or ammonium molybdate. Soil application of Mo has not been effective, especially when soil pH is low (Leonard, 1952). A supply of 0.025 ppm Mo was found to be sufficient for normal growth of sweet orange seedlings while Valencia orange tree responded at 0.5 ppm of Mo.

5. Guava

5.1. Introduction

Guava (*Psidium guajava*) is a native of tropical America, widely grown in all over the tropics and subtropics from Mexico to Peru (Chadha and Pandey, 1986). Recently, the guava has been introduced as sub-tropical fruit in Israel. The major guava producing countries are South Asian countries, the Hawaiian Islands, Cuba and India. The genus *Psidium* belongs to family Myrtaceae, contains about 150 species (Hayes, 1970). The Brazilian or Guinea guava (*P. guajava*) produces small-sized fruits while *P. pomiferum* is round shaped and *P. pyriferum* is pear shaped in nature. Fruits of Chinese guava, *P. friedrichsthalianum* are small and globose in shape with high acid content. The strawberry guava (*P. cattleianum*) is a relatively hardy sub-tropical species with round red fruit (Normand, 1994). The fruit of *P. cattleieanum* var. Lucidium are sulphur yellow in color and the trees are comparatively large in size. Besides, other promising species grown in different countries of the world are *P. acutangulum*, *P. firmum* (wild species), *P. araca*, *P. littorale*, *P. molle*, *P. coriaceum* (Mitra and Bose, 1999). Nathani and Srivastava (1965) tested 14 guava cultivars and identified Sofeda as a suitable variety for its whitish flesh with firm pulp, while Chittidar possess red dots on the body of the fruit. Golberg and Levy (1941) observed that pink-fleshed cultivar of South Africa were usually poor in vitamin C content as compared to white-fleshed fruits. In seedless cultivar, though, the fruit is of good quality and has few seeds, the bearing is very poor with small fruit size and less commercial value. Singh et al. (1979) recommended L 49 as an ideal cultivar due to its production, fruit weight and content of sugar, acidity and vitamin-C. In South Africa, Selection No 1 showed highest ascorbic acid content (Pensburg and Preez, 1985). In Brazil, the cultivars Rica found suitable for processing, while, another cultivar Paluma is high yielding with large size fruit and red pulp. The suitability of guava hybrids for the preparation of nectar, a process juice product, was evaluated and it was observed that the hybrid 3-22, 12-34 and 5-27 were suitable for preparation of nectar (Baramanray et al., 1995).

Guava trees can be grown in a variety of soil. However, best soil should be deep friable and well drained. It can be grown in pH range between 4.5 and 8.2. Among the tropical fruits, guava can tolerate certain extent of salinity. However, fruit weight, fruit size and ascorbic acid content decreased at high salinity value.

Guava is successfully grown under tropical and subtropical climate and has high capacity of tolerance to environmental stress (Marler, 1994). The optimum temperature lies between 23-28°C (Samson, 1980). Although guava can tolerate some low temperature but exposure to freezing temperature may cause shoot dieback and the trees may die at −2 to −3°C (Malo and Cambell, 1986a). However, Cattleya guava can survive temperature below −5°C (Morton, 1987). Guava is fairly tolerant to drought (Ogden et al., 1981).

5.2. Composition and Uses

Guava is rich source of ascorbic acid and pectin. The ripe fruits contain moisture (77.9–86.9%), dry matter (12.3–26.3%), ash (0.51-1.02%), crude fat (0.01–0.70%), crude protein (0.82–1.45%) and crude fiber (2.0–7.2%). The physico-chemical composition of guava fruit varies widely with cultivars, stage of maturity and season (Kundu et al., 1995; Ghosh and Chattopadhyay, 1996). The total soluble solids content in fruit varies from 8.2–10.5°C Brix (Kundu et al., 1995). The total sugar content ranges between 4.9–10.1%. The composition of guava fruit is presented in Table 4 below.

In ripe guava fruits, the predominant sugars are fructose (59%), glucose (36%) and sucrose (5%). Fructose is the principal sugar in green ripe fruits, while sucrose is the main sugar in fully-ripe fruits (Arenas de Moreno et al., 1995). Guava fruits are also a source of vitamin A (about 250 IU/100 gm) and an appreciable amount of thiamine, niacin and riboflavin. The ascorbic acid content varies from 75 – 260 mg per 100 gm, depending upon the type of cultivar and stage of maturity (Ghosh and Chattopadhyay, 1996). Cultivars with pink flesh are poor

Table 4. Composition of guava fruits

Constituents		Range
Moisture	per cent	77.9–86. 9
Dry matter		12.3–26.3
Ash		0.51–1.02
Crude fat		0.10–0.70
Crude protein		0.82–1.45
Crude fibre	g/100 g pulp	2.0–7.2
Sugars		
Reducing		2.4–5.2
Non-reducing		2.5–3.8
Total		4.9–10.1
Acidity		0.22–0.39
Ascorbic acid		75.2–234.3
Thiamine		0.03–0.07
Riboflavin		0.02–0.04
Niacin	mg/100 g pulp	0.20–2.32
Calcium		10.0–30.0
Phosphorus		22.5–40.0
Iron		0.60–1.39

Wilson (1980), Singh (1988), Das et al. (1995), Ghosh and Chattopadhyay (1996)

in ascorbic acid content than the white flesh one (Mitra et al., 1984). The guava fruit contains considerable Ca, P and Fe. However, Fe remains in the seed. Guava fruits are a good source of pectin, which ranges between 0.5 – 1.8% (Adsule and Kadam 1995). Citric and malic are the predominant organic acids present in guava fruit besides glycolic, tartaric and lactic acids. The flavor of the guava fruit is due to the presence of volatile compounds like hydrocarbons, alcohol and carbonyls. The decrease in astringence with the advancement of maturity is due to polymerization of leucoanthocyanidins. The pink flesh color of some cultivars is due to lycopene (Adsule and Kadam, 1995).

Guava is consumed fresh when mature. The guava jelly, jam and juice is processed from the flesh. Each individual cultivar has a characteristic taste. High quality nector can be prepared from guava fruit. The ripe fruits are also used for manufacturing of ice cream, cheese and toffee. Guava can be dehydrated and pulvarized into a powder form, which is a good source of vitamin C. The frozen product tastes almost similar to the fresh fruit (Ruehle, 1948). Wines are also prepared from guava fruits, namely guava juice wine and guava pulp wine (Bardiya et al., 1974). The guava seeds contain 5-13% oil, which is rich in essential fatty acid and can be used in salad dressing (Adsule and Kadam 1995).

5.3. Nutrient-Deficiency Symptoms of Guava

In guava, deficiency of nitrogen leads to stunted growth and purple color patches on both sides of the midrib and principal veins. The deficiency of K resembles similar symptoms to that of N but the necrotic patches were concentrated more towards the margin and base of the leaf. In case of P deficiency, the lower leaves show purple color patches, which gradually enlarge and form a complete band of deep brown color around the leaf. The chlorosis in guava leaves was interveinal for magnesium and the leaves become molted for sulphur deficiency. Murray (1960) recorded a reduction in growth rate of guava tree under deficiency of N, P, K, Ca and Mg.

Guava sometimes suffers from deficiency of Zn, exhibiting symptoms such as interveinal chlorosis, leatheriness of leaves, reduced leaf size, dieback of branches, production of less flower, drying and cracking of fruits and less fruit production. In case of Cu deficiency, the plants exhibit chlorosis and marginal necrosis of young leaves, brown pigmentation at both sides of the midrib and dieback of main and auxiliary shoots. Deficiencies of N, Zn and B may be the major cause of the physiological diseases. Due to physiological disorder, 'Fatio' disease in guava occurs in which the affected leaves show red spot and subsequently dry up and the branches develop cracks and finally cause dieback.

5.4. Nutrient Management on Growth, Yield and Fruit Quality

Macronutrients: The necessity of manuring guava tree at a regular intervals for increasing fruit production was emphasized by Hayes (1970). The fruiting occurs based on current seasons growth and manuring encourages vegetative growth and fruiting. Manuring improved both growth and quality of guava (Singh and Singh, 1970). In Sao Paulo, Brazil, Natale et al. (1994) observed that the yield of guava increased with N dose, attaining maximum production in the third year. The optimum dose of N applied at economic level was 131 kg and 199 kg/ha. in the second and third year, respectively. Spray of urea brought about significant improvement in fruit yield, sugar, vitamin C and TSS content (Singh and Rajput, 1977). There was increase in reducing and non-reducing sugars and pectin content by 3.61, 2.40 and 6.89%, respectively, and no change in fruit acidity due to spray of urea at 2% level.

Application of phosphorus showed a beneficial effect on the growth, yield and quality of guava (Kumar et al., 1995). An optimum dose of phosphorus @ 500 gm P_2O_5/tree/year improved the fruit quality. The fruit quality in terms of total soluble solids, sugar and ascorbic acid contents showed marked improvement due to application of P fertilizer.

Application of potassium significantly increased the fruit yield (Natale et al., 1996). Growth, fruit weight and yield were significantly increased with application of K @ 400 gm K_2O/tree/year. Increases in TSS, ascorbic acid, reducing sugar and pectin content of fruit juice were also increased when higher dose of potassium was applied @ 500 gm K_2O/ha. Kumar et al. (1996) demonstrated that maximum fruit yield and net return per tree was obtained when the crop received 682 gm K_2O/tree/year. Foliar application of muriate of potash at 4% concentration improved the number of flower buds, physico–chemical characteristics of fruits and yield.

Chonkar and Singh (1981) revealed that spraying of 3% urea along with a spray of calcium phosphate and muriate of potash, each at 1%, brought significant improvement in the yield of guava. Wagh and Mahajan (1985) obtained the highest yield (46 kg/tree) from 5-year-old guava trees by application of 600 g N and 300 g each of P_2O_5 and K_2O along with 25 kg of FYM as basal application in pits. A combined application of N, P and K dose of 225 g, 300 g and 255 g/tree, respectively, resulted in increase in yield, TSS and total sugar content. Application of N and K improved TSS, reducing sugar, acidity and pulp: seed ratio while potassium alone plays an important role in improving sugar amount in fruits. Foliar spray with 3% N, 1% P and 1% K in combination increased fruit setting, fruit retention and yield (Sharma and Sharma, 1992). Spraying of urea at 4% in combination with single super phosphate and GA_3 (100 ppm) improved the yield and ascorbic acid content of guava (Singh and Singh, 1995).

Micronutrients: Guava sometimes suffers from deficiency of zinc. Spraying 1lb zinc sulphate and 0.7 lb slaked lime corrected the deficiency and improved growth and yield of guava (Singh, 1969). Arora and Singh (1970) observed that application of 0.4% zinc sulphate improved vitamin C, total sugar and TSS content of fruits. Singh and Chhonkar, (1983) reported that application of zinc sulphate at 0.4% and boron 0.2% solution produced maximum TSS, reducing and non-reducing sugars of guava fruit in variety Allahabad Safeda. Pre-flowering spray with 0.4% boric acid increased the yield and fruit size (Rajput and Chand, 1976). Application of boric acid also has enhanced fruit maturity.

Copper deficiency was corrected by spraying copper sulphate at 0.2–0.4% and improved the yield of guava (Arora and Singh, 1971). Singh et al. (1983) reported that spraying of 3% urea and 0.3% boric acid thrice at an interval of 15 days from 1st October improved fruit size in guava variety L 49 with improvement in quality of total TSS (14.4%) and total sugar (8.2%).

6. Pineapple

6.1. Introduction

Pineapple (*Ananas comosus*) is a major fruit crop grown both in tropical and subtropical climates. The important pineapple-growing countries in the world are Java, Sumatra, Malaya, South Africa, Hawaiian Islands, Queensland, Singapore, Ceylon and India. However, the pineapple industry is developed maximum in Hawaii.

A mild tropical climate is best suited for pineapple cultivation. Bartholomew and Malezieux (1994) have reviewed the influence of environmental factors on growth and fruit production of pineapple. Among the climatic factors, temperature is considered to be the major one in distribution of pineapple in the world.

Table 5. Temperature variations in different pineapple growing regions of the world.

Region	Latitude	Temperature (°C)			References
		Max	*Min*	*Mean*	
Johore, Malayasia	1.22° N	35.0	18.9	26.9	Py (1965)
Sope, Brazil	7° 5′ S	31.2	20.6	25.9	Giacomelli and Py (1982)
Osorio, Brazil	29° 55′ S	23.8	15.4	19.6	
Brisbane, Australia	27° 28′ S	25.5	9.5	20.5	Tkatchenko (1947)
Wahiawa, Hawaii	21° 20′ N	30.1	14.3	22.6	Bartholomew and Kadzimin (1977)
Karenko, Taiwan	23° 58′ N	27.1	17.3	22.2	Tkatchenko (1947)
Port Elizabeth, South Africa	33° 58′ S	21.2	13.3	17.2	Tkatchenko (1947)
Malkerns Swaziland	26° 30′ S	28.0	4.0	16.8	Dodson (1968)

The suitable average temperature ranges between 16.8°C and 32.2°C (Neild and Boshell, 1976). Both frost and freezing injury become limiting factors in successful cultivation of pineapple, as reported by Broadley et al. (1993). Prolong exposure to low temperature without any frost protection measures may cause adverse effect for the aerial parts of the plant. Effect of temperature also influences the growth rate, duration and fruit quality. The maturity of the crop ranges from 140–300 days in different parts of the world (Malezieux and Lacoeuilhe, 1991). Fruits harvested during winter season are relatively sour. The variation in crown weight has a relation with prevailing temperature in the periods 2 – 4 months before harvest (Paull and Reyes, 1996). Leaves and roots grow best at 32°C and 29°C, respectively and their growth practically ceases when temperature remains below 20°C and above

36°C (Bartholomew and Kadzimin, 1977). Fruit shape and size are also regulated by temperature. Fruits that grow under low temperature (monthly mean temperature less than 20°C) or low irradiation (less than 12 MJm^{-2}) or both were generally smaller than the fruits developed under higher temperature and higher irradiation (Broadly et al., 1993). Temperature also plays a role in slip and sucker production. Flowering of pineapple is not dependent either on diurnal temperature or on day length.

The leaves of pineapples conditioned to 6:6 h light: dark (l:d) cycles with 32:24°C corresponding l:d temperature, had lower content of chlorophyll and malate at the beginning of the light period and higher malate concentration at the end of the light period than the leaves of pineapples conditioned to 6:18 h (l:d cycle) with 32:24°C temperatures (Olesen and Bailey, 1995). The increase in acidity in plants under high exposure to light accounted for 85% and 95% of the total increase in malate and citrate levels in pineapple varieties Brecheche and Spanish red, respectively, while in plants exposed to low lights, the corresponding values were 44% and 75% respectively. Being a xerophytic plant, it can produce high yield in dry environment and will be recognized for its drought tolerance. An annual rainfall of 150 cm is considered optimum for pineapple.

Pineapple grows in a wide range of soil groups from alluvial to lateritic soils. Water logging during root initiation and fruit growth reduce the yield (Barthlomew and Malezieux, 1994). Heavy clay soil with high water table is not conducive. Moisture stress causes yellowing of leaves and reduces the plant height (Py et al., 1987). The optimum pH for pineapple cultivation ranges from 5.5 to 6.0.

6.2. Composition and Uses

Pineapple is rich source of sugars, minerals (calcium, iron), organic acids and fairly rich in vitamins (A, B and C). Sucrose is the predominant sugar, followed by glucose and fructose (Gortner et al., 1967). This fruit contains a protein digesting enzymes bromelain.

The chemical constituents of a ripe pineapple fruit are given below.

The composition of pineapple, however, varies with geographical location, stage of maturity and management practices. Fruits from Ivory Coast contain higher acidity, soluble protein and pectin contents of pulp and organoleptic quality, whereas fruits from Costa Rica have higher crowns and lower Brix: acidity ratio (Lopez Lago et al., 1996). The contents of glycine, alanine, methionine and leucine are relatively higher than lysine, proline, histidine and arginine (Gortner et al., 1967). A large number of volatile constituents (total 157 numbers) have been identified from green and ripe fruits of which ethyl acetate and butane

Table 6. Composition of ripe pineapple fruit*.

Constituents		Content
	(% fresh weight)	
Moisture		80.0–85.0
Total soluble solids		12.7–15.8
Sucrose		5.9–12.0
Glucose		1.0–3.2
Fructose		0.6–2.3
Cellulose		0.43–0.54
Fat		0.1
Pectin		0.06–0.16
Titrable acid (as citric acid)		0.50–1.6
Citric acid		0.32–1.22
Malic acid		0.1–0.47
Oxalic acid		0.005
Ash		0.30–0.42
Calcium		0.02
Phosphorus		0.01
Iron		0.09
Fibre		0.30–0.61
Nitrogen		0.045–0.115
Carotene (mg)		0.13–0.29
Xanthophyll (mg)		0.03
Esters (ppm)		0.2–2.5
Vitamins	(μg/100 g fresh weight)	
Aminobenzoic acid		17–22
Folic acid		2.5–4.8
Niacin		200–280
Pantothenic acid		75–163
Thiamine		69–125
Riboflavin		20–88
Vitamin B_6		10–140
Vitamin A (I.U)		60
Ascorbic acid (mg/100 g)		30.0–50.2

*Akamine (1976); Samson (1980); Sen (1990); Toshi (1992)

2-3 diol diacetate are the major ones. The 3-methylpropionate esters comprise a significant fraction of the pineapple volatile components, which have been used for pineapple flavors. Several odorous lactones have also been found in pineapple, particularly the gamma and delta–octalactones and gamma-nonalactones (Flath, 1980).

Pineapple is consumed fresh or used for manufacturing juice, jam and squash. Slices are canned in sugar solution, fruit core is used for preparing candy, and pineapple powder is prepared from juice by freeze-drying. Coagulated milk products could be manufactured by using the non-traditional pineapple enzymic complex (Cattaneo et al., 1994). The pineapple leaves yield 2-3% of strong silky fibres, 38–90 cm in length, which are used for making a fine fabric called pina cloth in Philippines and Taiwan.

The pineapples have been classified according to their characteristics into Spanish, Queen, Cayenne, Abacaxi and Maipure group (Knight, 1980). The Spanish variety is spicy–acid in taste and fibrous; queen variety is sweeter, less acid and low in fibre; Abacaxi is sweet, tender and juicy; Cayenne is sweet, mildly acid, low tender juicy; and Maipure is fibrous, tender and very juicy.

6.3. Nutrient-Deficiency Symptoms of Pineapple

Nitrogen deficiency symptoms in pineapple showed stunted growth with smaller leaves of pale green color, delayed and deformed fruiting, reduction in fruit size and inhibition of sucker production. In case of phosphorus deficiency, the plant produces darker green leaves and the fruit tastes watery (Kanapathy, 1959). Collins (1960) recorded a reduced vigor of plants and inhibition of sucker production due to phosphorus deficiency. In case of K deficiency, the leaves usually become narrow and shorter with brown spots (Cannon, 1957). Soft, thick, dark green leaves and tasteless fruits generally develop on calcium-deficient plants, while pale greenish yellow leaves and poor fruit firmness are the symptom of magnesium deficiency (Kanapathy, 1959).

Burn and Py (1952) described the zinc deficiency symptom as an initial development of small yellowish spot on the upper surface of the adult leaves, which subsequently enlarge and coalesced causing complete destruction of the leaf tissue. Dunsmor (1957) reported 'crookneck' disease, which was caused due to zinc deficiency. Aldridge (1960) explained that deficiency of copper in soil had an unfavorable influence of the absorption of zinc even when adequate reserve available zinc was present in the soil. In copper deficiency, the leaves become dark green and waxy. Chlorosis is the characteristic symptom of Fe deficiency in pineapple. Under severe Fe deficiency, the plant develops stunted growth with necrosis in young leaves and mortality of stem apex (Mohammad and Mohammad, 1983). Tay (1974) observed death of apical region of the plant due to boron deficiency.

6.4. Nutrient Management on Growth, Yield and Fruit Quality

Macronutrients: Pineapple plants respond to specific nutrient elements. A pineapple crop produce yield of 40 t/ha removes 123 kg of N, 33 kg of P and 308 kg of K. It, therefore, requires abundant supply of N and K in order to maintain a desired level of yield and quality. Pineapple respond well to nitrogenous fertilizer and flowering was found to be closely related to C:N ratio. Increasing levels of nitrogen were found to retard flower formation due to imbalance in C:N ratio (Nightingale, 1942). Toshi (1992) reported that increase in N application from 200 kg to 360 kg/ha. resulted in a sharp rise in fruit weight, yield and improved fruit quality in variety Kew. Tay (1975) established a positive quadratic response in fruit weight due to N application upto 672 kg N/ha. Application of nitrogen one month before flower differentiation @ 350–430 kg per hectare produced same yield as that of 600 kg/ha. applied throughout the growing period (Langenegger, 1976). The higher doses of nitrogen increase the production of slips and suckers, early flowering but delayed in ripening. However, higher doses of nitrogen could not influence sugar content of fruits and also reduced both TSS and acidity of fruit (Singh et al., 1977). The pineapple plants receiving the recommended dose of nitrogen produced fruits which showed better storability at 30°C and also retention of Brix: acid ratio as compared to the crop received low dose of fertilizer (Balakrishnan et al., 1980). Singh et al. (1981) recommended application of 16 gm N/plant for optimum growth, yield and quality of pineapple. They also found out that both soil application plus foliar spray of N proved to be superior to either soil application or foliar spray. Foliar spray of urea 2 or 4 times during the growing season showed significant increase in fruit size and fruit weight of variety Caynne (Py, 1963). The application of urea during fruiting is not recommended due to reduction in firmness of fruits and, thereby, poor export quality.

Phosphorus did not show much influence on yield and quality of pineapple (Selamat and Ramlah, 1992). However, in ratoon crop, application of phosphorus @ 4 gm P_2O_5/plant increased the fruit weight and yield (Reddy and Prakash, 1982). Nightingale (1942) considered that sufficient quantity of phosphorous is needed at the time of differentiation of inflorescence and at flowering to avoid loss as in the yield. Su (1965) found acceleration in flowering and fruit maturity due to application of phosphorus at recommended dose @ 2-4 gm/plant. Khatua et al. (1980) identified rock phosphate as better phosphatic fertilizer than bone meal and super phosphate. The acidity of pineapple reduced considerably due to application of rock phosphate.

Pineapple plant requires more of K fertilizer than N or P. Sideris and Young (1945) applied 200, 400, 800, 1600 and 3200 lb K_2O/acre and observed that the weight of plants progressively increased with increase in doses of potassium. In an experiment on the effect of potassium on fruit quality, Su (1958) noticed that the sugar contents of fruits did not increase significantly by increasing the levels of potash but the acidity of fruits did increase. The sugar:acid ratio decreased with high dose of potassium. Selamat and Ramlah (1992) reported that increase in doses of K significantly increased plant height, fruit diameter, core diameter and fruit acidity but decreased fruit weight and sugar content. Smuels and Gandia-Diaz (1960) observed that potassium sulphate was superior to potassium chloride in respect of yield and quality. Potassium sulphate enhanced fruit maturity. Dalldorf (1975) noted a vigorous growth of plant by application of 200–400 kg K_2O/ha. in the form of potassium sulphate, potassium nitrate and potassium chloride.

Pineapple plants respond well to both organic and inorganic sources of fertilizer. Godfrey–Sam–Aggrey (1977) studied the response of N, P, and K on pineapple and found that 57 g/plant of N, P, K mixture was a profitable dose as compared to control. Roy et al. (1986) obtained the highest yield of pineapple with N, P and K applied @ 600, 400 and 600 kg/ha., respectively. In a trial in Mexico, the pineapple applied with 12-5-24 g N, P, K/plant produced high yield (Orona Moreno and Ovando Cruz, 1994). Neog et al. (1995) identified the commercial dose of N, P, and K as 7.2:1.2:7.2 g/plant for achieving best growth and maximum fruit size and yield. Fruit ripening was accelerated by the application of N and K.

Table 7. Recommended fertilizer dose of pineapple in different countries.

	N	P_2O_5	K_2O	*References*
Australia	10	2	10 g/plant	Canon (1960)
Brazil	120	60	210 kg/ha.	Vas Concelos (1952)
Sri Lanka	10	6	10 g/plant	Richards (1952)
Guinea	15–20*	4.5**	8.3–16.6*** g/plant	Philippe (1960)
Hawaii	440-670	220–460	170–280 kg/ha.	Collins (1968)
India	600	400	600 kg/ha.	Roy et al. (1986)
Malaysia	896	224	896 kg/ha.	Tay (1972)
South Africa	10	6	10g/plant	Le Roux (1951)
Taiwan	700*	300**	200**** kg/ha.	Chen et al. (1957)

*Ammonium sulphate, **Single superphosphate ***Potassium sulphate, ****Potassium chloride.

On the basis of yield and quality of pineapple fruit, Martin Prevel (1961) determined the optimum proportion of three nutrients, i.e. K: Mg:Ca at 42.5:42.5:15.0. Su (1965) reported that the critical value of magnesium was 0.22% in leaf tissue and 60 ppm of exchangeable magnesium in the soil.

Micronutrients: Application of zinc at 2 ppm was beneficial to overcome chlorosis and improved fruit quality (Srivastava, 1969). High level of micronutrient application significantly increased fruit yield (Verawudh, 1992). Two foliar sprays of micronutrients (Cu, Zn, Mn, B, Na and Fe) applied at the flower bud differentiation and pre-flowering stages caused significant increase in yield of pineapple variety Kew.

7. Litchi

7.1. Introduction

The litchi (*Litchi chinensis*) is a delicious, juicy table fruit, distributed in tropical and subtropical countries of China, Taiwan, Vietnam, Thailand, India, Madagascar, South Africa and Reunion Islands. It is also grown in limited areas in Australia, New Zealand, Indonesia, Israel, Spain, Mexico and the USA (Menzel and Simpson, 1994).

Climate is most important limiting factor in the area coverage of this crop. It requires a moist subtropical climate without heavy frost or hot dry winds. The four essential climatic factors for successful cultivation of this crop are frost-free weather, high humidity, cool and dry winter and abundant moisture under high rainfall. In areas where litchi grows successfully, the summer weathers are moist with high humidity and winters are cool but non-freezing. Poor productions are experienced in many places because the winter season is not cool and dry enough to induce a growth check prior to flowering. The ideal climatic conditions for litchi in relation to annual growth cycle in Australia are as follows (Menzel et al., 1986a).

Growth stage	Climate
Flushing	28–30°C, high RH, heavy rainfall
Dormancy	Minimum 15°C, 50 mm rainfall/month
Flowering	16–20°C, light rainfall
Fruit set	18–24°C, moderate RH
Harvest	25–28°C, even rainfall, high sunlight, high RH

Pollination is optimum between 19–22°C but fruit setting can be satisfactory at lower temperature of 15°C or higher temperature of 27°C (Menzel and Simpson, 1994). Low humidity may be detrimental to fruit setting of litchi (Batten, 1986). Extended overcast weather after fruit setting (10% of full sun) has been reported to cause poor harvest (Yuan and Huang, 1988). Zhuang et al. (1983) conducted long duration trial and confirmed that early flowering was not conducive for satisfactory fruit setting because of low temperature in association with overcast weather.

Litchi can be grown in a variety of soils but develops best in deep, well-drained loamy soil rich in organic matter in the cooler sub-tropical areas. However, it can also be grown well in clay soil in warmer areas. In China, the best litchi plants are found in alluvial sandy soils close to the rivers with good drainage (Chapman, 1984). Litchi grows well in slightly acidic soil with pH range from 5.5 to 6.5. The plant cannot sustain under excess soil moisture stress.

7.2. Composition and Uses

The major composition of litchi fruit is sugar and acid, which varies due to climate and the types of cultivars grown. The sugar content ranges from 6.74 to 18.86% in India (Singh and Singh, 1964), 12–15% in Florida and 11.8–20.6% in Hawaii (Miller and Bazore, 1945). Besides sugar, litchi contains 1.1% protein and considerable amount of Ca and P and vitamin C, B1 and B2.

Table 8. Composition of litchi fruits (per 100 gm fresh weight)*

Composition	Contents
Moisture	80.6%
Carbohydrate	17.5 gm
Protein	1.1 gm
Fat	0.1 gm
Fibre	0.1 gm**
Calcium	7 mg**
Phosphorus	41.0 mg**
Iron	1.3 mg**
Thiamine	0.05 mg
Riboflavin	0.07 mg
Niacin	0.5 mg
Ascorbic acid	49 mg

*Wills et al. (1986); **Menzel and Simpson (1993)

However, some cultivars contain a degree of niacin (Galan Sauco, 1989). The acidity of litchi fruit varies from 0.2-0.64% (Mathew and Pushpa, 1964), decreasing during the process of ripening and storage. The predominant non-volatile acid in litchi is malic acids. However, the ascorbic acid content varies widely from 40.2 mg/100 g in cultivar Kwai-Mi to 80.8 mg/100 g in Brewster (Wenkam and Miller, 1965), 90 mg/100g in unspecified variety (Thompson, 1955), 44 mg/100 g in Calcutta late (Chadha and Rajpoot, 1969).

The litchi is popular as excellent canned fruit in China, Taiwan and Thailand. About 20% of the total Chinese production of this fruit is canned (Menzel, 1991). The preservation of fruit in the form of syrup is also a common practice. In China, litchi is preserved in honey. A highly flavored squash is also prepared from the litchi fruit. Other products like ice cream, wine are also made from litchi in China. Dried litchi, commonly known as litchi nut, is also very popular in China.

Table 9. The Physico-chemical composition of some litchi cultivars.

| Cultivar | Fruit weight (gm) | Percentage of | | | Total soluble soild (%) | Acidity (%) | References |
		Peel	Seed	Aril			
Shahi	21.0	15.6	16.7	67.7	20.8	0.35	Sharma and
Rose Scented	19.2	15.0	18.3	66.7	19.6	0.48	Ray (1987)
Purbi	18.5	16.9	22.9	60.2	19.0	0.44	
China	19.8	24.9	20.8	54.3	20.5	0.57	
Bedana	14.9	18.5	13.4	68.1	19.0	0.59	
Bengal	20.8	26.1	17.2	56.7	17.3	-	Menzel et al.
Gee Kee	14.6	21.1	6.3	72.6	18.9	-	(1986b)
Tai So	18.7	19.0	15.5	65.5	17.0	-	
Wai Chee	17.1	23.7	8.1	68.2	18.6	-	
Bombai	19.0	17.9	19.9	62.2	17.7	0.42	Ghosh et al.
Deshi	16.6	13.3	21.1	65.6	17.1	0.86	(1987)
Elaichi	15.7	13.8	12.6	73.6	17.5	0.45	
Kasba	15.9	17.6	19.5	62.9	16.8	1.14	
Mclean	15.2	19.0	21.8	59.2	16.8	1.24	
Muzaffarpur	18.2	13.7	16.5	69.8	17.7	0.79	

7.3. Nutrient-Deficiency Symptoms of Litchi

The deficiency of different mineral elements showed the following symptom in litchi as described by Menzel and Simpson (1986).

Nitrogen: Yellowing of old leaves stunted growth, and resulted in poor flowering and small fruit.

Phosphorus: Tip and marginal necrosis of older leaves, leaf curling, desiccation and leaf fall.

Potassium: Leaf yellowing, necrotic leaf tips and margins, leaf fall, poor fruit set, stunted growth.

Calcium: Drying of growing points.

Magnesium: Small leaves, leaf necrosis, leaf drop, poor flowering.

Zinc: Bronzing of leaflets, small leaflets, and small fruits.

Iron: Leaf yellowing, dieback.

Copper: Dieback, small fruit, reduced pulp recovery.

Boron: Small fruit.

7.4. Nutrient Management on Growth, Yield and Fruit Quality

Macronutrients: Fertilizer management of litchi is essential for achieving a satisfactory growth and quality. Nitrogen is considered to be the most important nutrient besides other nutrients like phosphorus, potassium, calcium and magnesium. The fertilizer dose in litchi plant depends on cultivars, tree size, vigor and soil type. In West Bengal, India, six-year-old litchi trees fertilized with 600 g each of N and K_2O and 200 gm P_2O_5/year were found optimum for better growth and yield of crop. According to crop age, fertilizer doses for high yielding varieties of litchi in Australia as suggested by Menzel and Simpson (1986), are given as follows:

Table 10. N, P and K doses for litchi trees*

Tree age (Years)	Canopy Diameter (m)	N (g/tree)	P_2O_5 (g/tree)	K_2O (g/tree)
4 –5	1.0 –1.5	200	80	300
6 –7	2.0 –2.5	300	100	450
8 –9	3.0 –3.5	400	130	550
10 –11	4.0 -4.5	500	170	700
12 –13	5.0 –5.0	600	200	800
14 –15	6.0 –6.5	800	250	1200
15+	> 6.5	1000	300	1400

* Menzel and Simpson (1986).

Annual application of P and K may not be required in soil with a large amount of exchangeable P and K (Menzel *et al.*, 1995). However, Lal *et al.* (1996) reported that application of P @ 600 gm K/tree resulted increase in fruit weight and improved pulp quality of fruit. Yamdagani *et al.* (1980) in an experiment with a 12-year-old litchi crop observed a marked increase in fruit set, fruit retention and length, diameter and weight of fruits by increasing the levels of nitrogen from 0.25 – 1.0 kg/tree/year. In subtropical Queensland (Paxton and Chapman, 1980), it was recommended that the litchi trees should be fertilized during spring as it promotes growth, flushes and prevails premature flower and fruit abscission. Koen (1977), in an experiment with three levels each of N, P and K, observed that high level of N application markedly increased the yield and sugar content of fruit, while P has shown no positive response and potash showed detrimental effect at a higher dose.

Micronutrient: The important micronutrients like Fe, B, Cu, Zn and Mn are also required for best performance of the litchi crop. Foliar spray of zinc sulphate at the concentration of 0.5, 1.0 or 1.5% proved to be effective in reducing fruit drop, increase fruit and pulp weight and TSS and decrease in fruit acidity (Awasthi *et al.*, 1975). In another experiment, Kumar *et al.* (1995) observed that foliar application with 1% zinc sulphate twice immediately after fruit setting and after a fortnight from first application produced best quality fruit in terms of TSS content, sugar : acid ratio and ascorbic acid content. Application of boron at 0.4% reduced fruit cracking, while at higher concentration (0.8%), the fruit ripening was advanced by 22 – 27 days (Mishra and Khan, 1981).

8. Papaya

8.1. Introduction

Papaya (*Carica papaya* L.) is widely distributed throughout the tropical and subtropical regions of the world. The papaya cultivation is confined to 32°N and S latitude of the equator on the globe (Ram, 1996). The major papaya-growing countries are Australia, Hawai, Taiwan, Puerto Rico, Peru and the USA (Florida, Texas and California), South Africa, Pakistan, Bangladesh and India. The optimum temperature for papaya cultivation is reported to be 21–33°C (Knight, 1980). High temperature tended to produce more male flowers in papaya (Malo and Campbell, 1986). The net CO_2 assimilation in papaya leaves lies between 16 and 30°C and declines linearly when the temperature rose beyond 30°C (Allan and de Jager, 1978). The freezing damage occurs at temperature –6°C (Maxwell et al., 1984). The incidence of malformation of fruits was high from flowers during low winter temperature (less than 16°C) (Sippel, 1988). A relative humidity of 60% or more was reported to be essential for optimum plant growth (Anonymous, 1986). The flavor of the papaya fruit is correlated with the temperature at which the fruit matures. Low temperature reduces the flavor of fruit and makes it insipid. Papaya is best grown in loamy soil with good drainage and rich in fertility. The plant thrives well in soil with pH ranging from 5.5 –7 (Maxwell et al., 1984). The crop is highly sensitive to water logging and salt stress (Raheja, 1966). Salinity delays the percentage of the emergence of papaya seedlings.

8.2. Composition and Uses

Papaya is a rich source of minerals and also vitamins A, B, B1, B12, C and D (Kapanadze and Khasaya, 1988). The vitamins are generally associated with carotene. The yellow pigment in papaya is caricaxanthin. In the orange-fleshed cultivar, β–cryptoxanthin (8.1 µgm/ g) is the major pigment representing 62% of the total carotenoid content. However, in red-fleshed cultivar, lycopene is the major pigment ranging from 56–66%. In the warmer climate of Saou Paulo, Brazil, β-carotene, β-cryptoxanthin and lycopene contents are higher in papaya. Lowest vitamin A values were found in Formosa from Saou Paulo and highest value from Bahia. Geographic effect has a greater influence in quality of papaya particularly on vitamin A than that of cultivars (Kimura et al., 1991).

The important volatile components in ripen fruits have been identified as linalool, trans-linalool oxide and cis-linalool oxide (Franco and Rodriguez- Amaya, 1993). Several enzymes have been isolated and purified from papaya fruits, viz., invertase, UDP-glucose phenol- B-D–glucosyltransferase (Keil and Schreier, 1989), proteinases (Redina et al., 1993) and pectinesterase (Fayyaz et al., 1994). Leaves contain flavonoids, tannins, organic acids and alkaloids, including the valuable carparine (Kapanadze and Khasaya, 1988). The air-dried papaya seed contains moisture (4.0%), oil (25.6%), ash (17.8%) and protein (26.6%) (Raie et al., 1992). Fatty acid and triacylglycerol composition of papaya seed oil is similar to that of olive oil (Nguyen and Tarandjiska, 1995).

Fresh papaya fruits are used for consumption and also for preparation of soft drinks, ice cream flavoring, crystallized fruit, jam and syrup. The seeds are used for their medicinal value. Papain, derived from dried latex of unripened fruit, is a proteolytic enzyme, which is similar in action to pepsin. Papain has various uses as meat tenderizing preparation, manufacturing chewing gum and cosmetics, drug for digestive ailments, for bating-hides in tannin industry, for degumming natrural silk and to give shrink resistance to wool (Purseglove, 1968). The dried and powdered latex had a proteolytic activity, which is marginally higher than that of fresh latex (Narinesingh and Mohammed–Maraj, 1988).

8.3. Nutrient Deficiency Symptoms of Papaya

Deficiency of N and P nutrient of the papaya plant showed a stunted growth of plant, while for K and Ca, the deficiency symptom was reflected in foliage. Deficiency of K, Mg, Ca and B produced necrotic lesions on the leaves. Boron deficiency resulted restricted growth of roots and shoots. Wang et al., (1975) reported a secretion of the latex and formation of tumors on the fruit due to boron deficiency.

8.4. Nutrient Management on Growth, Yield and Fruit Quality

Macronutrients: In view of quick growth, continuous fruiting habit and higher production of fruits, the nutrient requirement of papaya differs from other crops. The uptake of major nutrients, N, P, K, Ca and Mg was more between flowering and harvesting stages. The removal of major and minor nutrients per tonne of papaya fruits were 1170 g N,

200 g P, 2120 g K, 350 g Ca, 180 g Mg 200 g S, 989 mg B, 300 mg Cu, 3364 mg Fe, 1847 mg Mn, 8 mg Mo and 1385 mg Zn (Cunha and Haag, 1981).

Papaya responds well to both organic and inorganic sources of fertilizers. Supply of 50% of nutrient through organic source and 50% through inorganic source have shown better crop performance. Hoosain et al. (1990) reported that nitrogen applied @ 100 g N/ pit in the form of decomposed cowdung produced highest number of fruits/plant, while better quality of papaya fruits were obtained with application of decomposed cowdung and mustard cake to supply same amount of N/ pit in addition to 100 g triple super phosphate and 75 g Muriate of Potash. Ochse et al. (1961) observed that the productivity of papaya plant was maintained by a continuous supply of fertilizers and organic materials at a regular interval. In Southern Florida, an application of 0.1 kg of 4: 8: 5 fertilizer mixture of N: P: K to each plant at 15 days interval for the first 6 months and thereafter, 0.2 kg of a similar mixture resulted in significant increase in yield. In another experiment, application of 250 g N/plant/year in six split doses at bimonthly intervals was optimum for better fruit yield and papain content. Satyanarayana Rao (1971) noted that application of phosphorus reduced the acidity, while potassium increase the TSS, total sugars and ascorbic acid content of the fruit. Sulladmath et al. (1981) obtained highest yield of 79.9 fruits/plant weighing 25.76 kg (64.4 tonne/ha.) due to application of N and P each @ 250 g/ plant/ year and reduced the seed weight. They also found that application of K increased TSS and skin thickness, while application of higher doses of N, P and K reduced the seed weight. Werner (1993) reported that the yield of papaya variety Tainung-2 increased linearly with increase in N doses from 50 to 150 kg/ha, but application interval did not affect the yield. Petiole N content was increased with N application but fruit quality was not affected. Soil application of 200 g N, 100g P_2O_5/plant in 4 split doses at 90 days' interval along with foliar spray of 1% urea, 0.2% boron and IAA at 50 ppm produced the best yield and quality of fruit (Lokhande and Moghe, 1991). Increase in K application rate from 168g to 504g/plant increased the total fruit weight/plant and extended the shelf-life of the fruit but the proteolytic activity of crude papain tended to decrease. In general, $K_2Mg(SO_4)_2$ form of potassium showed better results than KCl in terms of both fruit and latex yield (Harjadi et al., 1995). Best response with respect to number of fruits and yield/plant and yield/ha was obtained with highest rates of P (300 g P_2O_5/plant/ year) and K (600 g K_2O/plant/year) and the medium rate (200 g N/ plant/year). Fruit TSS, sugar, acid, ascorbic acid and total carotene contents increased with optimum dose of K and P but these parameters were adversely affected with the highest N dose (Ghanta et al., 1995).

Micronutrients : The micronutrients also influence the growth, yield and fruit quality of papaya. Fruit quality was best with application of boron through 0.15% borax (Pant and Lavania, 1989). They also reported that spraying papaya with 0.15% $FeSO_4$ and 0.15% $ZnSO_4$ produced highest percentage of female plants (56.3%). Ghanta et al. (1992) observed that spraying of 0.1% B, 0.25% Mn and 0.25% Cu alone or in combinations at 60 or 90 days after transplanting improved the quality of fruit in papaya variety Ranchi. The activities of peroxidase, catalase and polyphenol oxidase increased with increase in Fe concentration in the nutrient solution (Lin et al., 1994).

9. Grapes

9.1. Introduction

The grape *Vitis vinifera* is an important subtropical fruit grown as a cultivated plant in Southern Europe, border countries of Eastern Mediterranean and North Africa. Recently, it has been introduced in tropical countries, e.g. the vinifera vines were introduced into Philippines from California. In India, the grapes might have been introduced in 1300 AD. The grape is one of the most delicious, refreshing and nourishing fruit usually consumed as a table fruit. The production of grape accounts for nearly half of the total world production of world fruits. Grape requires a warm dry climate and a cool rainy winter. In such climate, the grape vine shed leaves and become dormant during winter and the new leaves and fruit flushes emerge with the advent of spring. The fruit matures in summer during the rainless period. The grape does not grow well in regions of humid summer, whether temperate or tropical, as moisture in the air causes fungal disease on the leaves and fruits. Bright sunny days help in the development of sugars in the berries. Midday light rich in UV radiation had a beneficial effect on pigment synthesis, inhibited leaf and shoot growth, hastened tissue differentiation, lignifications, shoot maturation, ripening and sugar content of berries (Zhakole et al., 1979). Both light quality and quantity affect the composition of sugar, acidity, pH, phenol and anthocyanins in grape fruit (Smart, 1986). However, under high temperature, the skin of the berry becomes thicker. Temperature is considered to be a major factor influencing the composition and quality of grapes (Coombe, 1986). The total amount of heat (degree days) determines the ripening time. Each cultivar requires a specific heat unit for ripening. The early and late cultivars require about 1600 to 2000 and 3000 or more degree days for ripening, respectively (Jacob, 1950). The root stock and scion cultivars excelled in drought tolerance, while wine grape cultivars showed cold tolerance to –21°C. Both type of stresses affected the quantity and quality negatively. Moderately cool weather, under which ripening proceeds slowly, is favorable for the production of dry table wine of quality. Cool weather fosters a high degree of acidity, good color and for table wine varieties, the fruit attains optimum aroma and flavor. In warm climates, the aromatic qualities of the grape lose their delicacy and richness and other constituents of the fruit, hence the resultant table wines, even from the best grape varieties, cannot be compared with the best wines of cooler region. The ripening time of a particular variety mainly depends upon the availability of heat units along with certain growth regulators like abscissic acid, ethephon resulted in early accumulation of sugar and increased TSS (Kim et al., 1998).

The grape is adapted to a wide range of soil types. However, the grape vine does better on light friable soil with well drainage. This crop is relatively tolerant to salinity and alkalinity. Joolka (1972) reported that some grape cultivars can be grown upto 7 mmhos/cm EC.

9.2. Composition and Uses

The minerals substances in grape constitute 0.2 – 0.6% of the fresh fruit weight. Major constituents are carbonates, oxides, phosphate or sulphate of potassium, sodium, calcium or iron. Besides other trace elements such as bromide, iodine and fluoride are also found to be present in grapes. The levels of all the mineral constituents in grape depend on soil type, climatic factors and nutrient management practices. During maturation of grape, the cation content (K, Ca, Mg and Na) increases two to three times in skin, 1.2-1.9 times in pulp and 1.5–2.5 times in the peduncle (Peynaud and Ribereau–Gayon, 1970). The anions, particularly phosphates also increase, which are mostly concentrated in seeds. Puissant (1960) reported that potassium content as mg/l changes little during ripening, while calcium increases by 50% and considerable increase of magnesium content during ripening. Potassium accounted for 50–70 % of the cations in grape juice.

Grape contains vitamin A and vitamin B complex like thiamin, riboflavin, pyrodoxin, pantothenic acid, nicotinic acid, inositol, biotin and folic acid. In fresh grapes, the B complex vitamins are reasonably stable during cold storage. Grapes are not a rich source of ascorbic acid. Ournac and Poux (1966) found an increase in ascorbic acid for a few weeks after fruit setting, followed by a decrease until the fruit started to ripen. Then, another increase during ripening was followed by decline during maturity. The skin of the berries contains greater concentration of ascorbic acid than juice but the total quantity is more in juice than in skin. Most of the ascorbic acid in grapes is lost during wine clarification, filtration and drying processes for raisin.

Table 11. Composition of grapes (ranges in percentage of more organic and inorganic components of freshly extracted juice by volume)*

	Percent		*Percent*
Water	70 –80	Residual	0.01 –0.02
Carbohydrates	15 – 25	**Mineral compounds**	0.03 – 0.6
Dextrose (glucose)	8 – 13	Aluminium	0.003 (T)
Levulose (fructose)	7 – 12	Boron	0.007 (T)
Pentoses	0.01 – 0.10	Calcium	0.004 – 0.025
Pectin	0.01 – 0.10	Chloride	0.001 – 0.010
Inositol	0.02 – 0.08	Copper	0.0003 (T)
Organic acids	0.3 – 1.5	Iron	0.003 (T)
Tartaric	0.2 – 1.0	Magnesium	0.01 – 0.025
Malic	0.1 – 0.8	Manganese	0.0051 (T)
Citric	0.01 – 0.10	Potassium	0.15 – 0.25
Nitrogenous compounds	0.03 – 0.17	Phosphate	0.02 – 0.05
Protein	0.001 – 0.01	Rubidium	0.001 (T)
Amino acid	0.017 – 0.11	Silicic acid	0.0002 – 0.005
Humin	0.001 – 0.002	Sodium	0.020 (T)
Amide	0.001 – 0.004	Sulphate	0.003 – 0.035
Ammonia	0.001 – 0.012	**Tannins**	0.01 – 0.10

* Amerine *et al.* (1972), T: indicates traces

In grape, the pectin present in three forms, i.e. protopectin, pectin and pectic acid. Protopectin is found abundantly in the primary cell wall. Middle lamella of the cell wall consists of pectic compounds like calcium and magnesium pectate. During ripening, the protopectin is transformed into pectin and the berries get softened as a result of the removal of middle lamella pectate. The effect of the changes in pectates on the texture has an important influence on keeping the qualities of table grapes. Some varieties remain firm, whereas others get very soft. Vinifera grapes contain more amount of pectin as compared to American grapes.

The green colour of unripe grapes fades during ripening. White grapes lose the chlorophyll gradually during ripening and become a translucent straw color. Generally, the pigment in grape is found in the outer three to four layers of cells in the skin. There are a few varieties of grapes like Alicante Bouchet whose juice is colored in both skin and the pulp. There are five anthocyanins that make up the basic part of grape pigment. The yellow pigment of both white and red grape is quercetin, a flavonol and its glucosides. The pigments in the berries increase with ripening and continue until physiological maturity (Fernandez – Lopez et al., 1992). The level of pigments in grapes is governed by the environmental factors such as light, temperature, soil moisture and nutrition.

During ripening, the grapes develop various substances that give rise to aroma. The kind and intensity of these aromatic substances differ according to the varieties. The precursor materials for these substances are produced in leaves and the aroma is synthesized in the berries, which accumulate during the last stage of maturity.

Flavor is the complex reaction of taste. Although many chemical substances contribute to the flavor of grapes, the primary taste of sugar and acids dominate. Tannin also influences the astringency of the fruits and, thereby, its flavor. Flavor is determined by an intermixture of the amount of sugar, acids and tannins present in it. The satisfaction one gets from consumption of grapes is due to constituents of the aroma and flavor.

In grape, most of the enzymes are found in the skin and to a limited extent in the pulp. A direct relationship is found between internal browning of berries and the presence of high levels of poly phenol oxidase activity and low levels of dihydroxy phenolic substrate. Golodriga and Pu Chao (1963) determined the amount of peroxidase, polyphenol oxidase and catalase in early and late ripening varieties. Besides, the presence of other enzymes are catechol oxidase, invertase, hexokinase, glucose–6-phosphate dehydrogenase, sucrose synthetase and enzymes increased primarily with synthesis and breakdown of malic acid have also been reported.

Grape contains a large proportion of sugars and useful minerals. Some grape varieties called raisins are used as dried fruit and others called table grapes used as fresh fruit and juice and wine grapes used for manufacturing juices and wines. Grape juice is a nourishing thirst quencher, a stimulant to the kidney as also a laxative.

The principal product of grape is wine confined in European countries like France, Italy and Spain. Other uses of grapes are in the form of raisin, juice and canning, besides it being used as a table fruit. Based on the use the grapes are divided into following classes: -

Table grapes – The grapes must be attractive in appearance and good in taste and should have good transport and storage quality. Varieties with large uniform size berries, firm pulp, and tough skin are most suited for this purpose. The cluster should be well filled and arranged.

Raisin grapes – In this class, the seedless table grapes with high sugar content are chosen for raisin. The seeded cultivar should have few, soft tasteless seeds. The standard suitable raisins must be soft in texture and meaty to eat and should not stick together. Light yellow-coloured raisins are preferred. The fruits of raisin cultivars should get dry rapidly and have good flavor. Some of the promising varieties are

Thomson seedless, Fiesta and Black Corinth, Sultana of California, Dattier in Greece, Monukka in Russia and Cape Current in Australia and South Africa.

Juice grapes – In this class, the strong muscat-flavored cultivars, which retain their flavor under various steps of processing like pasteurization, germ-proof filtration, are preferred for juice purpose. The promising juice-making cultivars are Beauty Seedless, Early Muscat, Bangalore Blue, Champion and Black Champa. In juice processing technology, the clarifying and preserving procedure should not destroy the natural color and flavor.

Wine grapes – The grape fruit produced vineyards in Europe, South Africa, Australia and the USA is used for manufacturing wine. The wine grapes of high acidity and low sugar content are suitable for dry or table wines, while sweet or dessert wines are prepared from grapes high in sugar and moderately low in acid content. The wine grapes should retain flavor and color in the wines. The famous red wine cultivars are Cabernet, Sauvignon, Barbera, Gamay, Grenache, Mataro, Merlot, Pinot Noir, Ruby Cabernet and Saugiveto. For white wines, the promising varities are Aligot, Chardonnay, Chenin Blanc, French Colombard, Muscat Blanc, Gray Riesling, Semillon and Saint Emilion.

Canning grapes – only seedless grapes like Thompson seedless are canned. Seedless white grapes with large berries are preferred for canning in 20 to 24° Brix sugar syrup.

9.3. Nutrien-Deficiency Symptoms of Grapes

The major nutrient elements required in a higher quantity for grape vine are N, P, K, Ca, Mg, S and the micronutrients like B, Fe, Zn, Mg, Mo, Cu and Cl. Growth and fruitings of a vine becomes adversely affected if any of these elements becomes deficient or is in short supply. The plant subjected to nutrient deficiency shows symptoms on leaves, twigs or flowers. Nitrogen deficiency is most common and frequent. Decrease in green color, diffused chlorosis and stunted growth are common symptoms of nitrogen deficiency (Cook and Kishaba, 1956). A correlation exists between nitrogen deficiency and fruit yield, as reported by Beattie and Forshey (1954). Nitrogen content in petiole is considered important for the determination of N requirements.

The phosphorus deficiency symptoms appear first on older leaves and the margins of the leaves into golden yellow to light brown in color. With P deficiency, the plant shows reduced growth, premature defoliation and fruit ripening. A complex disorder known as acid injury was reported to occur in acid soils in Germany. The symptoms appear first on older leaves and the margin of the leaves turn golden yellow to light brown. The chlorotic areas of the leaves become necrotic and in extreme cases, the whole leaf blade dries up (Gartel 1965). Larsen (1959) reported average P content to be 0.24% for the best performance of vine.

In both *vinifera* and *Labrusca* varieties of grapes, the most common nutrient disorder is K deficiency. Potassium deficiency in grape vines have been reported from New York, (Shaulis and Kimball, 1956), Germany (Gartel, 1958), Switzerland (Peyer, 1955), California (Cook and Kishaba, 1956a) and France (Levy, 1965). The visual K deficiency symptom starts from leaf margin and the leaf color becomes pale or yellow green with white grapes or into reddish bronze with black cultivars. Fregoni and Scienza (1976) observed redness in leaves in Barbera and Bonarda cultivar and leaf browning in Cabernet Souvignon cultivars due to K deficiency. The symptom known as 'black leaf' in the USA is the second syndrome of K deficiencies, which are different from the leaf scorch or leaf age chlorosis. Woodbridge and Clore (1965) opined that black leaf occurs generally in heavy crop vines. Potassium deficiency was observed when the petiole K level was less than 1.5 % (Beattie, 1955). A significant correlation exists between K content in tissue and K deficiency.

Magnesium deficiency commonly occurs in vines grown in acid soil. In young leaves, the Mg deficiency shows development of necrosis, while mature leaves show chlorosis. Vanek et al. (1967) recorded interveinal chlorosis and necrotic spots on the green leaf tissue. The leaves drop prematurely, thus reducing the assimilating potential and quality of fruit (Bucher, 1978). Rotundifolia vines with 0.1% Mg content at blooming time showed severe Mg deficiency symptom at harvest. Lott (1952) observed appreciable chlorosis when Mg level in leaves was 0.15% at harvest. In Japan, Mg deficiency was recorded in acid sandy soils and deficient leaves showed Mg content between 0.05 – 0.25% as compared to normal leaves, i.e. 0.27– 0.50% (Sugiyama et al., 1952).

The symptom of boron deficiency are poor growth, weak shoots with short internode, double buds, death of growing points, increased growth of side shoots and failure of fruit development (Christensen, 1986). The vines sometimes bear seedless fruit (Scott, 1944).

The presence of berries of different sizes in the same bunch is a characteristic symptom of boron deficiency (Jardine, 1946) and in severe cases, there may be complete failure in fruit setting (Satish Kumar and Sharma, 1974). Gartel (1956) stated that leaves containing less than 10 ppm B showed B deficiency. A safe level of B content in leaf blades or petiole should be above 30 ppm during bloom time. Verona (1959) noted the deficiency threshold value of B to be 12 – 15 ppm and the normal value as 20 –25 ppm in leaf dry matter.

The Zn deficiency symptoms appear on the apical leaves of the main shoot and leaves on the lateral shoot remain small, which is referred to as 'little leaf'. Low level of Zn also causes poor berry set, resulting in formation of loose bunches. Cook and Kishaba (1956a) found Zn deficiency symptom in grape vine when Zn content ranges between 5 –51 ppm.

In case of Fe deficiency, the loss of green color and yellow or creamy white color in leaves are iron deficiency symptom found in grape vine (Satish Kumar and Sharma, 1974). The deficiency only occurs in alkaline soil in relatively high lime content. In severe cases, growth is stunted and fruit setting is poor (Cook and Kishaba, 1956).

The Cu deficiency symptom shows poor root development, roughness in bark, short internodes and reduced yield and poor fruit quality. The copper content in deficient vines varied from 2–4 ppm in December and 1-2 ppm in April as compared to 7.5–10 ppm, and 2–4 ppm, respectively, in normal plants (Teakle et al., 1943).

9.4. Nutrient Management on Growth, Yield and Fruit Quality

As grapes are grown in wide climatic condition in the world, the fertilizer requirement may vary greatly. The grape vine removes appreciable quantities of nutrients from soil. An average crop of grape removes 40 – 60 kg N, 10 – 15 kg P and 50 – 70 kg K/ha, (Nijjar, 1970). Fertilizer doses of grape vine depend on the climate of the region, fertility status of soil, cultivar, and the age of the vine.

Macronutrients: Application of nitrogen was found to improve the vegetative growth, berry weight and yield/plant and reduce the alcohol content of the must (Agulhon et al., 1964). Sparks and Larsen (1966) noted better growth and foliage density and low soluble solid content in fruits with increase in level of nitrogen. Bell et al. (1979) reported that the concentration of esters and wine quality was dependent on N fertilization. Application of N @ 112 kg/ha. was found to be sufficient for the juice to ferment rapidly and produce wine of good quality. El- Shourbagy and Ismail (1961) reported that application of N @ 120 gm/vine/year significantly improved the number of clusters and crop weight/vine. Further increase in N dose @ 150 g N/ vine increased the number of fruit buds, which developed into bearing shoots. However, Ough and Bell (1980) observed that application of N higher than 112 kg/ha. reduced the production of isobutyl, active amine and isoamyl alcohol production in the wine but had no effect in changes in composition of grapines. Juice derived from vines received no fertilization, fermented slowly and resulted in wines of poor quality. Fertilization experiment at Laimburg, Italy using five different N sources showed the foliar spray of urea to be best for meeting acute N deficiency (Raifer, 1999). He suggested that white wine grapes should always have sufficient nitrogen and over supply should be avoided in red wine grape since it raises the pH and consequently decreases color and tannin content.

Application of phosphorus or soil application through foliar spray increased the fruit size and weight, percentage of juice and acid/sugar ratio but decreased the percentage of fruit rind and citric acid content. Dragas (1957) observed a high yield and content of sugars and phosphoric acid by using granulated super phosphate. Hassan (1968) reported that yield, ripening and quality of grapes were favorably influenced with application of 3 lb potash/vine. Survival of winter buds in vine was higher by 35% than control when K was applied at a dose of 90 kg/ha. Split application of 2 kg of KCl/vine produced berries with least acidity and highest sugar: acid ratio. Seven sprayings of K_2SO_4 at 1% concentration at an interval of 7 days from berry set to fruit harvest significantly improved the average berry weight, juice percentage, reducing sugar, TSS and berry firmness (Singh et al., 1979). High level of K was found to reduce the amount of green fruit at harvest but had very little effect on soluble solid content of fruit (Morris et al., 1980).

For successful cultivation of grape vine, a balanced fertilization with N, P, K fertilizer mixture is required as suggested by Bulgarelli, (1962). For attaining high vine yield and good grape quality, Biblina (1968) recommended N, P, K ratio of 1: 1: 1 or 2: 1: 1 to be most effective. The peak nutrient requirement was noted during flowering stage, while K requirement during ripening stage. Georgieva (1976) revealed that application of 500 kg N, 300 kg P_2O_5 and 300 kg K_2O/ha. increased the percentage fruit-bearing shoot and fruiting co-efficient. The nutrients when applied singly produced a yield of 4800–6260 kg/ha. and with combined N, P, K, the production was 7810 kg/ha. However, the best result was obtained in cultivar Italian Rieslineg with a high dose of K and medium dose of N and P. The fertilizers recommended in some grape-growing countries of the world are presented below.

Table 12. Fertilizer recommendation of grape grown in different countries (kg/ha) (Cheema et al., 2001)

Countries	N	P_2O_5	K_2O
Bulgaria	120	100	120
India (Punjab)	0.4*	0.4*	0.8*
Romania	150	225	150
USA (Pensylvania)	114	150	95
USSR	120	90	45

* kg/vine

Micronutrient: Grape vine needs a small amount of zinc for its growth and development. Spraying with 6 lb of zinc sulphate was found to be effective in correcting zinc deficiency (Cook, 1962). Eficient method of nutrient applicantion to grape vine is through foliar feeding, particularly the micronutries. The important micronutrients supplied by foliar spray includes fe,Zn, B, Mg. Foliar application of boric acid at bloom and pre-bloom stages have been found to improve the yield of Thompson seedless cultivars of grape (Kumar and Bhusan, 1978). Gavrilov (1984) reported that foliar nutrition with MnSO4 at the 0.5% or ZnSO4 at 0.2% concentration improved the productivity of grape vines. Mestor (1971) recorded that soil application of B @ 12 Kg/ha. followed by a foilar spray @ 2.5 kg/ha. before flowering increased the number of reproductive buds per bush, stimulated shoot growth, reduced flower abscission, and increased the yield and sugar content of the fruit. Spraying of Iron. Zinc and Manganese on Roumi red grape vines before bloom, after fruit set, and/or during berry developement, increased fruit weight size and TSS.

10. Acknowledgement

We acknowledge with thanks to publisher Naya Udyog, for the book *Fruits: Tropical and Subtropical*, Vol–1 and Naya Prakash, for the book *Mineral Nutrition of Fruit Crops* for reproduction of some tables and scientific informations.

11. References

Abd – El – Al, A. A., El – Demerdash, A. M. and Ebd – El – Kader, A. M. M. 1994. *Ann. Agric. Sci., Moshtohor*, 32: 2029–2038.

Adsule, R. N. and Kadam, S. S. 1995. In: *Handbook of Fruit Science and Technology-Production, Composition, Storage and Processing* (Eds D. K. Salunkhe and S. S. Kadam), Marcel Dekker Inc., New York, pp. 419–433.

Agulhon, R. and Others. 1964. *Prog. Agric. Vitic.*, 162: 222–228.

Akamine, E. K. 1976. *HortScience*, 11: 586–588.

Aldridge, N. L. 1960. *Qd. Agric. J.*, 86: 97.

Ali, S. 1952. *Punjab Fruit J.*, 15: 1–12.

Allan, P and de Jager, J. 1978. *Crop Production*, 7: 125.

Allen, R. N., Dettman, E. B., Johns, G. G. and Turner, D. W. 1988. *Aust. J. Agric. Res.*, 39 : 53

Amerine, M. A., Beng, H. W. and Cruess, W. V. 1972. In: *Technology of Wine Making* (3rd ed.), AVI Publishing Co., West Port, Conn.

Anderson, C. A. 1971. *Proc. Soil Crop Sci. Soc. Fla.*, 30: 150–157.

Anjorin, H.O. and Obigbesan, G. O. 1983. *Fruits*, 38 : 300–302.

Anonymous 1962. *Wealth of India*, Raw Materials, C. S. I. R., New Delhi.

Anonymous 1969. *Indian Fmg.*, 11: 29.

Anonymous 1986. In: *Food and Fruit – bearing Forest Species 3. Examples from Latin America*, FAO Forestry Paper, 44/3, Chap. 17.

Anonymous. 1994. *Global Aam' s race, Times of India,* New Delhi, June 18, 1994.

Arenas De Moreno, L., Mann, M., Castro De Rincon, C. and Sandoval, L. 1995. *Revista de la Facultad de Agronomia,* Universidad del Zulia, 12: 467–483.

Arora, J. S. and Singh, J. R. 1970. *J. Jap. Soc. Hort. Sci.,* 39: 207.

Arora, J. S. and Singh, J. R. 1971. *Indian J. Hort.,* 28: 108–113.

Asoegwu, S. N. and Obiefuna, J. C. 1987. *Trop. Agric.,* 64: 139–143.

Augustin, M. A. and Ling, E. T. 1987. *Pertanika,* 10: 53–59.

Awasthi, R. P., Tripathi, B. R. and Singh, A. 1975. *Punjab Hort. J.,* 15: 14–16.

Bacha, M. A. A. 1977. *Indian J. Agric. Sci.,* 45: 189–193.

Bain, F. M. 1941. *Proc. Agric. Soc.,* Trin, Tob., 41: 657–659, 661.

Balakrishnan, S., Aravindakshan, M., Marykutty, K. C. and Mathew, V. 1980. *Agri. Res. J.* Kerala, 18: 33–39.

Banik, B. C., Sen, S. K. and Bose, T. K. 1997 a. *Environment & Ecology,* 15: 122–125.

Banik, B. C., Sen, S. K. and Bose, T. K. 1997b. *Environment & Ecology,* 15 : 269–271.

Bar-Akiva, A and Lavon, R 1967. *Israel J. Agric Res.,* 17: 16.

Bar-Akiva, A. 1965. *Phyton* (Argentina), 22: 131–136.

Bar-Akiva, A. 1975. *J. Hort. Sci.,* 50: 85–89.

Bar-Akiva, A. and Gotfried, A. 1972. *Agrochimica,* 16: 127–135.

Bar-Akiva, A. and Lavon, R. 1967. *Israel J. Agric. Res.,* 17: 16.

Baramanray, A. Gupta, O. P. and Dhawan, S. S. 1995. *Haryana J. Hort. Sci.,* 24: 102–109.

Bardiya, M. C., Kundy, B. S. and Tauro, P. 1974 *Haryana J. Hort. Sci.,* 3: 140.

Bartholomew, D. P. and Kadzimin, S. B. 1977. In: *Ecophysiology of Tropical Crops* (Eds. P. de T. Alvim and Kozlowski, T. T.) Acad. Press, New York, Chap. 5.

Bartholomew, D. P. and Malezieux, E. P. 1994. In: *Handbook of Environmental Physiology of Fruits Crops.* Vol. II. *Subtropical and Tropical Crops* (Eds. B. Schaffer and P. C. Anderson) CRC Press Inc. Boca Raton, FL, pp. 243–291.

Batten, D. J. 1986. *Acta Horticulturae,* 175: 79.

Beattie, J. M. 1955. *Ohio Fm. Home Res.,* 40: 101–102.

Beattie, J. M. and Forshey, C. G. 1954. *Proc. Amer. Soc. Hort. Sci.,* 64: 21–28.

Bell, A. A., Ough, C. S. and Kliewer, W. M. 1979. *Amer. J. Enol. Vitic.,* 30: 124–129.

Bester, D. H. 1975. In: *Congreso Mundial de Citricultura* 1973, Murcia, Spain, Vol. 1, 103–108.

Beyers, E. and Jonbert, G. F. 1952. *Fmg. S. Afr.,* 27: 508–510.

Bhatt, J. R. and Shah, J. J. 1985. *Indian J. Exp. Biol.,* 23: 330–339.

Bhattacharya, A., Singh, R. P. and Singh, A. R. 1973. *Progressive Hort.,* 5: 41–52.

Biblina, L. T. 1968. *Agrohimija,* 11: 138–141.

Blackman, F. F. and Parija, P. 1928. Analytical studies in plant respiration. I: The respiration of a population of senescent ripening apples. *Proceedings of the Royal Society B. Biological Science,* 103, 412–418.

Bouma, D. 1959. *Aust. J. Agric. Res.,* 10: 804–817.

Bradford, G. R., Harding, R. B. and Miller, M. P. 1962. *Calif. Agric.* 16: 6–9.

Broadley, R. H., Wassman, III, R. C. and Sinclair, E. R. 1993 *Pineapple Pests and Disorders,* Queensland Dept. Primary Ind., Brisbane, p. 63.

Bruinsma, J. 1983. Hormonal regulation of senescence, ageing, fading and ripening. In: *Postharvest physiology and crop preservation,* ed. M. Lieberman, NATO ASI Series A, Plenum Press, 141–164.

Bucher, R. 1978. *Weinberg U. Keller,* 25: 278–336.

Bulgarelli, G. C. (1962) *Inf. Ortofruttic.,* 3: 301–302.

Burn, J. and Py, C. 1952. *Fruits d' Outre Mer.*, 7: 62–64.

Calvert, D. V. 1969. *Proc. First Int. Citrus Symp.*, Univ. of Calif., Riverside, 3: 1587–1597.

Cannon, R. C. 1957. *Qd. J.Agric. Sci.*, 14: 93–110.

Canon, R. C. 1960. *Proc. 2nd Australian Weeds. Conf.*, 16: 3

Cattaneo, T. M. P., Nigro, F., Messina, G. and Giangiacomo, R. 1994. *Milchwissenschaft*, 49: 269–272.

Chadha, K. L. and Pandey, R. M. 1986. In: *CRC Handbook of Flowereing.* Vol. V (Ed. A. H. Halevy), CRC Press, Inc. Boca Raton, p, 287.

Chadha, K. L. and Rajpoot, M. S. 1969. *Indian J. Hort.*, 26: 124–129.

Chalker, F. C. and Turner, D. W. 1966. *Agric Gaz.* N.S.W., 77: 697.

Chang, J. W., Chang, H. M., Chao, C. N. and Wang, H. L. 1996. *Chinese J. Agrometeorology*, 3: 151–157.

Chanturiya, I. A. 1975. *Subtrop. Kul'tury*, No. 4. pp. 61–64.

Chapman, H. D. and Brown, S. M. 1943. *Potash in Relation to Citrus Nurition*, pp. 87–100.

Chapman, H. D., Brown, S. M. and Bayner, D. S. 1945. *Calif. Citrogr.*, 30: 162–163.

Chapman, J. C. 1982. *Australian. J. Expt. Agric. and Animal Husbandry*, 22: 331–336.

Chapman, K. R. 1984. *Rept. Qd. Dep. Primary Ind.*, pp. 1–123.

Chattopadhyay, P.K. and Bose, T. K. 1986. *Indian Agric.*, 30: 213–222.

Cheema, S. S. and Jindal, P. C. 2001. In: *Fruits: Tropical and Subtropical,* (Eds. T. K. Bose, S. K. Mitra and D. Sanyal), Naya Udyog, Calcutta, Vol. 1, India, p. 378.

Chen, C. C., Fang, T. T. and Ling, R. S. 1957. *Mem. Coll. Agril. Nat. Taiwan Univ.*, 5: 15–26.

Chen, D. Q., Chen, M. H., Ban, J. Q., Xu, X. J., Zhang, Q. X. and Hu, S. L. (1998) *South China Fruits*, 27: 10–11.

Chonker, V. S. and Singh, P. N. 1981. *Abst. Nat. Symp. Trop. and Subtrop. Fruit Crops*, Banglore, p. 53.

Chowdhury, S. 1954. *FAO Plant Protection Bull.*, 2: 151–153.

Christensen, P. 1986. *Calif. Agric.*, 40: 22–23.

Chu, C. C. 1963. *J. Agric. For.*, 12: 1–9.

Chu, C. K. 1968. *Taiwan Agric. Quart.*, 4: 26–44.

Coetzer, L.A., Robbertse, P. J. and Janse van Vuuren, B. P.H. 1992. *Yearb. South Afr. Mango Growers' Assoc.*, 12: 48–51

Collado Fernandez, M. 2000. *Alimentaria*, 37: 91–95.

Collins, J. L. 1960. *The Pineapple*, Leonard Hill Inter- Science Publ. Inc. London.

Collins, J. L. 1968. *The Pineapple*, Leonard Hill, London.

Cook, J. A. 1962. West Fruit Gr., 16: 17–18.

Cook, J. A. and Kishaba, T. 1956 a. *Plant Analysis and Fer. Prob.* IRHO. Paris, pp. 158–175.

Cook, J. A. and Kishaba, T. 1956. *Proc. Amer. Soc. Hort. Sci.*, 68: 131–140.

Coombe, B. G. (1986) *Proc. 22nd Int. Hort. Cong., California*, USA, Abst. No. 1598.

Cooper, W. C., Ascension, P., Furr, J. R., Hilgeman, R. H., Cahon, G. A. and Boswell, S. B. 1963. *Proc. Amer. Soc. Hort. Sci.*, 82: 180–182.

Crane, J. H. and Campbell, C. W. 1991. The Mango, Fla. Co-op. Extn. Ser., IFAS, Univ. Florida, Fact Sheet FC–2.

Cunha, R. J. P. and Haag, H. P. 1981. *Anais da Escola Superior de Agricultura*, 37: 169–178.

Dalldorf, D. (1975) *Inf. Bull. Citrus & Subtrop.* Fruit Res. Inst., No. 32, p. 2.

Das, B. C., Chakraborty, A., Chakraborty, P. K., Maiti, A., Mandal, S. and Ghosh, S. 1995. *The Hort J.*, 8: 141–146.

Dasberg, S., Bielorai, H. and Erner, J. 1983. *Plant and Soil.*, 75: 41–49.

Datuadze, O. 1976. *Referetivnye Zhurnal.*, 7: 55–805.

De Fossard, R. A. and Lenz, F. H. 1967. *Qual. Plant. Mater. Veg.*, 14: 289–304.

Dhingra, D. R. and Kanwar, J. S. 1963. *Punjab Hort. J.*, 3: 54–59.

Dodson, P. G. C. 1968. *Exp. Agric.*, 4: 103.

Dragas, M. 1957. Z. Born. Rad. Poljoprev Fak. Belgrade., 5: 75–84.

Dunsmor, J. R. 1957. *Malay Agric, J.*, 40: 159–187.

Durmanov, G. L. 1974. *Ser. Sci. Inst. Nat, Et. Agron Cong.* No. 106, p. 8.

Echeverri-Lopez, M. and Garcia–Reyes, F. 1976. *Cenicafe*, 24: 104–114.

El–Shourbagy, M. A. and Ismail, Z. 1961. *Agric. Res. Rev.* Cairo, 39: 219–226.

Elahi, M and Khan, N. 1973. *J. Agric. Food Chem.*, 21: 229–231.

Embleton, T. W. and Jones, W. W. 1966. *HortScience*, 1: 25–26.

Ethiraj, S and Suresh, E. R. 1992. *J. Food Sci. Tech.*, 29: 48–50.

Fawusi, M. O. A. and Ormrod, D. P. 1975. *Hort. Res.* 15:1–8.

Fayyaz, A., Asbi, B. A., Ghazali, H. M., Man, Y. B. C. and Jaip, S. 1994. *Food Chem.*, 49: 373–378.

Felizardo, B. C., Galvez, N. L. and Davide, P. S. 1961. *Philipp. Agric.*, 45: 95–102.

Fernandez–Lopez, J. A., Hidalgo, V., Almela, L. and Lopez Roca, J. M. 1992. *J. Sci. Food Agric.*, 58: 153–155.

Flath, R. A. 1980. In: *Tropical and Subtropical Fruits: Composition, Properties and Uses* (Eds. S. Nagy and P. E. Shaw.), pp. 157–183. AVI. Pub., Westport, Connecticut.

Franco, M. R. B. and Rodriguez–Amaya, D. B. 1993. *Arquivos de Biologia e Tecnologia*, 36: 613–632.

Fregoni, M. and Scienza, A. 1976. *Vignevini*, 3: 7–11

Freiberg, S. R. 1955. *Annual Report*, United Fruit Co., Dept. of Res. Boston, Mass.

Freiberg, S. R. and Steward, F. C. 1960. *Ann. Bot.*, 24: 147–157.

Galan Sauco, V. 1989. *Lychee Cultivation.* FAO Plant Production and Protection Paper 83, FAO, Rome.

Ganje, T. J., Page, A. L. and Martin, J. P. 1966. *Calif. Citrogr.*, 51: 493–496.

Gartel, W. 1956. *Weinberg U. Keller*, 3: 132–139 and 233–241.

Gartel, W. 1958. *Weinberg U. Keller*, 5: 267–287.

Gartel, W. 1965. *Vignes et Vins* No, 138, pp. 25–29.

Gartia, C. R. and Jaramillo, C. R. 1984. *Informe Mensual*, UPEB, 8: 50.

Gavrilov, T. K. (1984) In: *Pitanie I Productvnost' Rastenii. Kishinev*, Moldovian SSR, pp. 121–124.

Georgieva, S. 1976. *Lozarstvo i Vinarstvo*, 25: 22–24.

Ghanta, P. K., Dhua, R. S. and Mitra, S. K. 1992. *The Hort. J.*, 5: 43–48.

Ghanta, P. K., Dhua, R. S. and Mitra, S. K. 1995. *Ann. Agric. Res.*, 16: 405–408.

Ghosh, B., Biswas, B., Mitra, S. K. and Bose, T. K. 1987. *Indian Fd. Pack.*, 41: 34–37.

Ghosh, S. 1960. *Sci. and Cult.*, 26: 287–288.

Ghosh, S. N. and Chattapadhyay, N. 1996. *The Hort. J.*, 9: 121–127.

Giacomelli, E. J. and Py, C. 1982. *Fruits*, 36: 645.

Godfrey-Sam-Aggrey, W. 1977. *Zeitschrift fur Acker und Pflanzenbau*, 145: 199–206.

Gofur, M.A., Shafique, M. Z., Helali, O. H., Ibrahim, M., Rahman, M. M. and Hakim, A. 1994. *Bangladesh J. scientific & Industrial Res.*, 29: 163–171.

Golberg, L. and Levy, L. 1941. *Nature*, 148: 286.

Golodriga, P. Ya and Pu Chao, K. 1963. *J. Nauchu Issled Inst. Vinodeliya i Vinogradorstva Magarach*, 12: 74–83.

Goodenough, P. W., Tucker, G. A., Grierson, D. and Thomas, T. 1982. Changes in colour, polygalacturonase, monosaccharides and organic acids during storage of tomatoes. *Phytochemistry*, 21, 281–284.

Gortner, W. A., Dull, G. G. and Krauss, B. H. 1967. *HortScience*, 2: 141.

Gowda, I. N. D. and Ramanjaneya, K. H. 1995. *J. Food Sci. Tech.*, 32: 323–325.

Haas, A. R. C. 1936. *Bot. Gaz.*, 98: 65–86.

Haas, A. R. C. and Brusca, J. N. 1954. *Calif. Agric.*, 8: 13–14.

Harjadi, S. S., Pribadi, F. I. and Koswara, S. 1995. *Acta Horticulturae*, 379: 83–88.

Hassan, S. I. 1968. *M. Sc. Thesis,*Punjab Agricultural University, Ludhiana.

Hasselo, H. N. 1961. *Trop. Agric.*, Trinidad, 38: 29–34.

Hayes, W. B. 1970. *Fruit Growing in India*, Kitabistan, Allahabad.

Hemavathy, J., Prabhakar, J. V. and Sen, D. P. 1987. *J. Food Sci.*, 52: 833–834.

Hernandez, J. 1983. *Proc. Int. Soc. Citriculture*, 1981, Vol. 2, Shimiju, Japan, pp. 564–556.

Holder, G. D. and Gumbs, F. A. 1983. *Trop. Agric.*, 60: 25–30.

Hoosain, M. M., Khan, L. A., Paul, N. K., Hoosain, M and Joarder, O. I. 1990. *Crop Res.*, 3: 222–233.

Huang, S. P., Wu, G. L. and Li, S. Y. 1993. *China Citrus*, 22: 3–5.

Ishihara, M., Knonna, S., Yokomijo, H. and Sato, K. 1965. *Bull. Hort. Res. Stat.*, Hiratsuka, No. 4, pp. 45–66.

Ishihara, M., Nishiba, S., Futai, S., Sakaguchi, S. Hashimoto, T., Yanase, M., Shibuya, H., Teraoka., Y., Hamaoka, S., Yokomizo, H., Konno, S. and Takatusji, T. 1974. *Bull. Fruit Tree Res. Stat.*, Hiratsuka, No. 1, pp. 33–56.

Iwasaki, T. and Owada, A. 1960. J. Hort. Asso. Japan, 29: 101–116.

Jacob, H. E. 1950. *Grape Growing in California*, Circular 116, Calif. Agric. Exten. Service Univ., California, Berkely.

Jagirdar, S. A. P. and Sheikh, M. R. 1970. *Agric. Pakistan*, 20: 175–184.

Jambulingam, A. R., Ramaswamy, N. and Muthukrishnan, C. R. 1975. *Potash Review*, 27: 4–6.

Janse, van Vuuren, B. P.H., Robbertse, P. J., Coetzer, L.A. and Hudson–Lamb, D.C. 1992. *Yearb. South Afr. Mango Growers' Assoc.*, 12: 14–19.

Jardine, F. A. L. 1946. *Qd. Agric. J.*, 62: 74–78

Jones, W. W. and Embleton, T. W. 1954. *Calif. Citrogr.*, 39: 252.

Jones, W. W. and Embleton, T. W. 1959. *Proc. Amer. Soc. Hort. Sci.*, 73: 234–236.

Jones, W. W. and Others. 1963. Calif. Citrogr., 48: 107, 128–129.

Jones, W. W. and Smith, P. F. 1964. In: *Hunger Signs in Crops* (Ed. H. B. Sprague), David Mckay Co. Inc. Washington.

Jones, W. W. and Steinacker, M. L. 1955. *Hort. News Letter*, Univ. Calif. Citrus Exp. Stat., 7:3.

Jones, W. W., Embleton, T. W., Foott, J. H. and Platt, R. G. 1973. *J. Amer. Soc. Hort. Sci.*, 98: 414–416.

Joolka, N. K. 1972. *M. Sc Thesis*, Haryana Agril. Univ., Hissar.

Jordine, O. G. 1962. *Nature*, 134: 1160–1163.

Kadman, A. and Gazit, S. 1984. *J. Pl. Nutrition*, 7: 283–290.

Kakde, J. R. 1956. *Agric. Soil Mag.*, 30: 19–24.

Kallarackal, J., Milburn, J. A. and Baker, D. A. 1990. *Aust. J. Pl. Physiol.*, 17: 79.

Kanapathy, K. 1959. *Malay Agric. J.*, 42: 157–160

Kanwar, J. S. and Dhingra, D. R. 1962. *Indian J. Agric. Sci.*, 32: 309–314.

Kanwar, J. S., Nijjar, G. S. and Kahlon, G. S. 1987. *Jour. Res., PAU*, 24: 411–422.

Kapanadze, I. S. and Khasaya, G. S. 1988. *Subtropicheski Kultury*, 1: 136–140.

Keil, U. and Schreier, P. 1989. *Phytochemistry*, 28: 2281–2284.

Kesterson, J. W., Braddock, R. J., Koo, R. C. J. and Reese, R. L. 1977. *J. Amer. Soc. Hort. Sci.*, 102: 3–4.

Khan, M. A., Nizami, S. S., Khan, M. N. I. and Azeem, S. W. 1994. *Pakistan J. Sci & Ind. Res.*, 37: 213–214.

Khatua, N., Sen, S. K. and Bose, T. K. 1980. *Nat. Symp. Pineapple Prod. & Utilization*, Jadavpur, pp. 27–37.

Kim, SeonKyu, Kim, JinTaek, Jeon, SeongHo, Nam, YangSuk and Kim, SeungHeui 1998. *J. Korean Soc. Hort. Sci.*, 39: 547–554.

Kimura, M., Rodriguez–Amaya, D. B. and Yokoyama, S. M. 1991. *Lebensmittel–Wissenschaft und Technologie*, 24: 415–418.

Knight, R., Jr. 1980. In: *Tropical and Subtropical Fruits: Composition, Properties and Uses* (Eds. S. Nagy and P. E. Shaw), pp. 1–120. AVI Pub., Westport, Connecticut.

Koen, J. 1977. *Inf. Bull. Citrus and Subtrop.* Fruit Res. Inst., No. 57, p. 7.

Koen, T. 1976. *Information Bull.* Citrus and Subtrop. Fr. Res. Inst., No. 46, p.4.

Koto, M. and Takeshita, S. 1957. *Bull. Kanagawa Agric. Exp. Stat.*, Hort. Branch, No. 5, pp. 13–22.

Koto, M., Takeshita, S. and Nakajima, T. 1963. *Bull. Kanagawa Hort. Expt. Stat.*, No. 11, pp. 11–19.

Kumar, R., Kotur, S. C. and Singh, H. P. 1995 *Indian J. Hort.*, 52: 254–258.

Kumar, R., Kotur, S. C. and Singh, H. P. 1996. *J. Potassium. Res.*, 12: 59–64.

Kumar, S., Sharma, R. K. and Choudhary, R. 1995. *The Hort. J.*, 8: 33–37.

Kumar, Satish and Bhushan, S. 1978. *J. Res. Punjab Agric. Univ.*, Ludhiana, 15: 43–48.

Kundu, S., Ghosh, S. N. and Mitra, S. K. 1995. *Indian Fd Packer*, pp. 11–16.

Kuznetsov, A. V. and Treshchov, A. G. 1979. Sbornik Nauch. Tr. Un-t Druzhby Narodov im Patrisa Lumuby, 89: 68–72.

Labanauskas, C. K., Jones, W. W. and Embleton, T. W. 1963. *Proc. Amer. Soc. Hort. Sci.*, 82: 142–153.

Lal, R. L., Tewari, J. P. and Mishra, K. K. 1996. *Recent Horticulture*, 3: 18–20.

Langenegger, W and Reynolds, R. 1986. *Information Bull.*, 166: 10.

Langenegger, W. 1976. *Inf. Bull. Citrus & Subtrop. Fruit Res. Inst.*, 44: 6

Larsen, R. P. and Others 1959. *Quart. Bull.* Mich. Agric. Exp. Stat., 41: 812–819

Le Roux, J. C. 1951. *Fmg. South. Afr.*, 26: 236–240.

Leigh, D. S. 1969. *Agric. Gaz. N.S.W.*, 80: 369–372.

Lenz, F. 1967. *Gartenbauwiss*, 32: 181–192.

Leonard, C. D. 1952. *Citrus Fruit*, Farm Chemicals, 130.

Levy, J. F. 1965. *Vignes et Vins*, No., 138: 18–24.

Lin, H. C., Lin, H. L. and Lee, K. C. 1994. *Proc. Symp. on Practical Aspects of Some Economically Imp. Fruit trees in Taiwan* (Eds. H. S. Lin and L.R. Chang) Special publication–Taichung Dist. Agric. Improvement Station, 33: 61–67.

Lokhande, N. M. and Moghe, P. G. 1991. *South Indian Hort.*, 39: 23–26.

Lopez, Lago, I., Diaz Varela, J. and Merino de Caceres, F. 1996. *Alimentaria*, 34: 59–64.

Lott, W. L. 1952. *Proc. Amer. Soc. Hort. Sci.*, 60: 123–131.

194 B.C. Ghosh and S. Palit

Lynch, S. J. and Ruehle, G. D. 1940. *Proc. Fla. Sta. Hort. Soc.*, 53: 167–169.

MacLeod, A. J., MacLeod, G. and Snyder, C. H. 1988. *Phytochemistry*, 27: 2189–2193.

Majorana, G. 1960. *Tech. Agric.*, 12: 645–661.

Majumder, P. K., Sharma, D. K. and Sanyal, D. 2001. In: *Fruits: Tropical and Subtropical,* (Eds. T. K. Bose, S. K. Mitra and D. Sanyal), Naya Udyog, Calcutta, India, Vol. 1, p. 21.

Malezieux, E. and Lacoeuilhe, J. J. 1991. *Fruits*, 46: 227.

Malhi, G. S. and Nijjar, G. S. 1985. *Abst. Int. Symp. On Mango*, Bangalore, p. 29.

Mallik, P. C. and De, B, N. 1951. *Indian J. Agric Sci.*, 21: 32–37.

Mallik, P.C. and Singh, D.L. 1959. *Indian J. Hort.*, 16: 228–232.

Mallik, S. K., Mitra, S. K., Banik, B. C., Maity, S. C. and Sen, S. K. 1985. 2nd *Int. Symp. on Mango*, India, Abst. No. 4.19, p. 37.

Malo, S. E. and Cambell, C. W. 1986a. *The Guava*. Univ. Fla. Institute of Food and Agricultural Sciences, Fruit Crops Fact Sheet FC- 4.

Malo, S. E. and Campbell, C. W. 1986. *The Papaya*, Institute of Food and Agricultural Sciences, Fruit Crops Fact Sheet FC–11.

Marler, T. K. 1994. In : *Handbook of Environmental Physiology of Fruit Crops*. Vol. II *Sub-Tropical and Tropical Crops* (Eds. B. Schaffer and P. C. Anderson), CRC Press, Inc., Boca Raton, FL, pp, 213–216.

Martin- Prevel, P. 1961. *Fruits*, 16: 341–51.

Martin- Prevel, P. and Charpentier, J. M. 1963. *Fruits d' Outre Mer.*, 18: 221–247.

Martin, J. P. and Page, A. L. 1965. *Plant and Soil*, 22: 65–80.

Martin-Prevel, P. and Montagut, G. 1966. *Fruits d' Outre Mer.*, 21: 283–294.

Mataa, M., Tominaga, S and Kozaki, I. 1998. *J. Japanese Soc. Hort. Sci.*, 67: 28–34.

Mathew, A. B. and Pushpa, M. C. 1964. *J. Food Sci, Technol.*, 1: 71.

Maxwell, L. S., Krezdorn, A. H., Will, A. A. and Golloy, E. V. 1984. *Florida Fruit*, Maxwell, Tampa. FL.

McKenzie, C. B. 1994. *Yearb. South Afr. Mango Growers' Assoc.*, 14: 26–28.

Menzel, C. M. 1991. In: *Plant Resources of South- East Asia*. No. 2. *Edible Fruit and Nuts.* (Eds. E. W. M. Verheij and R. E. Coronel), Pudoc, Wageningen., p. 191.

Menzel, C. M. and Simpson, D. R. 1986. *Proc. 1st Nt. Lychee Sem.*, Australia, pp. 22–36.

Menzel, C. M. and Simpson, D. R. 1993. In: *Encyclopedia of Food Science, Food Technology and Nutrition* (Eds. R. Macf, R. K. Robinson and M. J. Sadler), Academic Press, London, pp. 2108.

Menzel, C. M., and Simpson, D. R. 1994. In: *Handbook of Environmental Physiology of Fruit Crops.* Vol II: *Sub- tropical and Tropical Crops* (Eds. B. Schaffer and P. C. Anderson), CRC Press, Inc. Florida, pp. 123–145.

Menzel, C. M., Chapman, K. R., Paxton, B. F. and Simpson, D. R. 1986b. *Australian J. Exp. Agric.*, 26: 261–265.

Menzel, C. M., Piccone, M. F. and Simpson, D. R. 1986a. *Qd. Fruit Veg. News*, October, pp. 15–18.

Menzel, C. M., Simpson, D. R., Haydon, G. F. and Doogan, V. J. 1995. *J. South Afr. Soc. Hort. Sci.*, 5: 97–99.

Mester, I. M. 1971. *Referativnyi Zhurnal*, 11: 915.

Miller, C. D. and Bazore, K. 1945. *Hawaii Expt. Sta. Bull.*, p. 96.

Minessy, F. A. 1971. *Sudan. Agric. J.*, 6:71–74.

Misra, R. S. and Khan, I. 1981. *Progressive Hort.*, 13: 87–90.

Mitcham, E. J., Gross, K. C. and Ng, T. J. 1989. Tomato fruit cell wall synthesis during development and senescence. Plant Physiology, 29, 477–481.

Mitra, S. K. and Bose, T. K. 1999. In: *Tropical Horticulture*, Vol. I, (Eds.: T. K. Bose, S. K. Mitra, A. A. Farooqui and M. K. Sadhu), Naya Prakash, Calcutta, pp. 297–307.

Mitra, S. K., Ghosh, S.K. and Dhua, R. S. 1984. *Sci. and Cult.* 50: 235–236.

Mohammad, R. and Mohammad, A. 1983. Res. Bull., 11: 237–239.

Moreira, R. S. and Hiroce, R. 1978. *Bragantia*, 37: 59–63.

Morris, J. R. Cawthon, D. L and Fleming, J. W. 1980. *Amer. J. Enol. Vitic.*, 31: 323–325.

Morton, J. F. 1987. *Fruits of Warm Climates, Creative Resources Systems.* Inc., Winterville, NC.

Moss, G. I. 1972. *Aust. J. Expt. Agric.* Animal Hus., 12: 195–202.

Mungomery, W. V., Jorgensen, K. R. and Barnes, J. A. 1980. *Proc. Int. Soc. Citriculture,* Griffith, NSW, Australia, pp. 285–288.

Murray, D. B. 1959. *Trop. Agric.*, Trinidad, 36: 100–107.

Murray, D. B. 1960. *Trop. Agric. Trinidad*, 37: 97–106.

Mustaffa, M. M. 1983. *South Indian Hort.*, 31: 270–273.

Narashima Char, B. L., Azeemoddin, G., Atchyuta, R.D. and Thirumala, R. S. D. 1989. *Acta Horticulturae*, 231: 749–750.

Narasimha Char, B. L. and Azeemoddin, G. 1989. *Acta Horticulturae*, 231: 744–748.

Narinesingh, D. and Mohammed–Maraj, R. 1988. *J. Sci. Food Agric.*, 46: 175–186.

Natale, W., Coutinho, E. L. M., Boaretto, A. E. and Banzatto, D. A. 1994. *Revista de Agricultura*, 69: 247–255.

Natale, W., Coutinho, E. L.M., Boaretto, A. E., Pereira, F. M., Oioli, A. A. P. and Sales, L. 1996. *Revista Brasileira de Ciencia do Solo*, 20: 247–250.

Nathani, S. P. and Srivastava, H. C. (1965) *Allahabad Fmr.*, 41: 65–68.

Neild, R. E. and Boshell, F. 1976. *Agric. Meteorol.*, 17: 81.

Neog, M., Chakrabarty, B. K., Gogoi, D. K. and Baruah, K 1995. *The Hort. J.*, 8: 25–31.

Nguyen, H and Tarandjiska, R. (1995) *Fett Wissenschaft technologie*, 97: 20–23.

Nightingale, G. T. 1942. *Bot. Gaz.*, 104: 191–223.

Nijjar, G. S. 1970. *Progressive Fmg.*, 6: 7–8.

Nokrashy, M. A. El., Zorkani, S. El., Shorbagey, M. A. El., Abdel Baski, T. and Guindy, L. F. 1977. *Agric. Res. Rev.*, 55: 39–45.

Normand, F. 1994. *Fruits*, 49: 217–227.

Norton, K. R. 1963. *Annual Report*, United Fruit Co. Dept. of Research.

Norton, K. R. 1965. *Trop. Agric Trinidad*, 42: 361–365.

Nursten, H. E. 1970. Organic acids. In *The Biochemistry of fruits and their products.* (ed. A. Hulme), Vol. 2, Academic Press, pp. 89–113.

Obiefuna, J. C. 1984. *Fertilizer Res.*, 5: 315–319.

Ochse, J. J., Soule, M. J., Dijkman, M. J. and Wehlburg, C. 1961. *Tropical and Subtropical Agriculture.* The Mac Millan Company, New York.

Ogata, J. N., Kawani, Y., Bevenue, A. and Casarett, L. J. 1972. *J. Agric. Food Chem.*, 20: 113–115.

Ogden, M. A. H., Jackson, L. K. and Campbell, C. W. 1981. *Proc. Fla. St. Hort. Soc.*, 94: 222.

Olesen, T. D. and Bailey, G. J. 1995. *Australian J. Plant Physiol.*, 22: 387–390.

Oppenheimer, C. and Gazit, S. 1961. *Hort. Advance*, 5: 1–12.

Orona Moreno, J. D. and Ovando, Cruz, M. E. 1994. *Agricultura Technica en Medico*, 20: 33.

Orozco- Romero, J. and Sepulveda-Torres, J. L. 1983. *Proc. Int. Soc. Citriculture*, 1981, Vol. 2, Shimiju, Japan, pp. 542–543.

Ough, C. S. and Bell, A. A. 1980. *Amer. J. Enol. Vitic.*, 31: 122–123.

Ournac, A. and Poux, C. 1966. *Ann. Technol. Agr.*, 15: 341–347.

Ouyang ChoChun 1998. *South China Fruits*, 27: 21.

Palaniswamy, K. P., Muthukrishnan, C. R. and Shanmugavelu, K. G. 1974. *Indian Fd. Packer*, 28: 12–19.

Pant, V and Lavania, M. L. 1989. *Progressive Hort.*, 21: 165–167.

Parisot, E. 1988. *Fruits*, 43: 293–312.

Pathak, S. R. and Sarada, J. R. 1974. *Curr. Sci.*, 43: 716.

Paull, R. E. and Reyes, M. E. Q. 1996. *Scientia Hort.*, 66: 59–67.

Paxton, B. F. and Chapman, K. R. 1980. *Bienn. Rept. Maroochy Hort. Res. Stn.*, 2: 34–37.

Pensburg, N. van. and Preez., R. Du 1985. *Information Bull. Citrus and Subtrop. Fr. Res. Inst.*, No. 153, p. 8.

Peyer, E. (1955) *Obst. Und Weinbau.*, 64: 67–70

Peynaud, E. and Ribereau – Gayon, P. 1970. In: *Biochemistry of Fruits and Their Products*, Vol. 2 (Ed. A. C. Hulme), London, Academic Press.

Philippe, J. M. 1960. Bull. Agric. Congobelg, 51: 27–60

Pire, R. and Rojas, E. 1999. *Fruits*, 54: 177–182.

Plessis, S. F. Du. and Smart, G. 1982. *Infor. Bull.* Citrus and Subtrop. Fruit. Res. Inst., No. 123. pp. 15–19.

Prasad, A., Singh, H. and Shukla, T. N. 1965. *Indian J. Hort.*, 22: 254–265.

Puissant, A. 1960. *Ann. Technol. Agr.*, 9: 321–330.

Purseglove, J. W. 1968. *Tropical Crops – Dicotyledons*, The English Language Book Society.

Py, C. 1963. *Fruits d' Outre Mer.*, 18: 75–77.

Py, C. 1965. *Fruits*, 20 : 315.

Py, C., Lacoeuilhe, J. J. and Teisson, C. 1987. *The Pineapple Cultivation and uses*, Maisonneuve et Larose, Paris, p. 568.

Raheja, P. C. 1966. In: *Salinity and Aridity–New Approaches to Old Problems*, (Ed. H. Boyko and W. Junk), The Hauge, 43.

Raie, M. Y., Kamrah, S., Manzoor, A. and Qureshi, E. E. 1992. *Pakistan J. Scientific and Industrial Res.*, 35: 43–45.

Raifer, B. 1999. *Obst und Weinbau*, 36: 155–157.

Rajput, C. B. S and Chand, S. 1976. *Bangladesh Hort.*, 3: 22–27.

Rajput, C. B. S. and Tiwari, J. P. 1979. Res. Rep. Mango Workers Meeting, Panaji, Goa, p. 158.

Rajput, C. B. S., Singh, B. P. and Mishra, H. P. 1976. *Scientia Hort.*, 5: 311–313.

Ram, M. 1996. In: *A Text Book on Pomology*, Vol. II (Ed. T. K. Chattopadhyay), Kalyani Publishers, Ludhiana, pp. 113–140.

Ramaswamy, N. 1976. *Annamalai*, 6: 30–37.

Rao, D. P. and Mukherjee, S. .K. 1989. *Acta Horticulturae*, 231: 286–295.

Rath, S., Singh, R. L., Singh, B. and Singh, D. B. 1980. *Punjab Hort. J.*, 20: 33–35.

Reddy, B. M. C. and Prakash, G. S. 1982. *Ann. Rep.*, IIHR, Banglore, p. 18.

Reddy, S. E. 1984. *South Indian Hort.*, 32: 155.

Redina, E. F., Mezhlumyan, L. G., Kasymova, T. D. and Yuldashev, P. K. H. 1993. *Chemistry of Natural Compounds*, 29: 781–783.

Reese, R. L. and Koo, R. C. J. 1974. *Proc. Fla. St. Hort. Soc.*, 87: 1–5.

Reese, R. L. and Koo, R. C. J. 1975. *J. Amer. Soc. Hort. Sci.*, 100: 195–198.

Reig-Felin, A., Perez-Nievas, C. and Albert Bernal, A. 1963. *An. Inst. Nac. Invest. Agron.* Madrid, 12: 69–118

Reitz, H. J. 1984. *Outlook on Agriculture*, 13: 140–146.

Reitz, H. J. and Koo, R. C. J. 1960. *Proc. Amer. Soc. Hort. Sci.*, 75: 244–252.

Reuther, W and Smith, P. F. 1954. *Proc. Fla. St. Hort. Soc.*, 67: 20–26.

Riascos, R. G., Reyes, J. G. and Aguirre, J. O. 1996. *Agronomica Colombiana*, 13: 43–49.

Richards, A. V. 1952. *Trop. Agriculturist*, 108: 242–245.

Richmond, A and Biale, J. B. 1967. Protein and nucleic acid metabolism in fruit. II: RNA synthesis during respiratory rise of the avocado. *Biochimica et Biophysica Acta*, 138, 625 627.

Robinson, J. C. and Alberts, A. J. 1989. *Scientia Hort.*, 40: 215–225.

Robinson, J. C. and Anderson, T. 1991. *Newsletter of the International Group on Horticultural Physiology of Banana*, 14: 37.

Rosselet, F., Hefer, S. V., Helff, K. A. W., Langenegger, W. and Le Roux, F. H. 1962. *S. Afr. J. Agric. Sci.*, 5: 351–372.

Roy, A., Hossain, M., Mitra, S. K. and Bose, T. K. 1986. *Maharashtra J. Hort.*, 3: 38–43.

Roy, R. S., Mallik, P. C. and De, B. N. 1951. *Proc. Amer. Soc. Hort. Sci*, 57: 9–16.

Ruehle, C. D. 1948. *Econ. Bot.*, 2: 306–325.

Russo, F. and Raciti, G. 1955. *Rev. Argumic.*, 1: 29–42.

Sakamoto, T. and Okuchi, S. 1963. *J. Jap. Soc. Hort. Sci.*, 32: 256–264

Samoladas, T. H. 1964. *Bot. Zhurnal*, 49: 428–432.

Samson, J. A. 1980. *Tropical Fruits*, Longman, London.

Samuels, G. and Gandia-Diaz, H. 1960. *J. Agric. Univ. Puerto Rico*, 44: 16–20.

Sankar, K. P. 1963. *Sci and Cult.*, 29: 51

Sanyal, D., Roychoudhury, R., Dhua, R. S. and Mitra, S. K. 1991. *Indian Fd. Packer*, 45: 29–33.

Satish Kumar and Sharma, R. C. 1974. *Prog. Fmg.*, 10: 20

Sato, K. and Others 1962. *Bull. Hort. Res. Sta.*, Hiratsuka, Ser., No. 1, pp. 37–64.

Sato, K., Ishihara, M. and Kurihara, A. 1958. *Bull. Nat. Inst. Agric. Sci.*, Hiratsuka No. 6, pp. 109–144.

Satyanarayana Rao, D. V. 1971. *M. Sc (Ag) Thesis* submitted to the University of Madras.

Scott, L.E. 1944. *Soil Sci.*, 57: 55–65.

Selamat, M. M. and Ramlah, M. 1992. *First International Pineapple Symp.* Nov. 2–5, Honolulu, Hawaii, Abst. No. 66.

Sen, P. K., Roy, P. K. and De, B. N. 1947. *Indian J. Hort.*, 1: 35–45.

Sen, S. K. (1990). In: *Fruits: Tropical and Subtropical* (Eds. T. K. Bose and S. K. Mitra), Naya Prakash, Calcutta, India, pp. 252–279.

Sharma, D. P. and Sharma, R. G. 1992. *Adv. Plant Sci.*, 5: 313–315.

Sharma, R. C., Mahajan, B.V.C and Azad, A. S. 1993. *The Hort. J.*, 6 : 83–87.

Sharma, S. B. 1984. *Punjab Hort. J.*, 24: 89–91.

Shaulis, N. and Kimball, K. 1956. *Proc. Amer. Soc. Hort. Sci.*, 68: 141–156.

Shawky, I., Zidan, Z. El Tomi, A. and Dahshan, D. 1978. *Egyptian J. Hort.*, 5: 133–142.

Shawky, J., El-Tomi, A. L. and Nasr, A. F. 1979. *Egyptian J. Hort.*, 6: 1–12.

Shrama, S. B. and Ray, P. K. 1987. *Indian Fd Packer*, 41: 17–20.

Sideris, C. P. and Young, H. Y. 1945. Plant Physiol., 20: 609–630

Simmonds, N. W. 1959. *Bananas*, Longmans, London.

Singh, B. P., Singh, V. B., Srivastava, P. K.and Singh, R. R. 1973. Balwant Vidyapeet *J. Agric. Sc. Res.*, 15: 54–58.

Singh, D. S. and Singh, S. P. 1995. *Res. and Dev. Repoter*, 12: 35–39.

Singh, H. P. and Uma, S. 1996. *Banana Cultivation in India*, Farm Information Unit, Ministry of Agriculture, Govt. of India.

Singh, H. P., Dass, H. C. and Ganapathy, K. M. 1981. *Nat. Symp. Trop. & Subtrop. Fruit Crops*, Banglore, p. 57.

Singh, H. P., Dass, H. C., Ganapathy, K. M. and Subramanium, T. R. 1977. *Indian J. Hort.*, 34: 377–384

Singh, J. and Singh, M. P. 1970. *Fertilizer News*, 15: 45.

Singh, K. K. 1967. *The Mango: A Handbook*, ICAR, New Delhi, pp. 70–90

Singh, K. K. and Chadha, K. L. 1961. *Punjab Hort.* J., 1: 171–179.

Singh, L. B. 1960. *The Mango: Botany, Cultivation and Utilisation*, Leonard Hill, London .

Singh, L. B. and Singh, U. P. 1964. *The Litchi*, Supdt. Printing and Stationary, Uttar Pradesh, Allahabad.

Singh, M. 1988. *Punjab Hort*. J., 28 : 50.

Singh, M. P. 1957. *Hort. Adv.*, 1: 48–54.

Singh, M. P. and Agrawal, K. C. 1960. *A. R. Hort. Res. Inst.*, Sharanpur, pp. 57–70.

Singh, N. P. and Rajput, C. B. S. 1977. *Indian J. Hort.*, 34: 120–125.

Singh, P. N. and Chhonkar, V. S. 1983. *Punjab Hort.* J., 23: 34–37.

Singh, R. 1969. *Fruits*, National Book Trust, New Delhi, pp. 86–90.

Singh, R. P., Joon, M. S. and Daulta, B. S. 1983. *Haryana J. Hort. Sci.*, 12: 68–70

Singh, R. R. 1975. *South Indian Hort.*, 23: 126–129.

Singh, S. S., Kalyanasundaran, P. and Muthukrishnan, C. R. 1974. *Annamalai University Agric. Res.* 4/5, 23–33.

Singh, T. P., Yamdagni, R. and Jindal, P. C. 1979. Haryana J. Hort. Sci., 8: 207–208.

Singh, U. R., Pandey, I. C., Upadhyay, N. P. and Tripathi, B. M. 1979. *Punjab Hort. J.*, 19: 111–113.

Singha, H. N. 1961. *A. R. Hort. Res. Inst.*, Saharanpur, pp. 102–106.

Sippel, A. 1988. *Information Bull.*, Citrus and Subtropical Fruit Res. Inst., South Africa, 190: 2, 4.

Sites, J. W. and Deszyck, E. J. 1952. *Proc. Fla. St. Hort. Soc.*, 65: 92–98.

Smart, R. 1986. *Proc. 22nd Int. Hort. Cong.*, California, USA, Abst. No. 1599.

Smith, B. L. 1989. *Inf. Bll., Citrus & Subtrop. Fr. Res. Inst., S. Afr.*, 197: 4

Smith, P. F. 1959. *Citrus Mag.*, 22: 16–17, 23.

Smith, P. F. 1966. *HortScience*, 1: 101–102.

Smith. P. F. and Scudder, G. K. 1951. *Proc. Fla. Sta. Hort. Soc.*, 64: 243–248.

Sparks, D. and Larsen, R. P. 1966. *Quart. Bull.* Mich. Agric. Exp. Stat., 48: 506–513.

Srivastava, R. P. 1964. *Fertilizer News*, 9: 13–23.

Srivastava, S. S. 1969. *Indian J. Hort.*, 26: 146–150.

Stannard, M. C. 1973. *Agric. Gaz.* N. S. W., 84: 162–165.

Stover, R. H. and Simmonds, N. W. 1987. *Bananas*, 3rd Ed. Longman, London.

Su, N. R. 1958. *J. Agric. Ass. China*, 22: 27–50.

Su, N. R. 1965. *J. Agric. Ass. China*, 51: 26–30.

Su, N. R. 1965. *Mem, Coll. Agric. Nat.* Taiwan Univ., 8, pp. 31–56.

Sugiyama, T. Iwata, M. and Yashiro, H. 1952. *J. Hort. Ass.* Japan, 21: 161–164.

Sukonthasing, S., Wongrakpanich, M. and Verheij, E. W. M. 1991. In: *Plant Resources of Southeast Asia. 2. Edible Fruits and Nuts*, (Eds. E. W. M. Verheij and R.E. Coronel), Pudoc–DLO, Wageningen, The Netherlands, p. 221.

Sulladmath, U. V., Narayana Gowda, J. V. and Ravi, S. V. 1981. *Abst. Nat. Symp. Trop. and Subtrop. Fruit Crops*, Banglore, p. 54.

Summerville, W. A. T. 1944. *Lambest Qd. J. Agric Sci.*, 1: 1–127.

Suriyapananont, V. 1992. *Acta Horticulturae*, 321: 529–534.

Suzuki, T., Arakawa, Y., Takeshita, K. and Okamoto, S. 1977. *Bull. Fac. Agric.* Shiynoka Univ., 27: 33–38.

Talakvadze, K. B. 1973. *Subtrop. Kul'tury*, No. 3. pp. 58–60.

Tandon, D. K. and Kalra, S. K. 1983. *Plant Physiol. Biochem.*, 10: 113.

Tay, T. H. 1972. *Trop. Agric. Trinidad*, 49: 51–59.

Tay, T. H. 1974. *Mardi Res. Bull.*, 2: 43–45.

Tay, T. H. 1975. *Mardi Res. Bull.*, 3: 1–14.

Taylor, S. E. and Sexton, O. J. 1972. *Ecology*, 53: 143.

Teakle, L. J. H., Johns, H. K. and Truton, A. G. 1943. *J. Dept. Agric. W. Aust*, 20: 171–84.

Thompson, B. D. 1955. *Proc. Fla. Lychee Growers Assoc.*, 2: 27.

Tiwari, J. P. and Rajput, C. B. S. 1975. *Bangladesh Hort.*, 3: 31–36.

Tkatchenko, B. 1947. *Fruits*, 2 : 206.

Toshi, N. I. 1992. *M. Sc. (Ag) Thesis*, North Eastern Hill Univ., India.

Tucker, G. A. and Grierson, D. 1987. Fruit ripening. In *The Biochemistry of plants–A comprehensive treatise* (ed. D.D. Davies), Vol. 12, Academic Press, pp. 265–318.

Turner, D. W. and Lahav, E. 1983. *Aust. J. Plant Physiol.*, 10: 43.

Turner, D. W. and Lahav, E. 1985. *Scientia Hort.*, 26: 311.

Twyford, I. T. 1965. *Winban News*, 1: 43–44, 47.

Ulrich, R. 1970. Organic acids. In *The Biochemistry of fruits and their products*. (ed. A. Hulme), Vol. 1, Academic Press, pp. 89–118.

Vas Conceles, D. M. 1952. *Bol. Sect. Agric. Ind. Com. Est*. Pernambuco, 207.

Veerannah, L., Balakrishnan, R., Raman, K. R. and Alagiamanavalan, R. S. 1974. *Indian J. Hort.*, 31: 135.

Vega, E. and Molina, E. 1999. *Agronomia Costarricense*, 23: 37–44.

Venek, G., Mosny, V. and Kolonyova, V. 1967. *Pol' mohospodarstvo*, 13: 185–192.

Verawudh, J. 1992. *First International Pineapple Symp.* Nov. 2-5, 1992, Honolulu, Hawaii, Abst. No. 16.

Verona, O. 1959. *Inf. Filopat.*, 9: 296–299.

Wagh, A. N. and Mahajan, P. R. 1985. *Curr. Res. Rep.*, 1: 124–126.

Wali, Y. A. and Hassan, Y. M. 1965. *Proc. Amer. Soc. Hort. Sci.*, 87: 264–269.

Wallihan, E. F. and Garber, M. J. 1966. *Calif. Citrogr.*, 51: 450–459.

Wang, D. N., Chang, K. L. and Chow, M. 1975. *Crung Hua Nung Yeh Yen Chiu*, 24: 43–54.

Wardlaw, C. W. 1940. *Trop. Agric.*, Trinidad, 17: 124–127.

Weir, C. C. 1967. *Bull. Citrus. Res.*, Univ. W. Indies, 6: 3 and *J. Agric. Soc.*, Trin., Tob., 67: 469–72.

Weir, R. G., Bevington, K. B., Duncan, J. H and Cradock, F. W. 1978. *Proc. Int. Soc. Citriculture*, Griffith, N. S. W., Australia, pp. 292–295.

Wenkam, N. S. and Miller, C. D. 1965. *Hawaii Agr. Exp. Sta. Bull.*, 135.

Werner, H. 1993. *Proc, Interamerican Soc. Trop. Hort.* (Ed. R. J. Campbell), 37: 94–98.

Whiley, A. W. 1984. In: *Tropical Tree Fruits for Australia* (complied by P.E. Page), Qd. Dept. Prim. Indus. Inform. Series Q183018, 25.

Whiley, A. W., Rasmussen, T. S., Saranath, J. B. and Wolsetenholme. B. N. 1989. *J. Hort. Sci.*, 64: 753.

Whiting, G.C. 1970. Sugars. In *The Biochemistry of fruits and their products.* (ed. A. Hulme), Vol. 1, Academic Press, pp. 1–31.

Wills, R.B.H., Lim, J. S. K. and Greenfield, H. 1986. *Food Tech. Aust.*, 38: 118.

Wilson, C.W., III, Shaw, P. E. and Knight, R.J. Jr. 1990. *J. Agric Food Chem.*, 38: 1556–1559.

Wilson, W. C. 1980. In : *Tropical and Subtropical Fruits* (Eds. S. Nagy and P. E. Shaw) Composition, Properties and Uses. AVI Publ. Inc., Westport, Connecticut.

Wong, K. C. and Siew, S. S. 1994. *Flavour & Fragrance J.*, 9: 173–178.

Woodbridge, C. G. and Clore, W. J. 1965. *Proc. Amer. Soc. Hort. Sci.*, 86: 313–320.

Yamdagani, R., Balyan, D. S. and Zindal, P. C. 1980. *Haryana J. Hort. Sci.*, 9: 141–143.

Yokomizo, H. 1974. *Bull. Fruit Tree Res. Stat.*, Hiratsuka, No. 1, pp. 57–77.

Young, T. W. and Miner, J. T. 1960. *Proc. Fla. St. Hort. Soc.*, 75: 334–336.

Young, T. W. and Miner, J. T. 1961. *Proc. Amer. Soc. Hort. Sci.*, 74: 201–208.

Young, T. W., Koo, R. C. J. and Miner, J. T. 1965. *Proc. Fla. St. Hort. Soc.*, 78: 369–375.

Yuan, R. C. and Huang, H. B. 1988. *Scientia Hort.*, 36: 281–292.

Zhakole, A. G., Negru, P. V. and Gangesh, M. V. 1979. *Referetivni Zhurnal*, 10: 785.

Zhaong, H. Q., Luo, J. T., Lu, D. and Yuan, J. 1994. *Guangdong Agric. Sci.*, 5: 19–22.

Zhuang, W., Yan, J., Lin, X., Chen, R. and Chen, J. 1983. *J. Fujian Agric. College*, 12: 297.

7

Soil Fertility Management with Wood Ash

J.C. Voundi Nkana

Institut für Bodenkunde, Universität Bonn, Nussallee 13, D-53115 Bonn, Germany; IRAD, B. P. 2123 Yaoundé, Cameroon

1. Introduction

The challenge for soil fertility management is to restore, maintain and increase the soil productivity. Agriculture and forestry remove plant nutrients from the soil. Consequently, if the production system is to be sustainable, these nutrients have to be replaced by whatever sources that are available. In many developing countries, the loss of soil fertility by crop removal, without adequate replenishment, poses a serious threat to agricultural production. In Europe, atmospheric acidic deposition has led to a drastic depletion of available soil nutrient cations. The use of external sources of nutrients such as mineral fertilizers is, therefore, essential to meet the plant requirements as well as increase the production. However, intensification of agriculture and forestry by increasing the use of fertilizers is being regulated today by economic constraints and environmental hazards. In developing countries, the high cost of fertilizers and their inadequate availability restricts intensification. In developed countries, heavy metal contamination and water pollution are the current limiting factors for further intensification.

Considering the importance of nutrient supply for agricultural and forestry production, scientific developments during the past two decades have led to more analytical approaches, projecting the feasibility of using wood ash as a soil amendment. The material is produced in large amounts by wood and paper industries and considered to be waste. Recycling of wood ash in agricultural and forest soils is important to minimize the waste accumulation in the environment instead of the expensive land-filling option of disposal.

This chapter addresses most of the results on research aspects related to fertilization of soils with wood ash. After presentation of some generalities on wood ash, several parts are devoted to the effects on soil properties, plant nutrition and growth and nutrient availability for plants. Next, there is a discussion about the factors affecting the efficiency of wood ash as a soil additive. The latter parts of the text deal with proper management and its environmental impact. The chapter is assigned to assist everyone concerned with wood ash in working towards the efficient and sustainable management of soil fertility. The author has attempted to include some results of his own research work. In assembling the material for this chapter, he has drawn liberally from the works of other investigators.

2. Generalities on Wood Ash

2.1. Definition

Wood ash is the inorganic and organic residue remaining after the combustion of wood (Campbell, 1990). For many authors (Naylor and Schmidt, 1986; Unger and Fernandez, 1990; Etiegni et al., 1991; Büttner et al., 1998), wood ash is a by-product of steam and/or energy production by wood-burning utility plants.

2.2. Amounts of Wood Ash Produced

Large amounts of wood ash are produced as waste in the wood-processing industry (paper mills and sawmills) (Table 1), where huge quantities of paper mill sludge, waste wood and bark are burned either for waste reduction or to produce steam and electricity (wood-fired electrical generating plants). According to Unger and Fernandez (1990), a single plant generating 24 MW of electricity can produce annually 4000 to 5000 tons of wood ash. On site, wood ash can be obtained by prescribed burning under forests or by bush fire used as a clearing method in slash and burn agriculture.

The proportion of ash produced depends on the tree species (temperate-woods, or tropical and subtropical woods) and/or the plant

Table 1. Wood ash production in the USA and Sweden

Country	Annual ash production (million tons)	Source
USA	1.5-3.0	Campbell (1990)
		Etiegni et al., (1991)
	1.2-1.7	Someshwar (1996)
Sweden	000	Clarholm (1994)

part (bark, wood, leaves). Temperate-climate woods are reported to yield 0.1-1.0% ash, while tropical and subtropical woods yield up to 5%. In general, hardwoods contain more ash than softwoods (Campbell, 1990), and bark and leaves produce more ash than the inner woody parts of the tree.

2.3. Characteristics

Characteristics of wood ash depend on many factors such as: temperature of combustion, cleanliness of the fuel wood, type of wood processing, type and part of the plant combusted (bark, wood, leaves), type of waste (wood, pulp or paper residue), type of soil and climate from which wood is originated from, collection and storage conditions (Someshwar, 1996).

2.3.1. Physical Properties

Wood ash has a small particle size. Sieving of dry ash samples yields more than 80% of particles <1.00 mm, the remainder consisting of non-incinerated wood (Etiegni et al., 1991; Etiegni and Campbell, 1991). The density of wood ash is low and varies from 0.27 g cm^{-3} for ash originating from wood (Huang et al., 1992) to 0.51 g cm^{-3} for ash deriving from pulp and paper waste (Muse and Mitchell, 1995). The higher density of the latter is probably due to the addition of clays and salts to pulp during the production of paper (Demeyer et al., 2001). Density of wood ash also depends on its carbon content; the greater the carbon content, the lower the density (Campbell, 1990). Etiegni and Campbell (1991) found that wood ash expands 4.5% after 1 day, 8.9% after 2 weeks and 12.5% after 4 weeks, proving that ash particles can absorb water and swell in wet conditions.

2.3.2. Mineralogical Properties

During the combustion of wood, organic compounds are mineralized and the basic cations are transformed to their oxides, which are then slowly hydrated and subsequently, carbonated under atmospheric conditions. Crystalline compounds that can be found in wood ash include calcium oxide, calcite, portlandite, arcanite, periclase, and a small amount of sylvite. Additionally, trace amounts of merwinite, anorthoclase, bredigite and larnite occur. Other crystalline compounds may also comprise riebeckite, calcium chloride hydrate, calcium silicate, hydrotalcite and serandite (Table 2). Hence, wood ash appears as a complex and heterogeneous material.

Table 2. Crytalline compounds found in wood ash (Etiegni and Campbell, 1991; Holmberg and Claesson, 2001).

Name	Chemical formula
Quartz	SiO_2
Lime*	CaO
Calcite	$CaCO_3$
Portlandite	$Ca(OH)_2$
Anhydrite*	$CaSO_4$
Gypsum**	$CaSO_4.H_2O$
Calcium aluminate*	$Ca_3Al_2O_6$
Ettringite**	$Ca_6Al_2(SO_4)3(OH)12.26H_2O$
Aluminohydrocalcite**	$CaAl_2(CO_3)2(OH)4.6H_2O$
Gehlinite	$Ca_2Al_2SiO_7$
Anorthite*	$CaAl_2SiO_8$
Periclase	MgO
Hydrotalcite**	$Mg_6Al_{12}CO_3(OH)16.4H_2O$
Hydrated calcite**	$CaCO_3.H_2O$
Merwinite*	$Ca_3Mg(SiO_4)_2$
Calcium magnesium silicate*	$Ca_2MgSi_2O_7$
Calcium silicate*	Ca_2SiO_4
Hydrated calcium silicate**	$Ca_3(SiO_3OH)2.2H_2O$
Calcium phosphate*	$Ca_5(PO_4)_3OH$
Hydrated calcium hypochlorite**	$Ca(ClO)2.3H_2O$
Microcline	$KAlSi_3O_8$
Sodium calcium aluminosilicate	$(Na,Ca)Al(Si.Al)_3O_8$
Arcanite*	K_2SO_4
Potassium sodium sulfate*	$K_3Na(SO_4)_2$
Potassium calcium carbonate*	$K_3Na(CO_3)_2$
Syngenite**	$K_2Ca(SO_4)_2.H_2O$
Sylvite*	KCl
Halite*	$NaCl$
Potassium sodium chloride*	$K_{0.4}Na_{0.6}Cl$
Serandite**	$Na(MnCa)_2Si3O_8(OH)$
Hematite	Fe_2O_3
Rutile*	TiO_2
Manesioferrite*	$MgFe_2O_4$

*only found in dry wood ash, **only found in hardened wood ash. The others are found in both forms

2.3.3. Chemical Properties

2.3.3.1. Alkalinity The alkalinity or acid-neutralizing capacity of wood ash is high and depends on its carbonate, bicarbonate and hydroxide content. It can be measured by titration with hydrochloric acid of a water extract. The titration curve often presents two points of equivalence and can be used to calculate the hydroxide/carbonate/bicarbonate ratio. The calcium carbonate equivalence (CCE) of wood ash ranges between 13.2% and 92% (Vance, 1996). The soluble alkalinity is due mainly to hydroxides (92%) and a small amount of carbonate (8%) (Etiegni and Campbell, 1991).

The alkalinity of wood ash, which is largely influenced by the temperature of combustion and the period of storage, decreases with increasing temperature and time. According to Etiegni and Campbell (1991), carbonates and bicarbonates predominate at combustion temperatures below 500°C, while oxides become prevalent above 1000°C.

2.3.3.2. Macro-elements The concentrations of major elements of wood ash measured by total analysis are extremely variable (Table 3). Mineralogical composition shows that the alkali and alkaline earth elements are present particularly as oxides, hydroxides or carbonates, explaining the superb capacity of wood ash to capture sulphur. Sulphur can be present in the order of 1.0-5.2% in the ash (Holmberg and Claesson, 2001).

Since carbon and nitrogen are generally oxidized and transformed into gaseous constituents during combustion, they are mostly present in ash in negligible quantities or even absent. Nevertheless, C and N may still exist, especially in pulp mill bark boiler ashes, as a result of incomplete combustion (Muse and Mitchell, 1995; Someshwar, 1996).

Solubility tests carried out in water and different acids by many investigators (Ulery et al., 1993; Meiwes, 1995; Voundi Nkana, 1998) showed that K is by far the most soluble element in wood ash. Calcium, Mg and K are the most acid-soluble elements, while Si and Al are least soluble. This suggests, according to Ohno (1992) and Ohno and Erich (1993), that Si and Al are structural components of ash particles.

A comparison of different types of wood ash has shown that ashes from direct wood combustion generally have higher contents of major elements than those originating from pulp and paper (Someshwar, 1996). This is primarily due to modifications brought about during manufacturing of pulp and paper. Wood ash generally has higher Ca and K contents and a lower Al content than ashes of coal. Calcium and Mg contents in wood ash are inferior to those found in liming agents currently used in agriculture.

2.3.3.3. Micro-elements The concentrations of micro-elements in wood ash are as variable as the major elements (Table 3). Iron and Mn are the most abundantly available micro-elements. Like Si and Al, Fe is barely soluble in acid solutions as it is probably part of a structural framework

Table 3. Range in elemental composition of wood ash.

Element	Holmberg and Claesson (2001)[1]	Risse and Harris (2001)[2]
	Concentration in %	
Si	3.4 (1.8-5.7)	nd
Ca	15 (6.8-23.1)	15 (2.5-33)
K	12 (2:5-25)	2.6 (0.1-13)
Al	0.84 (0.26-1.9)	1.6 (0.5-3.2)
Mg	3.6 (1.9-5.4)	1.0 (0.1-2.5)
Fe	0.36 (0.27-0.55)	0.84 (0.2-2.1)
P	0.82 (0.32-1.3)	0.53 (0.1-1.4)
S	3.3 (1.0-5.2)	nd
Cl	0.72 (0.23-1.4)	nd
Mn	0.90 (0.35-1.5)	0.41 (0-1.3)
Na	0.95 (0.27-3.7)	0.19 (0-0.54)
N	0.43 (0:4-0:59)	0.15 (0.02-0.77)
	Concentration in mg kg^{-1}	
As	26 (23-29)	6 (3-10)
B	750	123 (14-290)
Cd	16 (4.8-31)	3(0.2-26)
Cr	80 (16-213)	57 (7-368)
Cu	184 (83-244)	70 (37-207)
Pb	45 (14-62)	65 (16-137)
Hg	0.43 (0.06-1.16)	1.9 (0-5)
Mo	4 (3-5)	19 (0-123)
Ni	38 (23-69)	20 (0-63)
Zn	935 (260-1400)	233 (35-1250)

1. Mean and (range) taken from analysis of 3-9 samples
2. Mean and (range) taken from analysis of 37 samples nd= non determined

of ash particles. Consequently, its reactivity in acid soils will be very low (Ohno and Erich, 1993).

Compared to ashes of coal, the Mn, Zn and Cd contents appear to be higher while As, Se and Cr contents are lower in wood ash (Someshwar, 1996). However, due to the broad ranges, comparisons are difficult for these trace elements. Compared to liming agents currently used in agriculture, B and Mn are greater in wood ash. In addition, wood ash may contain some organic compounds, notably polyaromatic hydrocarbons, chlorobenzenes, chlorophenols, etc., but in negligible amounts.

2.4. Wood Ash as a Beneficial Material for Agriculture and Forestry

Wood ash is an excellent source of potassium, lime and other plant nutrients. Ash may have a low fertilizer equivalent (NPK 0:1:2)

(Campbell, 1990) and a low acid neutralizing capacity, when compared to commercial NPK fertilizer and lime, respectively. However, it can be used as a substitute for lime and limestone and a source of calcium, potassium, magnesium and other nutrients.

The economical benefit of using wood ash instead of lime and fertilizers is important. For example, in Germany, between 1983 and 1995, about 350 Millions € have been allocated to lime 1.8 Million ha. which represents 17% of the forested area (Meiwes, 1995). By using wood ash, which is cheaper due to the absence of production costs, money can be saved by those countries that utilize lime and fertilizers.

3. Wood Ash and Soil Properties

3.1. Microbial Activity

In general, wood ash application affects the microbial community, microbial respiration, microbial biomass C and fungal biomass measured as the soil ergosterol content. Some field investigations demonstrate increases in microbial respiration rate and metabolic quotient qCO_2 (a measure of microbial respiration per unit biomass) with wood ash fertilization. But, no change has been noticed either in microbial biomass C (fumigation-extraction method) or in fungal ergosterol, which is an indicator of fungal biomass (Fritze et al., 1994; Baath and Arnebrant, 1994; Baath et al., 1995).

Furthermore, wood ash fertilization increases soil respiration (Weber et al., 1985). The increase in microbial activity in ash-amended soils, as indicated by increased soil respiration, is often coupled with an increased growth rate of soil bacteria (Baath and Arnebrant, 1994). Sometimes, the bacterial biomass can remain unchanged while there is an increase in respiration rate, suggesting an increase in the fraction of the microbial biomass that is active over the rest of the soil community (Baath and Arnebrant, 1994). An altered community composition appeared, therefore, evident. Using phospholipid fatty acid (PLFA) patterns, Baath et al., (1995) showed that fungi are more seriously reduced by wood ash application rather than bacteria. Wood ash fertilization seems, thus, to increase the turnover of nutrients without immobilization into the microbial biomass.

3.2. Physical and Mineralogical Properties

Little is known about the effects of wood ash on the physical and mineralogical properties of the soil. However, wood ash application may affect the texture, aeration and salinity. Since wood ash is essentially composed of fine particles, its application to the soil may alter the texture of soils (Demeyer et al., 2001). Ash particles swell in contact with water and can clog the soil pores (Etiegni and Campbell, 1991), with the result of decreased aeration and water drainage in soils, as observed in prescribed burning of forest lands (Ralston and Hatchell, 1971). The electrical conductivity of the soil solution increases linearly with ash dose and may cause salinity problems (Clapham and Zibilske, 1992).

Since the dominant minerals identified in wood ash are very soluble (Ulery et al., 1993), application of wood ash has only a minor effect with respect to addition of minerals to the soils. However, the dissolution and leaching of wood ash may alter the mineral composition of the soil. Processes such as clay dispersion and dissolution of alumino-silicates are expected to occur in wood ash-amended soils.

3.3. Chemical Properties

3.3.1. Soil Acidity

Various studies (Bramryd and Fransman, 1995; Kahl et al., 1996; Voundi Nkana et al., 1998a; Mozaffari et al., 2002) have shown that wood ash increases soil pH and decreases the exchangeable Al content of acid soils. Increases in pH are more strong in soils with low pH and low organic matter content (Ohno, 1992; Saarsalmi et al., 2001). According to Vance (1996), the liming ability of wood ash is attributed to the neutralizing capacity from hydroxides and carbonates of Ca, Mg and of K. The lower exchangeable Al concentrations in the soil solution after wood ash application could be either a result of cation exchange reactions or a pH effect on Al solubility or even a combination of both (Voundi Nkana, 1998; Saarsalmi et al., 2001).

Compared to lime, wood ash reacts more quickly, resulting in a stronger pH increase, but only for a relative short period (Clapham and Zibilske, 1992; Muse and Mitchell, 1995; Voundi Nkana, 1998). This is due to the fact that the oxides, hydroxides and carbonates of K and Na, principally responsible for the acid neutralization capacity, are very soluble and do not persist for a long time in the soil (Ulery et al., 1993). According to Ohno (1992), the rate of acid neutralization decreases with time due to increasing pH of the soil and the complex composition of wood ash, consisting of highly-reactive fractions, such as oxides and hydroxides and more slowly-reacting fractions such as carbonates.

3.3.2. Major Elements

Since wood ash contains virtually no carbon and nitrogen, no significant effect on the soil organic matter can be observed. However, its application to the soil may reduce the total contents in C and N through leaching, by increasing the solubility of organic matter (Kahl et al., 1996) and the nitrification rate (Weber et al., 1985; Meiwes, 1995; Pietikäinen and Fritze, 1995), without any effect on C/N ratio (Saarsalmi et al., 2001).

Wood ash is essentially a direct source of other major elements, notably P, Ca, Mg and especially K in soils (Meiwes, 1995; Kahl et al., 1996; Williams et al., 1996; Saarsalmi et al., 2001). The dissolution of wood ash in soil and the rate at which nutrients become available to the plant is more complicated than that of lime. Firstly, wood ash contains several cations and each of these cations forms oxides, hydroxides, carbonates and bicarbonates (Ohno, 1992; Erich and Ohno, 1992a; Ulery et al., 1993) which dissolve at different rates (Meiwes, 1995). Secondly, the change in soil nutrient availability is a combination of three factors: (i) nutrient addition from the ash; (ii) shifts in pH-dependent soil equilibria; and (iii) changes (mostly increases) in microbial activity and dissolution of organic matter (Vance, 1996; Voundi Nkana, 1998).

In an ammonium acetate extract, which is generally considered as an indication of nutrient availability, Meiwes (1995) has found 81% of total Ca, 57% of total Mg, 34% of total K and 20% of total P at pH 4.2. In general, P is the least available major nutrient in wood ash. In short-period incubations, Ohno (1992) found in the soil suspension less than 1% of the P added with wood ash. Nevertheless, this was higher than the percentage water soluble P, indicating that the concentration of P in the soil solution is also a function of the desorption of adsorbed P, following the increase of soil pH. Wood ash P is most probably occluded in alumino-silicates or in the form of weakly-soluble aluminium phosphate (Ohno and Erich, 1990; Erich, 1991; Erich and Ohno, 1992b).

As already mentioned for P, the adsorption capacity of the soil will also determine the availability of nutrient elements from wood ash. Certain fractions of elements found in the soil solution after wood ash treatment originated from the soil either due to interactions of Ca, Mg or K with the exchange complex (Ohno, 1992).

3.3.3. Trace Elements

Wood ash contains significant amounts of micro-elements and one may expect that soil amendment with wood ash will remarkably increase their concentrations in the soil solution. However, application of wood ash to the soil initially results in a reduction of the solubility of Fe, Mn, Zn and Cu (Clapham and Zibilske, 1992; Krejsl and Scanlon, 1996) due to the increasing soil pH. As soil pH decreases repeatedly over a period time, trace elements from the ash and soil become more mobile and plant available. Zhan et al. (1996) have studied the release of trace elements from wood ash as a function of pH by adding a range of nitric acid concentrations to the ash. Cadmium and Zn were readily released from wood ash at pH values of 6.5 and below. Chromium release behaviour varied between wood ash samples. Chromium (VI) level in water extracts was a good indication for Cr solubility. The solubility of Cu and Pb were low and very low, respectively.

3.3.4. Cation Exchange Capacity and Base Saturation

Addition of wood ash obviously increased the cation exchange capacity of the soil (Kahl et al., 1996; Eriksson 1998; Voundi Nkana et al., 1998a; Saarsalmi et al., 2001). This increase can be mainly attributed to the dissociation of pH-dependent exchange sites induced by an increase in soil pH. Wood ash fertilization significantly increased the degree of base saturation, but it levelled out the differences between soils with different fertility (Kahl et al., 1996; Saarsalmi et al., 2001). There was an indication of an increase in exchangeable bases Ca, Mg, K and Na, probably reflecting the solubility of the oxides.

3.3.5. Soil Solution

Numerous authors (Kahl et al., 1996; Lehnardt, 1998; Ludwig et al., 1999; Rumpf et al., 2001) have observed that the application of wood ash significantly increased soil solution pH and EC and this increase is often associated with increased concentrations of DOC, inorganic C, Cl, NO_3, SO_4, Ca, Mg, and K. Wood ash application affects the soil solution chemistry in two ways, as a liming agent and as a supplier of nutrients (Voundi Nkana, 1998).

As a liming agent, wood ash application induced increases in soil solution pH, Ca, Mg, inorganic C, SO_4 and DOC, as suggested by many authors in the case of lime application (Voundi Nkana, 1998; Bakker et al., 1999; Derome and Saarsalmi, 1999; Hildebrand and Schack-Kirchner, 2000). The increase in SO_4 concentrations could be attributed to the reduction in the number of positive charged sites in the soil induced by increased pH and consequent desorption of these anions from soil exchange sites (Curtin and Smillie, 1983; Su and Evans, 1996). The increased DOC was probably due to desorption or dispersion associated with the increases in pH (Curtin and Smillie, 1983).

As a supplier of elements, the increase in the soil solution pH with wood ash was partly due to ligand exchange between wood ash SO_4^{2-} and OH^- ions (Alva and Summer, 1990; Alva et al., 1990). This explains the relatively higher soil solution pH values of wood ash treatments compared to that of lime treatments (Voundi Nkana, 1998). Large increases in concentrations of inorganic C, SO_4, Ca and Mg with wood ash, relative to lime and especially increases in K, reflect the supply of these elements by wood ash and its high solubility (Ohno, 1992). The SO_4 additions with wood ash probably resulted in higher DOC concentrations; the sorbing SO_4 seems to displace organic polyelectrolyte molecules from the soils (Kaiser and Kaupenjohann, 1998). The relative higher NO_3 levels in soil solution from wood ash-amended soils suggest that this by-product enhanced nitrification more efficiently than lime (Voundi Nkana, 1998). The improvement of nitrification seemed to be the result of induced output of DOC, since according to Hildebrand and Schack-Kirchner (2000), this output is a potential energy source for heterotrophic nitrifiers. According to Ohno (1992), wood ash amendment can result in increased soil solution P. The weak effect of wood ash on P concentrations in soil solution can be explained by the low release rate of P from wood ash to soil (Fransson et al., 1999) and/or soil sorption that even exceeds release in those soils which exhibit high P adsorption capacity (Voundi Nkana, 1998).

Data on soil solution trace elements have been scarcely reported. For essential trace elements, due to the effect of increase in soil solution pH and depending on their initial content in the soil, one could expect a general decrease in soil solution concentrations after application of wood ash. For non-essential trace elements, a recent study (Rumpf et al., 2001) mentioned that the concentrations of Zn and Cd increased significantly, while the concentrations of Pb and Cr were less affected. Levels found were not a concern in terms of soil conservation and environmental quality.

There is evidence that wood ash application could represent the increased inavailability of the nutrient base cations, SO_4 and NO_3. However, large concentrations of basic cations, SO_4 and NO_3, resulting from higher application rates, could be a concern because of potential solute transport to surface and ground waters.

4. Effect of Wood Ash on Plant Growth and Nutrient Availability

4.1. Plant Growth and Crop Yield

After application of wood ash to the soil, several cultivated plants showing an increased growth and/or yield include oat (*Avena sativa* L.), winter wheat (*Triticum aestivum* L.), fescue (*Festuca arundinacea* Schreb.), spinach (*Spinacia oleracea* L.), bean (*Phaseolus vulgaris* L.), corn (*Zea mays* L.), poplar (*Populus* sp.), soy (*Glycine max* (L.) Merr.), and ray grass (*Lolium* sp.) (Erich, 1991; Etiegni et al., 1991; Huang et al., 1992; Clapham and Zibilske, 1992; Erich and Ohno, 1992b; Muse and Mitchell, 1995; Krejsl and Scanlon, 1996; Voundi Nkana et al., 1998a; Mozzafari et al., 2000). In some cases, plant dry matter production can decrease or remain unchanged following wood ash application, depending on factors such as ash type and rate of ash applied, plant species and soil properties. Oat (*Avena sativa* L.) grown in soil amended with 50 t ha^{-1}. of wood ash showed a decline in dry matter yield when compared to yields at lower ash rates (Krejsl and Scanlon, 1996). Wood ash application had no effect on the production of spinach (*Spinacia oleracea* L.) either in the greenhouse (Clapham and Zibilske, 1992) or on wheat yield in a field study (Huang et al., 1992). The specific cause for slower growth and non-growth at higher ash rates could be due to pH, which exceeded the threshold suitable for growth and/or nutrient uptake (Etiegni et al., 1991).

With regards to tree growth, Ferm et al., (1992) have observed for Scots pine grown on peatland (C/N ratio: 17-24) a four to five fold increase in volume of the growing stock 13 years after addition of 20 t ha^{-1}. wood bark ash. This was mostly due to the increased availability of K and B, as evidenced by the concentrations in the needles. Scots pine on poor Scandinavian sites (C/N >30), usually reacts with a slight growth decrease during the first 10-20 years. According to Bramryd and Fransman (1995), this can be caused by a decrease in N-mineralization or by N-immobilization in microorganisms, because of the high C/N ratio of the soil. McDonald et al., (1994) have studied the efficacy of different organic wastes on the growth of western red cedar (*Thuja plicata* Donn ex D. Donn) and observed that wood ash alone did not affect tree growth; combination with N-rich waste had a positive effect on the growth.

However, once the adequate fertility is established in the root zone, the extra fertility imparted by the higher ash concentrations provides little increase in growth. Higher ash concentrations may provide benefits to yields over the longer term, but the existing data could not substantiate this consideration.

4.2. Plant Nutrient Content and Uptake

Modification of soil chemical properties by wood ash application is likely to affect elemental composition of plants. Since wood ash contains virtually no N, its application normally will not affect the total N in plants (Voundi Nkana et al., 1998a) but in some cases, an increase in N uptake can be observed. Increase in N uptake could be attributed to enhanced nitrification as a result of the liming effect (Meiwes, 1995; Pietikäinen and Fritze, 1995), subsequent to marked increase in microbial activity (Weber et al., 1985). Combination of wood ash application with additional N may be necessary if a balanced fertilization is required (Voundi Nkana, 1998).

Phosphorus, Ca, K and Mg contents are remarkably affected by applications of wood ash (Krejsl and Scanlon, 1996; Voundi Nkana et al., 1998a; Mozaffari et al., 2000; Arvidsson and Lundkvist, 2002). All existing studies indicate that P contents in plants are not increased substantially and could restrict the crop yields. Comparative studies on P uptake by corn have shown that ash is significantly less efficient than P fertilizers (Erich and Ohno, 1992b). However, wood ash is a source of P (Ohno and Erich, 1990) and its application should increase nutrient soil test values for P (Ohno, 1992; Krejsl and Scanlon, 1996). Some cansative factors include low P solubility in wood ash due to high pH (Ohno and Erich, 1990; Erich and Ohno, 1992b), P immobilization in the soil mainly by fixation and/or to a limited extent, by precipitation with Ca, and low P desorption from the soil as a result of increase in pH (Erich, 1991; Ohno, 1992; Voundi Nkana et al., 1998a).

Calcium and K contents in plants increase noticeably following wood ash application. The increase in plant K levels with wood ash application is the result of the high K content, combined with high solubility of K in wood ash (Meiwes, 1995). The average concentrations found by Voundi Nkana et al., (1998a) in ray grass were generally above the normal range for *Lolium* spp (25000-35000 mg kg^{-1} DM) but were not that high to become toxic. In other studies, wood ash increased the K concentration in corn (*Zea mays* L.) and winter wheat (*Triticum aestivum* L.) in greenhouse studies (Erich, 1991; Etiegni et al., 1991) and alfalfa (*Medicago sativa* L.) in field studies (Meyers and Kopecky, 1998). Since there is a large input of basic cations with wood ash application, these elements are more absorbed by plants (Erich and Ohno, 1990). The uptake of K is most spectacular in such a way that Erich and Ohno (1992b) proved that the availability of wood ash K is similar to fertilizer K. According to Etiegni et al. (1991) K toxicity can be the cause of yield depression at high wood ash application rates. The inconsistencies in the plant content and uptake of Mg following wood ash application are probably the result of the interaction with Ca and/or K.

Application of wood ash decreases tissue Zn, Fe and Mn and increased that of B and Mo (Francis et al., 1985; Naylor and Schmidt, 1986). The concentrations of Mn and Cu in bean (*Phaseolus vulgaris* L.) plants decrease with wood ash application (Krejsl and Scanlon, 1996). Their uptake is affected accordingly. The decrease in minor elements (Fe, Mn, Zn and Cu) in plants and their uptake with the application of wood ash is often attributed to alkalinity brought about with wood ash, which reduced their solubility (Krejsl and Scanlon, 1996; Voundi Nkana et al., 1998a). Uncontrolled fertilization with high amounts of wood ash may even induce micronutrient deficiencies (Krejsl and Scanlon, 1996).

The agronomic effectiveness of nutrients was estimated in pot experiments by the coefficient of utilization (Voundi Nkana et al., 1998b). The coefficient of utilization is calculated as the percentage of a wood ash nutrient that has been taken up by the plant. Nutrient uptake from wood ash is calculated as the difference between nutrient uptake from the wood ash treatment and nutrient uptake from the lime treatment at the same target pH. Resulting data (Table 4) showed that wood ash is an important source of macronutrients P, Ca, Mg and K and micronutrients Mn, Zn and Cu for plant uptake, Fe being less taken up by plants. The low recovery of Fe by plants was caused by the low solubility of Fe in wood ash (Ohno, 1992; Ohno and Erich, 1993). The low or negative utilization coefficient for Ca was due to a higher Ca addition with lime. Coefficients of utilization decreased with higher application rates, as expected. Unused fractions were probably assigned to other nutrient pools in the soil (e.g. solid phase).

In general, proportions of nutrients in land-applied wood ash taken up by plants reflect their concentrations in the ash, their relative solubility, any pH-induced changes in soil nutrient availability and the crop requirements (Vance, 1996).

Table 4. Coefficient of utilization of wood ash nutrients after three growth cycles of rye grass in three tropical acid soils (Voundi Nkana et al., 1998b)

Soil	Target		Coefficient of utilization (%)					Identification	
	pH	P	Ca	Mg	K	Fe	Mn	Zn	Cu
MBA	5,5	74,7	4,1	18,6	88,2	0,2	9,1	5,2	5,0
	6,0	48,8	0,9	9,9	83,3	0,2	10,1	12,7	3,5
	6,5	30,6	0,7	9,7	63,2	0,2	3,6	8,5	2,6
MEN	5,5	31,8	0,0	6,8	90,5	0,8	7,8	9,7	0,8
	6,0	24,2	0,0	5,9	70,4	<0.1	3,9	6,2	0,3
	6,5	22,6	0,0	7,5	52,3	<0.1	2,8	3,9	1,0
NKO	5,5	42,8	3,2	14,5	90,3	0,5	12,8	14,0	1,4
	6,0	53,4	2,7	15,7	108,8	<0.1	7,8	19,3	5,3
	6,5	36,7	1,8	12,2	88,3	0,3	4,2	12,7	5,4

4.3. Nutrient Availability

Wood ash application influences the availability of plant nutrients, either directly through additions of its nutrient constituents or indirectly, through the modification of soil properties.

Numerous soil extractions have been studied by various investigators in order to assess the availability of nutrients in wood ash-amended soils. Based on Bray, Olsen, NH_4OAc or NH_4-citrate extractable phosphorus, wood ash application increase the soil available P (Voundi Nkana et al. 1998a; Mozaffari et al., 2002). According to Mozaffari et al. (2002) the relationship between ash application rate and Olsen-extractable P was best described by a linear regression. For Erich and Ohno (1992), citrate-extractable P as a measure of plant-available P overestimates the availability of wood ash P. However, citrate-extractable P is an improvement over total P as an estimate of the amount of wood ash P available for plant uptake. An increase in the availability of P following wood ash application could be the result of increased soil solution P. This is, according to Erich and Ohno (1990), achievable in two pathways. First, the soluble fraction of wood ash P would be released to the soil solution as the wood ash dissolves in the neutralization reaction of soil acidity. On the other hand, the addition of wood ash raises the soil pH which will support a higher concentration of P in the solution. All authors found that ash application increases exchangeable Ca, Mg, K and Na determined in 1 M NH_4OAc or 1 M NH_4Cl or 0.1 M $BaCl_2$ extracts for all soils and the pattern of increase in exchangeable cations is consistent with their concentrations in the ash (Kahl et al., 1996; Voundi Nkana et al., 1998a, 1998b; Ludwig et al., 1999; Mozaffari et al., 2000; Saarsalmi et al., 2001; Mozaffari et al., 2002). Extractable micronutrients using DTPA or NH_4OAc were generally reduced by ash application (Voundi Nkana et al., 1998a; Mozaffari et al., 2002). Reduced extractability was probably caused by the liming effect of wood ash (Krejsl and Scanlon, 1996; Voundi Nkana et al., 1998b; Mozaffari et al., 2002).

Nutrient additions and extractable soil nutrients are generally positively correlated with plant uptake of P, Ca, Mg and K and negatively correlated with that of Mn and Zn. The positive correlations indicate the availability of these nutrients from wood ash (Voundi Nkana et al., 1998b). Negative correlations for Mn and Zn are due to the lower solubility and availability at increasing soil pH. The strength of relationships depends on the initial levels of nutrients in the soil (Voundi Nkana et al., 1998b). In general, both nutrient additions and nutrient extractions from wood ash-treated soils provided a reliable measure for plant available nutrients.

Soil pH, cation exchange capacity, soil organic matter and Fe, Al and Mn oxides have important influences on the solubility and availability of nutrients in wood ash amended soils (Vance, 1996). Voundi Nkana et al., (1998b) assessed the influence of pH and ECEC on nutrient availability. Correlations including pH and ECEC present similar trends since ECEC of the soils used is pH dependent. Soil pH and ECEC are positively correlated with the uptake of N and Ca and negatively correlated with uptake of K, Mn and Zn uptake (Table 5), indicating that wood ash-amendments increase soil pH and ECEC and, at the same time, increase the availability of N, P, Ca, Mg and K for plant uptake but decrease the availability of Mn and Zn.

Table 5. Correlation coefficients (r) between pH and ECEC of the soils and nutrient uptake by plants in three wood ash amended tropical acid soils (n = 4) (Voundi Nkana et al., 1998b).

Soil characteristic Identification		Plant uptake						
		N	P	Ca	Mg	K	Mn	Zn
MBA	pH	0 01	0 01	0 01	0 01	0.95*	− 0.99**	0 01
	ECEC	0 01	0 01	0 01	0 01	0 01	− 0.95*	000
MEN	pH	0 01	0 01	0.96*	0 01	0.96*	− 0 01	− 0 01
	ECEC	0 01	0 01	0 01	0 01	0 01	− 0 01	− 0.98*
NKO	pH	0.97*	0.97*	0.98*	0 01	0.98*	− 0.97*	− 000
	ECEC	0 01	0 01	0 01	0 01	0.98*	− 0 01	− 000

*, ** Significant at 0.05 and 0.01 probability levels, respectively.

The beneficial effects of soil pH on N uptake resulted from better growth conditions in general and from enhanced mineralization of organic matter subsequent to increase in pH due to liming effect (Doerge and Gardner, 1985). For P, increased soil pH led to a decrease in phosphate adsorption due to a corresponding decrease in positively charged soil exchange sites and thereby, an increase in its availability (Erich and Ohno, 1992b; Voundi Nkana et al., 1998b). An increase in pH is always associated with increase in the soil saturation rate and the levels of exchangeable Ca, Mg or K. It is widely recognized that raising soil pH markedly reduces the uptake of Mn and Zn and restricts their availability. A general reduced availability of micronutrients Mn and Zn with wood ash application is, therefore, the result of increase in soil pH (Krejsl and Scanlon, 1996; Voundi Nkana et al., 1998b).

4.4. Crop Quality

According to Naylor and Schmidt (1989), the quality of crops such as hay is generally improved by applying wood ash. This improvement is not uniform with the quantity applied, but in each case, ash increases the mineral and crude protein content of the plant over that of the untreated control. The only detrimental effect of wood ash application on crop quality could be the critical concentration of Mo in plants, which may induce a Cu deficiency. The possibility of feeding an animal with forage crops grown on soils amended with very high ash rates may present a potential problem of molybdenosis, if supplemental Cu is not included in the animal's diet.

5. Factors Affecting Wood Ash Efficiency

5.1. Composition

Wood ash generally contains reactive oxides in addition to significant amounts of soluble salts. These types of compounds can dissolve easily in the soil and could result in high alkalinity and salinity conditions. Effects such as pH shock, initial release of ample amounts of most nutrients accompanied with nutrient imbalances, disturbance of soil microbial activity, burning damage of plant tissues and salt effects can be observed. Large leaching losses could occur in the short term in wood ash-amended soils, when amounts of nutrients in soil solution exceed that which plant roots can take up. This could affect its efficiency with regards to plant growth.

5.2. Solubility

The composition of wood ash suggests that it contains degrees of very soluble, less soluble and insoluble constituents. The solubilities, distributions and potential solubility controlling solid phases of elements contained in wood ash have been less studied. The total of ash dissolving does not change significantly, as approximately 10% of the ash dissolves at 50 g L^{-1} and 9% at 390 g L^{-1} (Etiegni and Campbell, 1991). The mineralogical speciation of wood ash, determined by Steenari et al., (1999a) showed that Ca occurs in a variable compounds, among which dominant CaO and Ca(OH)$_2$ are very soluble, while CaCO$_3$, calcium silicates and calcium aluminium silicates have a low solubility. The speciation of K is dominated by salts such as sulphate, chloride and carbonate which are all soluble. The only K-compound which has a low solubility is the microcline KAlSi$_3$O$_8$. Results from XRD analysis of several wood ash materials have indicated that P occurs in apatite and other phosphates associated with Ca (Steenari and Lindqvist, 1997). The amounts of elements found in the dissolved fraction of the ash suggest that elements such as SO$_4$, Na and K are highly soluble, whereas Mg and Ca are less soluble (Ludwig et al., 1999). Based on the dissolution experiment with different ratios of ash to water, Khanna et al., (1994) found that K, B and S were rapidly dissolved, Ca, Mg, Si, Fe and Al dissolved with increasing dilution while P remained relatively insoluble, implying that elements such as Na, K, S, B as well as Ca found in wood ash will exhibit greater plant availability and mobility through the soil profile than the others.

5.3. Leaching

Leaching of elements contained in wood ash is likely to be related to element speciation. Elements present as soluble salts are subjected to an initial rapid leaching (Steenari et al., 1999a). According to Voundi Nkana et al. (2000) the dominant ions generally present in leachates are Ca^{2+}, K^+, Na^+, OH^-, Cl^- and SO_4^{2-} (Table 6). The leaching rate for Mg and P are low; Mg being present in the ash in the form of MgO and magnesium silicates and P in the form of apatite which have low solubility at high pH. On the other hand, in a solution with high level of Ca^{2+}, phosphate ions tend to associate with calcium ions and precipitate as solid phosphates (Gray and Schwab, 1993). The leaching of Si and Al from wood ash is low as a result of the low dissolution kinetics of Si and Al-bearing components (Steenari et al., 1999b). The leaching amounts of Fe, Mn, Zn and Cu are generally low and those of Pb, Cd, and Hg very low when present in the ash (Steenari et al., 1999a). Iron and Mn, which probably occur in an oxide form in wood ash, tend to form hydroxides in contact with aqueous solution at a high pH; the hydroxides are only slightly soluble (Morel and Hering, 1993).

Table 6. Cumulative leaching losses of nutrients from wood ash-amended tropical acid soils (Voundi Nkana et al., 2000).

Soil Identification	Rates $(g.kg^{-1})$	Ca	Mg	K	Na	NO_3	Cl	SO_4	IC	DOC
						$(mg.kg^{-1})$				
Mb	0	6	3	5	5	8	10	52	2	185
	4,8	37	8	17	5	8	8	89	23	195
	6,4	73	13	26	11	23	11	132	48	225
	12,8	364	53	118	13	69	10	210	267	398
Me	0	13	5	15	4	19	13	131	3	230
	3,5	54	11	35	13	116	15	205	19	358
	5	84	17	42	11	93	15	160	33	361
	10	230	39	100	13	174	32	248	110	544
Nk	0	13	7	6	6	94	25	68	8	116
	1	31	14	11	5	147	16	90	24	88
	2	42	17	17	6	85	17	93	40	128
	4	103	33	40	9	82	16	94	111	173

IC = inorganic carbon; DOC = dissolved organic carbon.

Furthermore, calcite, quartz, feldspars and other minerals occurring in the ash-water system obtain a negative surface charge in a surrounding with high pH and, thus, positively-charged metal ions such as Zn^{2+} and Cu^{2+} may be adsorbed onto such surfaces and they are leached out in low amounts. Leachates produced from wood ash could enhance increase pH, DOC and salt content of the soil solution, increase metal leachability, disturb the microorganisms in the soil and may affect nutrient availability and surface and ground water quality.

5.4. Stabilization

To avoid detrimental effects occasioned by leaching of wood ash components, modification of physical characteristics of wood ash by carbonation/granulation or self-hardening have been suggested.

Carbonation involves the transformation of CaO over $Ca(OH)_2$ to $CaCO_3$ to reduce the solubility and provide the liming effect, which is highly desirable (Steenari and Lindqvist, 1997). According to Ohlsson (2000), carbonation of wood ash includes: addition of water to dry and finely dispersed ash, formation of granules (approximately spherical) of diameters ranging between 1 and 10 mm and allowing the wet ash to react with atmospheric CO_2 for several days. This carbonation process produces a hardened (or stabilized) ash that is easier to handle; presents less problems with dust and dissolves more slowly when placed in contact with soil (Ericksson et al., 1998) compared with the dry untreated ash. However, carbonation results in a solid product that is more brittle than the product that is only hydrated. The so-called self-hardening process is a simple and cheap stabilization method based on the fact that ash materials have the ability to solidify or self-harden on addition of water. If the ash possesses a high content of combustible matter (>10%), solidification is hindered (Steenari and Lindqvist, 1997).

It will probably be most convenient to spread wood ash when it has reached a degree of carbonation where the granules have an appropriate particle size distribution. Additional carbonation of the particles will take place after spreading. Leaching of K, Na, SO_4 and Cl is not significantly lowered by hardening because Na and K and their counter ions Cl^- and SO_4^{2-} are retained in the pore solution but not incorporated in solid phase during hardening (Steenari et al., 1999a). However, leaching of K is significantly lower in large-sized particles rather than smaller sized ash granules (Steenari et al., 1998). The hardening process transforms the calcium of the ash through the following steps: $CaO \rightarrow Ca(OH)_2$ (portlandite; in wetted ash) $\rightarrow CaCO_3$ (Calcite; in carbonated ash) $\rightarrow CaSO_4.2H_2O$ (gypsum) $\rightarrow$ ettringite $\rightarrow$ hydrated silicate and aluminium silicate phases (Steenari and Lindqvist, 1997), where portlandite is considerably more soluble than calcite (Ohlsson, 2000). Reaction with atmospheric CO_2 and lowering of pore liquid pH caused by weathering results in decreased stability for the ettringite in some ash materials, leading to formation of $CaCO_3$ and soluble gypsum (Steenari and Lindqvist, 1997).

Other factors may also affect the ash Ca dissolution rate; notably, swelling and hydration characteristics of ash (Etiegni and Campbell, 1991), ash porosity and chemical inhibition of the dissolution process (Steenari et al., 1998). The transformation of Ca occurring during the hardening process is important since it decreases the calcium leaching rate significantly (Steenari et al., 1999a). Thus, the liming effect of the ash is extended and possible shock effects of high pH is avoided. In some ash materials, the formation of ettringite during hardening contributes to a decrease in Ca and S leaching. Hydration and solidification of ash result in a low leachability for most metal species, but when the alkalinity of the ash decreases, the trace metals will become mobile (Steenari and Lindqvist, 1997).

6. Wood Ash Application

6.1. Useful Considerations

Campbell (1990) and Risse and Harris (2001) evoked several considerations to take into account when planning for wood ash application. Prior to ash application, the soil nutrient level and pH of the field must be determined through proper soil testing. Soil testing laboratory should be consulted in order to obtain a bulletin outlining the proper procedures for soil sampling and determining the proper liming rates. If the fields contain variable soils, they should be divided into blocks with similar characteristics and sampled individually. It is also important to obtain a representative sample.

The liming ability of wood ash is usually estimated using a laboratory measured parameter called the calcium carbonate equivalent (CCE). With proper soil testing and considering the calcium carbonate equivalent of the wood ash, the application rate for wood ash can be calculated by dividing the recommended lime application rate by the CCE of wood ash, as illustrated by the formula:

$$AR = \frac{LR}{CCE/100}$$

Where *AR* is wood ash application rate in tons per hectare and *LR* is the recommended liming rate in tons per hectare.

During application of wood ash to the soil, special care should be taken to prevent the ash from entering any surface or ground water. A minimum distance should separate the wood ash from any farm ditches, wells or other bodies of waters. Buffer vegetation zones should be put up. Wood ash should not be applied to areas with water standing on the soil surface. Care should also be taken to avoid wood ash applications immediately preceding periods of prolonged rainfall or when large storms are expected.

Wood ash should be land applied as soon as possible in order to avoid the need for on-site storage. When conditions such as inclement weather do require on-site storage, wood ash should be stored in a manner that blocks the previous runoff particulate from entering the surface or ground water. Indoor storage is ideal. However, when it must be stored outdoors, it should be placed on packed soils or in surrounded by pads to prevent the surface water from entering or leaving the storage area. The storage area should also be located away from wells, surface water and animal watering areas and covered or shielded as much as possible to prevent nuisance conditions if it were moved or disturbed during dry or windy weather.

One of the major obstacles to land spreading of wood ash is the undesirable handling and spreading characteristics of ash. Most ash has a low density and small particle size and consequently, creates dust problems during transport and application. Wood ash should always be covered during transport to prevent losses on route to application sites. The handling characteristics of ash are generally improved with increasing humidity. Attempts should be made to avoid spreading in extremely dry days. Moisture can be added to improve the handling characteristics of ash. However, if too much moisture is added, the ash will cake and not spread uniformly.

Health considerations must be taken into account when dealing with ash so as to prevent both particle inhalation and contact with skin. Inhaling any small particle is dangerous. Masks should be worn during application or when dusty conditions warrant them. Ash is an alkaline material with a pH ranging from 9-13. Therefore, this material could irritate the skin. To prevent this, skin should be covered during application and transport and skin areas exposed to ash should be washed and thoroughly rinsed with water immediately following application.

6.2. Agricultural Land Application

Wood ash can be used to lime acidic agricultural soils and replace micro- and macro-elements removed during crop growth and harvesting. Liming with wood ash may also reduce the toxic effect of Al and Mn in acidic soils (Voundi Nkana et al., 1998a). The application rate should be limited to a level that maintains the soil pH within acceptable ranges for plant growth. If additional fertilizer applications are being used, they should be formulated to ensure that the plant requirements for these elements are not being exceeded, especially on crops that are known to be sensitive to particular nutrients such as P and K.

Application of wood ash to agricultural fields can be mechanically assisted, by spreading the ash with conventional manure spreading or lime broadcasting equipment. Ash may be either top dressed or incorporated, but these spreaders are not designed for the higher application rates required for spreading ash (Wilhoit and Ling, 1996). To get maximum benefit, incorporate the wood ash throughout the root zone whenever possible as the benefits only occur where the ash and soil are in contact. It is also essential to calibrate the spreader to meet the target application rate. Due to the physical characteristics of ash, it is often difficult to obtain a uniform application, but calibration and knowledge of the application distribution of the spreader can help to minimize any non-uniformity.

Ash should not be applied immediately preceding planting or during early emergence as it could cause short-term concentrated alkaline conditions that could interfere with plant growth. Ash may also absorb pesticides if it is not given time to neutralize in the soil; so chemical applications should be avoided for three to five days prior to or after wood ash application. Agricultural soils could routinely receive repeated applications of ash.

6.3. Forest Land Application

Utilization of wood ash on land by spreading over forest soils enables the replacement of nutrients removed by intensive forest growth, harvesting practices and losses of basic cations due to atmospheric acidic deposition, which can be locally intensified by excessive biomass removal. The consequence of atmospheric acidic deposition is the development of Al toxicity to tree roots, limiting their growth or even causing die-back. Additions of wood ash would decrease Al concentrations in the soil solution by raising the soil pH. Applying wood ash to forest soils would also help to reduce the soil acidity as well as improve the nutrient supply. One problem with forest land spreading is that the pH requirements for forest trees are not fully established and the ash could increase the soil pH, favouring the growth of undesirable hardwoods over softwoods.

It is more difficult to apply wood ash to forest land than to agricultural land. Access can be limited either by rough conditions or the presence of forest stands. Spreaders must be rugged and be able to throw the material at a long distance. Equipment for spreading wood ash must be able to operate reliably under rough conditions at the required application rates and swath widths.

Several different types of spreading mechanisms have been used to spread ash as also other similar materials. One of the main types used for spreading wood ash is the horizontal spinner, a centrifugal-type of spreader. This is the standard type of spreader used for spreading lime and poultry litter and has been used for several spreading operations also. The performance of this type of spreading mechanism broadcasting granular materials such as fertilizer has been investigated extensively (Broder, 1983). However, little has been reported about the performance of this or any other mechanism broadcasting non-granular materials. Wilhoit and Ling (1996) have tested a specialized tractor for hauling logs out from the forest to conduct distribution pattern trials of spreading wood ash. They observed that wood ash could be spread satisfactory at a lower application rate (7.1 cm s^{-1} conveyor speed) at swath widths up to about 12 m. Wider swath widths which will be desirable from productivity standpoint in many land application situations should be possible with this type of spreader, through the optimization of material conditions and spreader-operating parameters.

7. Environmental Considerations

Most authors stress that wood ash used as a soil amendment should originate solely from burning of forest residues or untreated wood (Meiwes, 1995; Bramryd and Fransman, 1995; Zollner and Remler, 1998). Such ashes generally contain few noxious elements. Heavy metal loads increase from the bottom ash to the boiler and fly ash. Therefore, application of bottom ash should be preferred (Zollner and Remler, 1998). Ashes from demolition wood, painted or impregnated wood can be highly contaminated with heavy metals, rendering them unsuitable. Campbell (1990) suggests that the strong increase in soil pH and available K, brought about by wood ash addition, is often the most limiting factor of its application instead of the contamination with noxious elements. The doses based on the lime requirement, estimated either on the actual base saturation and the CEC or on the exchangeable Al of the soil, represent a negligible risk for the environment.

Many organic compounds of environmental concern such as PAHs, PCBs, chlorbenzenes and chlorophenols appear in negligible quantities in wood-fired boiler-ashes and the contamination risks are minimal (Someshwar, 1996; Vance, 1996). However, dioxin levels may be elevated, especially in ash from boilers working at a lower temperature when high levels of chloride are present in the fuel mixture as from wood stored in ocean water or from bleached kraft mill sludge (Someshwar, 1996).

There is also a probability of nutrient losses and contamination of surface and ground waters after soil amendments with wood ash

(Ohno, 1992; Kahl et al., 1996; Williams et al., 1996). Two studies have shown that wood-derived boiler ashes applied to forest land would not threaten the surface water and groundwater quality due to metal or nutrient contamination (Steponkus, 1992; Williams et al., 1996). However, all authors acknowledge that, at reasonable wood ash rates, there is no risk for the environment.

8. Conclusions

Wood ash is a fine material with low density, containing crystalline compounds—most of them consisting of lime materials. In addition, ash is composed of many major and minor elements needed for plant growth. Soil fertility management with wood ash is a way of reducing waste accumulation in the environment and fertilization costs.

Applying wood ash to the soil resulted in the increase of microbial activity, thus, leading to enhanced mineralization of organic matter, soil pH and contents of most major nutrients and decreases Al and minor element mobility. Changes in soil solution composition of wood ash-amended soils and in plant response to wood ash application is principally the result of a different chemical balance in the soil and availability of the nutrients elements. The benefits on the growth of plants are the result of an increase in availability of mainly P, Ca, Mg, K, B and a decrease in Al and Mn toxicity. Wood ash fertilization appeared as a mean to counteract the natural and anthropogenic acidification of agricultural and forest soils, loss of nutrients resulting from crop and tree harvesting, maintenance and improvement of the nutrient balance. Wood ash is a valuable residual material appropriate for management of fertility of naturally-tropical acid soils and forest lands from the northern hemisphere that are suffering from acidification by acid deposition and clear-cutting. However, because of the absence of N in wood ash, this can only part of a total fertility management strategy. Depending on the nutritional requirements of the plant species, wood ash can be used either alone or in combination with other fertilizers. More research has to be undertaken in order to test ash-fertilizer combinations.

The high solubility of some of its constituents and their susceptibility to leaching under humid conditions can lower the efficiency of wood ash as a liming agent as well as a nutrient supplier. The stabilization of the material has been proposed to solve the problem. This also helps to facilitate the handling operations and land application of wood ash. For application in forest soils, more knowledge is required, especially with respect to pH requirements of forest trees. Spreading equipment for small scale farm have to be tested.

Utilization of wood ash in agriculture and forestry does not present any risk for the environment, provided that reasonable amounts are applied and ashes from untreated wood material are used.

Acknowledgements

The author wishes to thank Prof. Dr. Ir. F. M. G. Tack from the Laboratory of Analytical Chemistry and Applied Ecochemistry, University of Gent (Belgium) and Dr M. Hamer from Institut für Bodenkunde, Universität Bonn (Germany) for reviewing the manuscript and offering many helpful suggestions.

References

Alva, A. K. and M. E. Sumner. 1990. Amelioration of acid soil infertility by phosphogypsum. *Plant Soil* 128: 127-134.

Alva, A. K., M. E. Summer and W. P. Miller. 1990. Reactions of gypsum or phosphogypsum in highly weathered acid subsoils. *Soil Sci. Soc. Am. J.* 54: 993-998.

Arvidsson, H. and H. Lundkvist. 2002. Needle chemistry in young Norway spruce stands after application of crushed wood ash. *Plant Soil* 238: 159-174.

Baath, E. and K. Arnebrant. 1994. Growth rate and response of bacterial communities to pH in limed and ash treated forest soils. *Soil Biol. Biochem.* 26: 995-1001.

Baath, E., A. Frostegard, T. Pennanen and H. Fritze. 1995. Microbial community structure and pH response in relation to soil organic matter quality in wood-ash fertilized, clear-cut or burned coniferous forest soils. *Soil Biol. Biochem.* 27: 229-240.

Bakker, M. R., A. Dieffenbach and J. Ranger. 1999. Soil solution chemistry in the rhizosphere of roots of sessile oak (*Quercus petraea* L.) as influenced by lime. *Plant Soil* 209: 209-216.

Bramryd, T. and B. Fransman. 1995. Silvicultural uses of wood ashes-effects on the nutrient and heavy metal balance in a pine (*Pinus Sylvestris* L.) forest soil. *Water Air Soil Pollut.* 85: 1039-1044.

Broder, M. F. 1983. Performance testing of fertilizer application equipment. *ASAE Pap.* 83-1503. ASAE, St. Joseph, MI.

Büttner, V. G., C. Gering, U. Nell, S. Rumpf and K. V. Wilpert. 1998. Einsatz von Holzasche in Wäldern. *Forst und Holz* 53: 72-81.

Campbell, A. G. 1990. Recycling and disposing of wood ash. *Tappi J.* 73: 141-146.

Clapham, W. M. and L. M. Zibilske. 1992. Wood ash as a liming amendment. *Commun. Soil Sci. Plant Anal.* 23: 1209-1227.

Curtin, D. and G. W. Smillie. 1983. Soil solution composition as affected by liming and incubation. *Soil Sci. Soc. Am. J.* 47: 701-707.

Demeyer, A., J. C. Voundi Nkana and M. G. Verloo. 2001. Characteristics of wood ash and influence on soil properties and nutrient uptake: an overview. *Bioresour. Technol.* 77: 287-295.

Derome, J. and A. Saarsalmi. 1999. The effect of liming and correction fertilization on heavy metal and macronutrient concentrations in soil solution in heavy-metal polluted Scots pine stands. *Environ. Pollut.* 104: 249-259.

Doerge, T. A. and E. H. Gardner. 1985. Reacidification of two lime amended soils in western Oregon. *Soil Sci. Soc. Am. J.* 49: 680-685.

Erich, M. S. 1991. Agronomic effectiveness of wood ash as a source of phosphorus and potassium. *J. Environ. Qual.* 20: 576-581.

Erich, M. S. and T. Ohno. 1992a. Titrimetric determination of calcium carbonate equivalence of wood ash. *Analyst* 117: 993-995.

Erich, M. S. and T. Ohno. 1992b. Phosphorus availability to corn from wood ash amended soils. *Water Air Soil Pollut.* 64: 475-485.

Eriksson, H. M. 1998. Short-term effects of granulated wood ash on forest soil chemistry in SW and NE Sweden. *Scand. J. For. Res. Suppl.* 2: 43-55.

Eriksson, H. M., T. Nilsson. and A. Nordin. 1998. Early effects of lime and hardened and non-hardened ashes on pH and electrical conductivity of the forest floor, and relations to some ash and lime qualities. *Scand. J. For. Res. Suppl.* 2: 56-66.

Etiegni, L. and A. G. Campbell. 1991. Physical and chemical characteristics of wood ash. *Bioresour. Technol.* 37: 173-178.

Etiegni, L., A. G. Campbell and R. L. Mahler. 1991. Evaluation of wood ash disposal on agricultural land. I. Potential as a soil additive and liming agent. Commun. *Soil Sci. Plant Anal.* 22: 243-256.

Ferm, A., T. Hokkanen, M. Moilanen and J. Issakainen. 1992. Effect of wood bark ash on the growth and nutrition of Scots pine afforestation in central Finland. *Plant Soil* 147: 305-316.

Francis, C. W., E. C. Davis and J. G. Goyert. 1985. Plant uptake of trace elements from coal gasification ashes. *J. Environ. Qual.* 14: 561-569.

Fransson, A. M., B. Bergkvist and G. Tyler. 1999. Phosphorus solubility in an acid forest soil as influenced by form of applied phosphorus and liming. *J. For. Res.* 14: 538-544.

Fritze, H., A. Smolander, T. Levula, V. Kitunen and E. Mälkönen. 1994. Wood-ash fertilization and fire treatments in a Scots pine forest stand: Effects on the organic layer, microbial biomass, and microbial activity. *Biol. Fert. Soils* 17: 57-63.

Gray, C. A. and A. P. Schwab. 1993. Phosphorus-fixing ability of high pH, high calcium, coal-combustion, waste materials. *Water Air Soil Pollut.* 69: 309-320.

Hildebrand, E. E. and H. Schack-Kirchner. 2000. Initial effects of lime and rock powder application on soil solution chemistry in a dystric cambisol – results of model experiments. *Nutr. Cycl. Agroecosyst.* 56: 69-78.

Holmberg, S. L. and T. Claesson. 2001. Mineralogy of granulated wood ash from heating plant in Kalmar, Sweden. *Environmental Geology* 40: 820-828.

Huang, H., A. G. Campbell, R. Folk and R. L. Mahler. 1992. Wood ash as a soil additive and liming agent for wheat. Field studies. *Commun. Soil Sci. Plant Anal.* 23: 25-33.

Kahl, J. S., I. J. Fernandez, L. E. Rustad and J. Peckenham. 1996. Threshold application rates of wood ash to an acidic forest soil. *J. Environ. Qual.* 25: 220-227.

Kaiser, K. and M. Kaupenjohann. 1998. Influence of the soil solution composition on retention and release of sulphate in acid forest soils. *Water Air Soil Pollut.* 101: 363-376.

Khanna, P. K., R. J. Raison and R. A. Falkiner. 1994. Chemical properties of ash derived from Eucalyptus litter and its effects on forest soils. *For. Ecol. Mange.* 66: 107-125.

Krejsl, J. A. and T. M. Scanlon. 1996. Evaluation of beneficial use of wood-fired boiler ash on oat and bean growth. *J. Environ. Qual.* 25: 950-954.

Lehnardt, F. 1998. Einfluß der Kalkung und Düngungauf den Ionenaustaush und die chemische Zusammensetzung der Bodenlösung am Beispiel von vier Waldstandorten im Hessischen Bergland. *J. Plant Nutr. Soil Sci.* 161: 41-50.

Ludwig, B., P. K. Khanna, D. Hölscher and B. Anurrugsa. 1999. Modelling changes in cations in the topsoil of an Amazonian Acrisol in response to additions of wood ash. *Eur. J. Soil Sci.* 50: 717-726.

McDonald, M. A., B. J. Hawkins, C. E. Prescott and J. P. Kimmins. 1994. Growth and foliar nutrition of western red cedar fertilized with sewage sludge, pulp sludge, fish sludge, and wood ash on northern Vancouver Island. *Can. J. For. Res.* 24: 297-301.

Meiwes, K. J. 1995. Application of lime and wood ash to decrease acidification of forest soils. *Water Air Soil Pollut.* 85: 143-152.

Meyers, N. L. and M. J. Kopecky. 1998. Industrial wood ash as a soil amendment for crop production. *Tappi J.* 81: 123-130.

Morel, F. M. M. and J. G. Hering. 1993. *Principles and Applications of Aquatic Chemistry.* Wiley & Sons, New York.

Mozaffari, M., C. J. Rosen, M. P. Russelle and E. A. Nater. 2000. Corn soil response to ash application generated from gasified Alfalfa stems. *Soil Sci.* 165: 896-907.

Mozaffari, M., M. P. Russelle, C. J. Rosen and E. A. Nater. 2002. Nutrient supply and neutralizing value of Alfalfa stem gasification ash. *Soil Sci. Soc. Am. J.* 66: 171-178.

Muse, J. K. and C. C. Mitchell. 1995. Paper mill boiler ash and lime by-products as soil liming materials. *Agron. J.* 87: 432-438.

Naylor, L. M. and E. J. Schmidt. 1986. Agricultural use of wood ash as a fertilizer and liming material. *Tappi J.* 69: 114-119.

Naylor, L. M. and E. Schmidt. 1989. Paper mill wood ash as a fertilizer and liming material: field trials. *Tappi J.* 72: 199-206.

Ohlsson, K. E. A. 2000. Carbonation of wood ash recycled to a forest soil as measured by isotope ratio mass spectrometry. *Soil Sci. Soc. Am. J.* 64: 2155-2161.

Ohno, T. and M. S. Erich. 1990. Effect of wood ash application on soil pH and soil test nutrient levels. *Agric. Ecosyst. Environ.* 32: 223-239.

Ohno, T. 1992. Neutralization of soil acidity and release of phosphorus and K by wood ash. *J. Environ. Qual.* 21: 433-438.

Ohno, T. and M. S. Erich. 1993. Incubation-derived calcium carbonate equivalence of papermill boiler ashes derived from sludge and wood sources. *Environ. Pollut.* 79: 175-180.

Pietikäinen, J. and H. Fritze. 1995. Clear-cutting and prescribed burning in coniferous forest: comparison of effects on soil fungal and total microbial biomass, respiration activity and nitrification. *Soil Biol. Biochem.* 27: 101-109.

Ralston, C. W. and G. E. Hatchell. 1971. Effects of prescribed burning on physical properties of soil. In: Prescribed Burning. Proceedings of Symposium, 14-16 April, Charleston S. C., US Department of Agriculture, Forest Service, South Eastern Forest Experimental Station, pp. 68-81.

Risse, M. and G. Harris. 2001. Best management practices for wood ash used as an agricultural soil amendment. Soil Acidity and Liming. Internet In service Training.

Rumpf, S., B. Ludwig and M. Mindrop. 2001. Effect of wood ash on soil chemistry of a pine stand in Northern Germany. *J. Plant Nutr. Soil Sci.* 164: 569-575.

Saarsalmi, A., E. Mäkönen and S. Piirainen. 2001. Effects of wood ash fertilization on forest soil chemical properties. *Silva Fennica* 35: 355-368.

Someshwar, A. V. 1996. Wood ash and combination wood-fired boiler ash characterization. *J. Environ. Qual.* 25: 962-972.

Steenari, B-M. and Q. Linqvist. 1997. Stabilisation of biofuel ashes for recycling to forest soil. *Biomass Bioenergy* 13: 39-50.

Steenari, B-M., N. Marsic, L-G. Karlsson, A. Tomsic and O. Lindqvist. 1998. Long-term leaching of stabilized wood ash. *Scand. J. For. Res. Suppl.* 2: 3-16.

Steenari, B-M., L. G. Karlsson and O. Lindqvist. 1999a. Evaluation of the leaching characteristics of wood ash and the influence of ash agglomeration. *Biomass Bioenergy* 16: 119-136.

Steenari, B-M., S. Schelander and O. Lindqvist. 1999b. Chemical and leaching characteristics of ash from combustion coal, peat and wood in a 12 MW CFB – a comparative study. *Fuel* 78: 249-258.

Steponkus, P. C. 1992. Recycling wood ash in established loblolly pine plantations in eastern NORTH California. Proceedings of TAPPI Environmental Conference VA 12-15 April 1992. TAPPI Press, Book 1, Richmond, Atlanta, GA.

Su, C. and L. J. Evans. 1996. Soil solution chemistry and alfalfa response to $CaCO_3$ and $MgCO_3$ on an acidic gleysol. *Can. J. Soil Sci.* 76: 41-47.

Ulery, A. L., R. C. Graham and C. Amrhein. 1993. Wood-ash composition and soil pH following intense burning. *Soil Sci.* 156: 358-364.

Unger. Y. L. and I. J. Fernandez. 1990. The short-term effects of wood-ash amendment on forest soils. *Water Air Soil Pollut.* 49: 299-314.

Vance, E. D. 1996. Land application of wood-fired and combination boiler ashes: an overview. *J. Environ. Qual.* 25: 937-944.

Voundi Nkana, J. C. 1998. Utilisation des déchets de l'industrie du bois en vue de l'amélioration de la fertilité chimique des sols acides tropicaux. Ph.D Thesis, Univeriteit Gent. Belgium.

Voundi Nkana, J. C., A. Demeyer and M. G. Verloo. 1998a Chemical effects of wood ash on plant growth in tropical acid soils. *Bioresour. Technol.* 63: 251-260.

Voundi Nkana, J. C., A. Demeyer and M. G. Verloo. 1998b. Availability of nutrients in wood ash amended tropical acid soils. *Environ. Technol.* 19: 1213-1221.

Voundi Nkana, J. C., A. Demeyer. and M. G. Verloo. 2000. Nutrient dynamics in tropical acid soils amended with wood ash. AGROCHIMICA 44: 197-210.

Weber, A., M. Karsisto, R. Leppänen, V. Sundman. and J. Skujins. 1985. Microbial activities in a Histosol: effects of wood ash and NPK fertilizers. *Soil Biol. Biochem.* 17: 291-293.

Wilhoit, J. H. and Q. Ling. 1996. Spreader performance evaluation for forest land application of wood and fly ash. *J. Environ. Qual.* 25: 945-950.

Williams, T. M., C. A. Hollis and B. L. Smith. 1996. Forest soil chemistry following bark boiler bottom ash application. *J. Environ. Qual.* 25: 955-961.

Zhan, G., M. S. Erich and T. Ohno. 1996. Release of trace elements from wood ash by nitric acid. *Water Air Soil Pollut.* 88: 297-311.

Zollner, V. A. and N. Remler. 1998. Eigenschaffen von Holzaschen und Möglichkeiten der Wiederverwertung. *Forst und Holz* 53: 77-81.

8

Role of Grafting in Horticultural Plants

R.M. Rivero, J.M. Ruiz and Luis Romero[1]

Department of Plant Biology, Faculty of Sciences, University of Granada,
E-18071 Granada, Spain. E-mail: lromero@ugr.es

1. Introduction

The cultivation of grafted horticultural plants began in Korea and Japan towards the end of 1920 on grafting watermelon plants to squash rootstocks (Ashita, 1927; Yamakawa, 1983). After the first experiments, the cultivation of grafted plants gradually increased in these countries and currently, most watermelon, cucumber and various Solanaceae crops are grafted before being transplanted either in the greenhouse or in the field (Ryu et al., 1973; Lee, 1989; Ito, 1992; Kurata, 1992). Thus, this technique is widely used in many parts of Korea and Japan, as also throughout Asia and Europe for intensive crop systems (Hartmann and Kaster, 1975), reaching 81% of the cultivation in Korea and 54% in Japan.

Initially, the cultivation of grafted plants was intended to diminish the damage incurred by soil pathogens, primarily *Fusarium oxysporum*. However, as the use of this technique spread, the aims also expanded until today, when grafting serves a spectrum of purposes: (1) to increase the nutrient and mineral uptake to the shoot; (2) to boost plant growth and development; (3) to improve the quality and yield, (4) to control wilt caused by pathogens; (5) to reduce viral, fungal and bacterial infection; (6) to augment the production of endogenous hormones that promote greater development of the aerial part; (9) to strengthen the tolerance to thermal or saline stress, etc. Each of these reasons for the expansion of grafting will be discussed below in detail.

The fact that one rootstock can be more resistant than another against certain factors of biotic or abiotic stress is unquestionable. However, the mechanisms of such resistance has not yet been thoroughly investigated. For example, the fact that substances associated with tolerance to *Fusarium* are synthesized in the roots and translocated to the shoot through the xylem is still a controversial issue and requires extensive research.

The cultivation of grafted plants has expanded greatly in recent years due primarily to the discovery that the same variety can be grafted to different rootstocks, depending on the aim. Crops such as watermelon, melon, cucumber, tomato and eggplant are commonly grafted onto different rootstocks prior to being transplanted in the field or, predominantly, in the greenhouse. Table 1 lists the main rootstocks used for these crops, according to the objective pursued.

Table 1. Rootstock and purpose of grafting for some vegetables (Lee, 1994)

Vegetables	*Popular rootstock species*	*Purpose*[1]
Watermelon	Gourd (*Lagernaria siceraria* var. *hispida*)	1, 2
	Interspecific hybrids	1, 2, 3
	Wax goourd (*Benincasa hispida* Cong.)	1, 2
	Pumpkin (*Cucurbita pepo* L.)	1, 2, 3
	Squash (*Cucurbita moschata* L.)	1, 2, 3
	Sicyos angulatus	5
Cucumber	Figleaf gourd (*Cucurbita ficifolia*)	1, 2, 3
	Interspecific hybrids	1, 2, 3
	F$_1$ (*Cucurbita maxima* x *C. Moschata*)	1, 2, 4
	Cucumber (*Cucumis sativus*)	1, 2
	Sicyos angulatus	2, 5
Oriental melon	Interspecific hybrids	1, 2, 3
	Cucurbita moschata	1, 2, 3
	Cucumis melo	3, 4
Melons	*Cucumis melo*	1
Tomato	*Lycopersicon pimpinellifolium* (L.) Mill.	5
	Lycopersicon hirsutum Humb. & Bonpl.	5
	Lycopersicon esculentum	5
Eggplant	*Solanum integrifolium* Poir	6
	Solanum torvum Sw.	7

[1]Purpose of grafting: 1 = *Fusarium* wilt control; 2 = growth promotion; 3 = low-temperature tolerance; 4 = growth-period extension; 5 = nematode resistance; 6 = bacterial-wilt control; 7 = viral-infection reduction.

In this review, we seek to summarize the current role of grafted plants in horticulture, our main aim being to analyse the advantages and drawbacks of using such plants in cultivation today.

2. Types and Methods of Grafting

Grafting is the process by which two or more plants are joined to grow as an individual (Baldwin, 1996). To predict the result of a grafted plant is complex, although generally, it can be said that the success of a graft is closely linked to the botanic affinity of the parts grafted: on the one hand, the morphological and anatomical affinity of the tissues brought into contact with one another, and, on the other, physiological affinity in function as well as and similarity of sap quantity and constitution. All this can be summarized as the creation of a plant having roots that must grow and develop with the sap synthesized by the green organs of another plant, which in turn, grows and develops with the sap supplied by a root that is not its own (Baldwin, 1996). The grafting technique used—from the many established types—depends on the species to be grafted and the expected survival rate of the plant. Among the most-widely used types are: (1) cleft grafting; (2) tongue approach grafting; (3) pin grafting; and (4) Tongue approach grafting. Below, is described as follows:

(1) *Hole insertion grafting* (Fig. 1): This type of graft is often used for forest and garden species which have a woody stem. The scion must have a smaller diameter than the rootstock. In the rootstock, a vertical incision is made in the shape of a wedge from the centre of the stem outwards. Afterwards, the scion is separated from its roots at the height of the first green leaves and inserted into the wedge, and a clip is placed over the graft until the formation of scar or callus.

(2) *Tongue approach grafting* (Plate 1): In this type of graft, the size of the rootstock and scion seedlings can be same or different. In the rootstock, a small blade is used to make a slit downwards through the hypocotyl, while in the scion, the cut is upwards. Afterwards, the two hypocotyls are joined by grafting clips or aluminium or tin foil (grafting tin or aluminium bandage) so as to aid healing (see photos 1b and 1c). The clip is kept on 4 to 6 days and, finally, the hypocotyl of the rootstock plant is cut below the union of the graft (see photo 1e).

(3) *Pin grafting* (Plate 2): For this type of graft, young seedlings are a must, where the hypocotyl does not exceed 3 mm in diameter. To assure a similar size in the species selected for the graft, both should be sown at the same time. For the graft, the rootstock

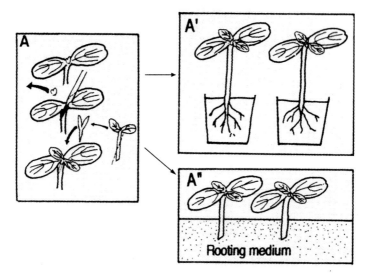

Fig. 1. Hole insertion grafting. (A) Method; (A') Grafted seedlings are transplanted in pot for futher growth; (A") Rootstocks are cut just above the root zone and he grafted as in (Lee, 1994).

seedling is cut cleanly and horizontally just above the cotyledons and the scion seedling cut horizontally below the cotyledons. In the centre of the hypocotyl of the rootstock, a pin—generally of porcelain—is inserted half its length (like a dowel) and the scion is inserted over the exposed half of the pin(see Plates 2b and 2c). The joint is reinforced with a grafting tin bandage surrounding the graft, this being removed after 5 to 7 days, when the scar or callus has completely formed. This type of graft is being used frequently more throughout the world because it is quick, easy and effective, and gives high percentages of seedling survival. In addition, it can be used with a wide variety of plants (tomato, eggplant, watermelon, pepper, cucumber, squash).

(4) *Cleft grafting* (Fig. 2): To perform this type of graft, the rootstock diameter should be greater than that of the scion. In the seedling to be used as the rootstock, a slit or hole is cut in the central stem at the height of the cotyledons. Afterwards, a cut is made in the hypocotyl of the seedling to be used as the scion, below the cotyledons. This is then inserted into the slit or hole in the rootstock. Finally, the grafted seedling is moved to a growth medium until the graft heals and forms a scar or callus.

Plate 1. Tongue approach grafting in watermelon plants. (a) Left plant: rootstock; right plant: scion; (b) Graft union; (c) Grafting tin bandage is used to adhere the graft union; (e) The scion's hypocotil is cut below the graft union 1 week after grafting; (f) callus formation.

Plate 2. Pin Grafting in tomato plants: (a) Rootstock; (b) y (c) Pin insertion; (d) Scion; (e) Graft union; (f) Callus formation

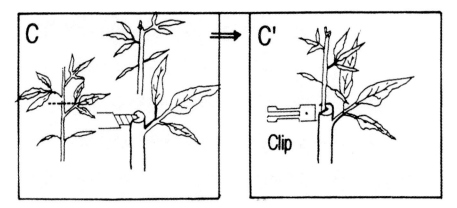

Fig. 2. Cleft grafting. (C) Method; (C′) Specifically devised grafting clips are used to hold the scion and stock together tightly. (Lee, 1994)

In the previous century, various methods of grafting have been developed and farmers select them on the basis of experience and preference. Curbitaceae crops are generally grafted when the seedlings are young—that is, just after the sprouting of the first green leaf. However, currently, the grafting methods vary considerably with the type of crop, and the growth of the species depends, in turn, on the type of graft used. For example, the cleft graft can be effectively used for watermelon because of the small size of the scion seedling compared with the stem of squash seedlings used as rootstocks. Cucumber, however, is usually grafted by the whip method, mainly for the large size of this seedling, the length and diameter of the hypocotyl and the ease of grafting.

3. Compatibility and Incompatibility during Formation of the Graft

It is generally accepted that the success of the graft is due to the morphological and physiological union of the two parts involved (root and shoot). The histological development of the union of the root and the shoot has been subject to numerous studies and appears to be similar in all herbaceous dicotyledons (Robert, 1949; McCully, 1983). While many authors initially based their studies on the development of the connection of the xylem between the shoot and root (Funk, 1929; Silberschmidt, 1935 and 1936; de Stigter, 1961), recent studies indicate the importance of the phloem, since the connection of the xylem occurs in most grafts (Monzer and Kollmann, 1986; Tiedemann, 1989; Kollmann and Glockmann, 1990). Generally, the vascular connection progresses in three phases:

(a) Phase I begins after the grafting and consists of the development of an isolation layer or a necrotic layer at the interface of the graft, similar to the protective layer that forms after tissue damage. This isolation layer is formed by the residues of the cell walls destroyed by the cut during the grafting procedure (Roberts, 1949; Moore and Walker, 1981a, 1981b, 1981c; Moore, 1982, 1984; McCully, 1983). Yeoman and Brown (1976) and Yeoman et al. (1978) found that the isolation layer developed both in compatible and incompatible grafts.

(b) During phase II, cell division of the parenchyma cells forms the scar or callus similar to that caused by injuries (Tiedemann, 1989).

(c) Finally, in phase III, differentiation occurs in the parenchyma layer and in the scar or callus cells in sieve tubes, bringing about a symplastic vascular connection—that is, through plasmodesms and sieve pores (Moore and Walker, 1981a).

There are five main elements to take into account in order to perform a successful graft (Parkinson et al., 1987; Camacho and Fernández, 1997; Miguel, 1997):

1. Compatibility between the scion and the rootstock, which must be capable of forming a union.

2. The vascular cambium of the scion must be located in close contact with that of the rootstock. The cut surfaces must be strongly held together and it is necessary to perform the graft quickly for the scion to receive the water and nutrients from the rootstock until the buds begin to sprout.

3. The graft must be done in the appropriate physiological state of the rootstock and the scion. This means that in general the sprouts of the scion are dormant and, at the same time, the tissues of the cut at the union of the graft are able to form the scar necessary for the graft to heal. The rootstock can be in active growth or in dormancy, depending on the method of grafting used.

4. Immediately after the grafting, all the cut surfaces should be protected against drying, with applications of tape, grafting wax, etc.

5. The grafts should receive appropriate care for a certain time until complete healing has taken place. Sometimes, the scion can grow vigorously and break, if a brace or splint is not provided or if the scion is not pruned.

The joint of the graft is often represented as a barrier to translocation between the shoot and root, resulting in an accumulation of solutes on the upper side of the connection zone (Silberschmidt, 1933, 1935 and

1936; de Stigter, 1961). These situations arise in incompatible grafts as well as in grafts in an early stage of growth, fundamentally because:

1. Most of the elements or sieve tubes in the section of the graft do not connect with the phloem bundles of the different parts (Kollmann and Glockmann, 1990).
2. The sieve pores can be blocked by callose or by pore protein, due to the reactions of incompatibility of the cell contents in the sieve tubes (Kollmann and Glockmann, 1990).

According to Miguel (1997), incompatibilities can be recognized by the following symptoms:

• Low percentage of grafting success.
• Yellowing of leaves, at times defoliation and lack of growth.
• Premature death of the grafted plant.
• Differences in the growth rate between the rootstock and scion.
• Formation of a girdle or knot around the graft scar.
• Breakage around the grafted zone.

Tiedemann (1986) demonstrated that in compatible grafts, the mean number of connections between sieve tubes was fourfold higher than in the incompatible grafts, reflecting that the translocation of assimilates from the shoot to the root was much greater in compatible grafts. Rachow-Brandt (1987) indicated that the transport of assimilates through the interface of the graft is generally parallel to the number of connections of sieve tubes, illustrating the importance of the regeneration of the phloem for the transfer of assimilates in grafted plants.

It can be deceptive to distinguish compatibility from incompatibility of a graft. From species that join together with ease, to others that are incapable of being joined, there is a wide range of intermediate possibilities which, even while joining together at first, soon exhibit symptoms of lack of affinity, either during the graft or in growth habits (Camacho and Fernández, 1997). At times, the appearance of these symptoms in isolated cases does not signify incompatibility, as these can appear under inappropriate environmental conditions. The incompatibility can be local, depending on the contact between the rootstock and scion, or translocational, involving phloem degeneration. The former can be corrected with a grafting bridge, but translocated incompatibility cannot, since it is fundamentally due to difficulties in the movement of carbohydrates and other compounds through the graft zone (Hartmann and Kester, 1991; Camacho and Fernández, 1997).

4. Response of Grafted Plants to Stress

4.1. Resistance to Soil Pathogens

Most horticultural corps are often exposed to a broad spectrum of adverse environmental factors, both biotic and abiotic. Some of the most dangerous and common biotic factors that most crops currently face are soil diseases caused by viruses, fungi and bacteria, as well as nematodes (Hain et al., 1993; Bais et al., 2000; Thévenot et al., 2001). The main injuries to roots caused by these soil pathogens, which are becoming more and more numerous and threatening in greenhouse cultivation, are smaller foliar size, as well as thin weak stems, wilt, depressed flowering, and poorer fruit quality—in short, a reduced life span of the plant (Hain et al., 1993; Bais et al., 2000).

The primary problem confronting farmers is that by the time the plants begin to present visual symptoms, the crop can be already lost. As the pathogens attack the roots, the first symptoms usually appear in the leaves when the plant is completely infected. The only feasible option for the farmer is to take preventive measures, involving soil treatments for the following crop season. Plant resistance offers another possibility, and, in this sense, the development and use of rootstocks resistant to different soil pathogens could offer one of the most promising advantages of grafted plants in present-day horticulture (Forner and Alcaide, 1993, 1994).

There is no doubt that a strong and vigorous rootstock has better tolerance against diseases caused by fungi, such as *Verticillum* or *Fusarium*; by bacteria such as *Pseudomonas*; by viruses such as TYLCV (tomato yelow leaf curl virus, transmited by *Bemisia tabaci*), although the degree of tolerance varies considerably from one plant to another, depending on the genotype of the rootstock (Schneider et al., 1995).

Despite its importance, this mechanism of resistance or tolerance has not been intensely investigated (Lee, 1994). Tolerance to these diseases in grafted plants may be due to the resistance of the rootstocks, as it is accepted that the root system synthesizes substances resistant to pathogen attack, and these are transported to the shoot through the xylem (Biles et al., 1989). The activity of these substances, related to disease resistance, can vary during the development stages of grafted plants (Padgett and Morrison, 1990; Heo, 1991).

One of the points to be studied in this sense, having received little or no research, is that rootstocks are resistant or tolerant to pathogen attack, since no work to date concludes whether the reduction of damage by these agents in grafted plants is due to the greater resistance to diseases for the rootstock selected. It has become accepted that the characteristics of disease susceptibility of the shoot are not translocated to the rootstock. This implies that the rootstocks govern the degree of infection of a disease (Lee, 1994).

On the other hand, chemical soil fumigation, being strongly limited by different governments for reasons of environmental pollution, has sharply declined. Also, given the economic expense involved, this practice offers little solution to diseases that attack plants in infected soils (Rattink, 1983; Schneider et al., 1995). Thus, the selection of rootstocks resistant to soil pathogens implies an advantage in terms of reduced use of chemical agents. Nevertheless, as indicated above, more research is needed in this regard to identify the mechanisms controlling disease tolerance.

4.2. Resistance to Low Root Temperatures

Injuries inflicted by low temperatures are defined as the damage caused by physiological and biochemical alterations induced by temperatures above the freezing point but below 12°C (Raison and Lyons, 1986; Salveit and Morris, 1990). Many of the most economically important crops, such as corn, tomato, squash, cucumber, watermelon and cotton are highly sensitive to cold temperatures during their vegetative development and reproduction (Markhart, 1986; Jackman et al., 1988; Wang, 1990). In addition, seed germination and seedling development are two critical stages in the survival of temperature-sensitive crops. Low soil temperature is one of the major factors affecting this survival, inflicting heavy economic losses in yield (Bradow, 1990a, 1990b), by reducing plant growth and development, causing wilt and necrosis, and retarding fruit ripening (Herner, 1990; Bolger, 1992; Reyes and Hennings, 1994; Ahn et al., 1999). All this results from a limitation in the uptake of water and essential mineral nutrients (Ahn et al., 1999), bringing about a serious fall in root conductance (McWilliam et al., 1982; Bolger et al., 1992) and an extrusion of endogenous solutes triggered by degraded membrane integrity (Mistrik et al., 1992).

On the other hand, as soil temperature tends to vary more slowly than aerial temperature, the roots can suffer consequences of cold over longer periods than does the shoot. It is a well known fact that some rootstocks are more resistant to low temperatures, implying an adaptive advantage of certain crops. However, the reason is to why these rootstocks offer more resistance than do others is still not clearly understood. Exudation of xylem sap has been proposed (Masuda and Gomi, 1982), as has high oxygen consumption (Masuda and Gomi, 1984), as possible mechanisms of resistance to low temperatures. In addition, Tachibani (1988) demonstrated that the roots of squash plants respond to the low temperatures by stimulating meristem activity and photosynthate translocation. This resistance could be explained by the mechanisms described above, although they have received little attention and their effect on the shoot is completely unknown (Ahn, 1999).

However, different species have been studied at low temperatures in order to select the most resistant rootstocks for use in grafted crops. For example, the cucumber (*Cucumis sativus* L.) is often grafted onto *Cucurbita ficifolia* rootstocks as well as different genotypes of *Sicyos angulatus* which are resistant to low temperatures, boosting growth and yield (den Nijs, 1980, 1984a, 1984b; Tachibani, 1982).

Criteria for selecting rootstocks tolerant to low temperatures are based on lipid differences in membranes (Smolenska and Kuiper, 1977; Lyons and Breidenbach, 1979; Horvath et al., 1980; Vigh et al., 1985). The desired profile includes an increase in the total of lipids per gram of fresh weight, an increase in the ratio fatty acid/total lipid, a high degree of unsaturation of fatty acids, and finally, a reduction of the ratio esterol/total lipids (Lyons and Breidenbach, 1979; Markhart et al., 1980; Horvath et al., 1983; Tachibani, 1986 and 1987; Bulder et al., 1990 and 1991). Horvath et al. (1987) demonstrated that the ratio esterol/phospholipids was the main indicator of low-temperature tolerance, since they found an inverse relationship between the ratio esterol/phospholipids and resistance to cold in the leaves of cereals. A high value of this ratio corresponds to a more cold-sensitive plasma membrane, prompting a strong esterol-esterol interaction in the lipid bilayer, in turn, causing disruptions in the membrane. These effects were also noted in membranes of roots subjected to low temperatures (Bulder et al., 1990).

Studies on cucumber by Horvath et al. (1983) and Bulder et al. (1990) demonstrated that some of the effects caused in roots by tolerance to low temperatures were found in the leaves of sensitive species, and thus the presence of these effects in the leaves is not indicative of low-temperature tolerance in the roots, since the foliar lipid composition differs from the lipid composition of the roots.

These are four examples that show that the use of grafted plants onto rootstocks tolerant of low soil temperatures would assure good development and optimal production of the crop with little or no economic loss from low environmental temperatures.

4.3. Resistance to Salinity

In the past, stress caused by high concentrations of salts in the environment had little importance because these situations arose only in areas near the coast or in particular environments with evaporation of salt-rich waters. However, the development of agricultural techniques in recent years has made salt stress one of the chief problems in agriculture today. The indiscriminate use of heavy quantities of chemical fertilizers and the overexploitation of aquifers has dramatically multiplied the surface area affected by salinity (Tudela and Tadeo, 1993). Currently, a third of all irrigated lands in the world is affected to a greater or lesser degree by salinity (Hasegawa et al., 1986). A heavy environmental concentration of salts unleashes various types of physical and chemical stress in plants (Kuiper, 1984; Pasternak, 1987), provoking complex responses that involve changes in plant morphology, physiology and metabolism (Hasegawa, 1986; Pasternak, 1987; Cheeseuman, 1988; Borochov-Neori and Borochov, 1991). It is commonly accepted that growth inhibition by salt stress is associated with alterations in the water relationships within the plant, caused by osmotic effects with specific ionic consequences (excesses or deficits) or energy availability related to carbohydrate concentrations (Munns, 1993; Lazof and Bernstein, 1998). Prior research on the possible mechanisms of growth inhibition and salt tolerance is often based on the comparative study of tolerant and sensitive lines. However, as with most types of environmental stress, the assessment of the tolerance level of the different processes of developing salt stress is a complex task, though it is fundamental to establish the criteria for action in such situations. For example, growth and reproduction can be altered in different ways in the same plant, as the shoot is usually more salt sensitive than the root (Deban et al., 1982; Weimberg et al., 1984).

Salt stress has two components negatively affecting plant growth: the osmotic component and the ionic component. A heavy salt concentration lowers the water potential in the soil, inducing water stress in plants. This is known as the osmotic component of salinity. On the other hand, certain ions are toxic for glycophytes (the immense majority of cultivated plants; Tudela and Tadeo, 1993), and this represents the ionic component. Among the most abundant toxic ions are Cl^- and Na^+, although other ions can also cause problems, such as NO_3^-, SO_4^{2-} and NH_4^+ (Kramer, 1984).

Damage from salinity has been attributed principally to an excess of Cl^- and Na^+ accumulation in the leaves (Cooper, 1961; Smith, 1962), provoking a nutritional imbalance, as these ions reduce the concentration of Ca, Mg and K (Jones et al., 1957; Patil and Bhambota, 1980; Zekri and Parsons, 1990 and 1992). The high concentration of Cl^- in the aerial parts of citrus can be the prime cause of physiological disturbance and eventual visible damage to the foliage (Cooper and Shull, 1953). The high foliar concentration of Na^+ interferes with photosynthesis and transpiration (Behboudian et al., 1986). Some of these adverse effects of salinity have been attributed to a K^+ deficiency, due to the antagonistic action by Na^+ (Levitt, 1980).

With specific reference to the behaviour of the root system, plants that tolerate salinity can be grouped into ion-exclusive and ion-inclusive plants (Kramer, 1984). Ion-exclusive plants have diverse adaptive mechanisms to restrict salt from reaching the aerial parts except in very small quantities. These mechanisms include:

1. The efficient uptake selectivity by the roots with respect to certain ions.
2. Accumulation of Cl^- and Na^+ in the roots.
3. Exclusion of ions from entering the roots from the exterior.

On the contrary, inclusive plants take up salt in great quantities and store it in the stems and leaves. In this case, the main adaptations are based on the elimination of Cl^- and Na^+ from the cytosol either by storage in the vacuoles or from the cell itself (Tudela and Tadeo, 1993).

If the ions Cl^- and Na^+ damage the plant at high concentrations in the leaves, the selection of exclusive rootstocks (salt tolerant) may increase resistance to salinity in grafted plants (Bernstein et al., 1956,1969). This has promoted research on the growth and yield with new salt-tolerant rootstocks (Furr and Ream, 1969; Sykes, 1985). In grafted citrus, Cooper (1961) demonstrated that the foliar Cl^--concentration was the most useful value to assess the damage caused by salinity and to classify salt tolerance. Later, Zekri and Parsons (1989) observed that the accumulation of heavy foliar concentrations of Cl^- caused leaf burn, notably reduced growth and upset water relationships. Meanwhile, no injury was noted in plants that accumulated Na^+ in the leaf, and neither growth nor the water relationship were not severely affected.

Zekri (1991) reported that the reduction in the foliar chlorophyll content was due more to Cl than Na^+ accumulation. Behboudian et al. (1986) observed that the aerial parts of citrus grafted to the rootstock 'Cleopatra mandarin' (*Citrus reshni* Hort. ex Tan.) accumulated less Cl^- in its stems and leaves than did the aerial parts of citrus grafted to the rootstock 'rough lemon' (*Citrus jambhiri* Lush), an evidence evident in the more severe injury in the aerial parts of the latter. Walker (1986) demonstrated that the rootstock 'Cleopatra mandarin' excludes Cl^- but not Na^+. This confirms the data reported by Sykes (1992) identifying grafted fruit trees with the capacity of excluding Cl^-. However, these trees were incapable of either excluding Na^+ or including it, suggesting that the ability to exclude these two ions stems from different mechanisms (Sykes, 1992). Picchioni et al. (1990) studying the effect of salinity on grafted *Pistacia* spp., found higher Cl^- concentrations in the roots and rootstock than in the aerial parts. This same study on *Pistacia* reported fivefold greater storage capacity of Na^+ in the roots than in the leaves.

With respect to the salt-tolerance mechanisms among inclusive plants, Zekri and Parsons (1992) determined that in citrus grafted onto sour orange (*Citrus aurantium* (L.) Cl⁻ was not excluded, but instead, it was accumulated in the aerial part, where no effects or severe injury was evident; therefore, the conclusion was that the leaves of the aerial part must have the capacity of partially excluding Cl⁻ from the cytoplasm towards the vacuoles, where it could exert a certain influence on metabolic functions.

One method of ascertaining whether the rootstock presents exclusive salt tolerance is to determine the proline in the leaves, since, as reported by Sánchez-Diaz and Aguirreolea (1993), when the salts do not accumulate in leaves, the plants utilize organic substances to lower the osmotic potential of the cytoplasm and the vacuole thereby, lowering the foliar water potential. Among these organic compounds that do not interfere in the cell metabolism at high concentrations is proline. The quantity of carbon used for the synthesis of these organic solutes can be high. Nevertheless, when ions are taken up by the leaves and accumulated in the vacuoles, the cell osmotic potential increases without damaging the salt-sensitive enzymes of the cytoplasm. In these leaves, the water balance is maintained between the cytoplasm and the vacuole, accumulating organic compounds such as proline in the cytoplasm. Because the volume of the cytoplasm in a adult vacuolated cell is small as compared with the volume of the vacuole, the quantity of organic compounds synthesized is lower.

All the above data leads to the conclusion that the selection of rootstocks that are exclusive or inclusive of ions could strengthen the resistance to high salt concentrations in the soil.

5. Nutritional Status in Grafted Plants

The fact that certain rootstocks show stronger resistance to soil pathogens (Ashwort, 1985; Schneider, 1995; Yakawa, 1996), greater tolerance of low soil temperatures (Reyes and Jennings, 1994; Ahn, 1999) as well as of salt stress (Borochov-Neori and Borochov, 1991; Bernstein et al., 2001) has been well documented in the past. Nevertheless, few works demonstrate the effect that the rootstock exerts on the foliar content in nutrients. Thus, the selection of rootstocks is rarely based on characteristics related to nutrient uptake, but rather, almost always on resistance to environmental stress (Ruiz et al., 1997).

Knowledge of the rootstock/scion nutritional relationship could be decisive in choosing rootstocks tolerant or resistant to soils that are deficient or toxic in one or more nutrients, as well as in preparing fertilization programmes after the grafted plant is transplanted to the field or greenhouse (Chapin and Westwood, 1980).

Initially, Chapin and Westwood (1980), working with fruit trees grafted to different rootstocks, found no clear evidence that the rootstock engendered a different nutritional composition in the leaves of those trees, concluding that the phenotype of the aerial part appeared to be more determinant of the foliar content in nutrients than was the rootstock. Later, Tagliavani et al. (1992) in a similar study, suggested that the differences in the uptake and translocation of nutrients depended on the vigour of the aerial part, and that the concentrations of the different nutrients in the xylem appeared to be more indicative of the vigour derived from the combination of the rootstock and the scion. Later, Brown et al. (1994) investigating grafted *Pistacia* spp., found that a change in the rootstock brought about variations in the foliar content of several essential nutrients of the plants. Ruiz et al. (1997) suggested that in melon plants grafted to different rootstocks, the foliar contents of N, Na and K were determined by the rootstock genotype, and also that the foliar content in N and Na found in these plants also led to differences in yield.

In an earlier work by Ruiz et al., (1996) on grafted melon plants, the rootstock was found to have a positive effect on the foliar levels of total P, reflected by the greater shoot vigour in these plants as well as a higher carbohydrate content (glucose, sucrose, fructose and starch). These authors concluded that with good P uptake by the roots, the concentration in carbohydrates falls, these components being transported from the root to the shoot, thereby increasing the vigour of the aerial part of the plant (Lee, 1994; Ruiz et al., 1996).

In addition, Pulgar et al. (1998) studying the concentration of different micronutrients in the leaves of watermelon grafted to different rootstocks, found that the free-Zn content was lower in grafted plants but presented greater foliar biomass than did the same species without grafting. Since Zn is directly involved in the synthesis of nitrogenous compounds of high molecular weight (Cakmak, 1988), Pulgar et al. (1998) concluded that the free-Zn levels were lower in grafted plants due to greater transport of this metal to the aerial part and higher efficiency in integrating this element into nitrogenous compounds that form chelates with Zn, thereby also explaining the greater foliar biomass found in these plants.

In short, there is no doubt that in grafted plants, the uptake of water and mineral nutrients is greater, giving rise to variations in the foliar concentration of these plants with respect to ungrafted conspecifics. The first and basic consequence is a more vigorous development of the plant and a proportional reduction in the susceptibility to different types of environmental stress.

6. Production of Endogenous Hormones in Grafted Plants

The predominant concerns in the cultivation today refer to the plant resistance to pesticides and diseases to and, as indicated above, other physiological aspects appear concern the farmer less. However, there is a growing demand by farmers to find varieties that boost yield and improve fruit quality.

The root system supplies the shoot with water and nutrients and, therefore, plays a crucial role in vegetative growth. On the other hand, the root is the main organ for the synthesis of cytoquinines (CQ), abscisic acid (ABA) and gibberellins (GB), for which its function as a hormonal growth regulator is very important (Itai and Birubaum, 1991; Zijlstra et al., 1994). The root system functions on the basis of closed interactions with the aerial part and roughly 305 of the total plant genes are expressed in the root (Zobel, 1975). Some works on the genetic variation of root systems of crops offer the morphological and anatomical descriptions without concluding the physiological reasons for achieving such characteristics (O'Toole and Bland, 1987).

Due to its economic importance, many researches worldwide have attempted to improve the rooting system of plants, taking into account the effects of some of the most important growth regulators (Haissig and Davis, 1993). In this way, putrescine appears to be involved in inducing adventitious roots, while phenolic compounds such as catecol, fluoroglucinol, or ferulic acid, act synergetically with auxins in rhizogenesis (Hess, 1969; James and Thurbon, 1981). On the other hand, work on grafted cucumber and eggplant reveal striking contrasts regarding root genotypes and their effects on the growth and production of the aerial part. In addition, it has been observed that the concentrations of CQ, GB, ABA, AIA and *t*-zeatine in xylem sap is higher in grafted than in ungrafted plants. It appears that these hormones fortify development in the aerial part thereby, augmenting yield in these plants (Carlson, 1963; Kato and Lou, 1989).

Therefore, in view of prior research on this subject, the selection of root genotypes that boost the concentration of endogenous plant hormones, such as CQ, GB, ABA and AIA could confer an advantage for the farmer in terms of enhanced vegetative and reproductive development, which translates as improved yield.

7. Grafting in Realtion to Horticultural Quality and Yield

Till date, we have seen how the use of grafted plants offers great advantages with respect to ungrafted ones in resistance to different types of environmental stress—either biotic or abiotic—in the uptake and translocation of water and nutrients as well as in the production of endogenous hormones. Together, these constitute the most important physiological processes to achieve optimal plant development.

According to Lee (1994) and Oda (1995), the use of grafted plants improves yield. However, depending on the root genotype selected, the yield index and fruit quality can oscillate considerably, since compatibility problems can arise between species, provoking a gradual decline in these indices. After the compatibility of the grafted species has been confirmed, a study should be made on the basis of prior research on production parameters (biomass production or growth) and the quality of the new plant, which could be resistant to different factors of environmental stress and but nevertheless, register serious deterioration in the indices of yield and fruit quality. Previous works on grafted cucumber plants have revealed sharp contrasts between the genotypes of rootstocks as well as their effects on growth and production of the aerial part, presenting a correlation between the root system, vegetative growth and fruit yield in soil-grown plants (Carlson, 1963).

Den Nijs (1985) observed that the rootstocks that promote vigorous growth of the shoot also induce high fruit yield. This was true when certain rootstocks such as *Cucurbita ficifolia* and *Sicyos angulatus* were grafted to cucumber scions, promoting growth and early fruit production. In fruit trees, different rootstocks also resulted in notable differences in size, earliness and yield (NC140 Cooperators, 1987 and 1991; Elfving and Mckibbon, 1991).

Zijlstra et al. (1994) studying grafted cucumber plants cultivated on vermiculite, noted that most of the rootstocks used induced a lengthening of the shoot, directly influencing the first flowering and, therefore, early production in these plants. Also, in contrast to the experimental findings of Hayase (1966) and Friedlander et al., (1977) Zijlstra et al., (1994) found that the rootstock had no effect on the sex expression of the aerial part of the plant.

Working with the aerial part of grafted melon hybrids, Leoni et al., (1990) reported that yield increased with the combinations RS-841 × Pacio, RS-841 × Paquito, RS-841 × Symphony y RS-841 × Supermarket by 310, 65, 62 and 52%, respectively, compared with control. On this basis, these authors drew the following conclusions:

1. The rootstock can spur vegetative development of the plant and thus yield.
2. The high yield in grafted plants is due to greater mean weight and number of fruits.

3. Grafting encourages early maturation of the crop (ripening of the fruits).

4. Rootstocks specific to hybrids and varieties do not deteriorate fruit quality, either morphologically nor organoelectrically.

5. There is a clear positive interaction between certain rootstocks and certain varieties selected, resulting in greater productivity in some combinations with respect to others.

Since the shoots of watermelons grafted to vigorous rootstocks often enlarges the fruit in comparison to ungrafted plants, grafting is practiced in many zones of the world. Nevertheless, other characteristics of quality such as fruit shape, colour, smoothness and thickness of the rind, colour and texture of the pulp, and concentration of soluble solids are known to be influenced by the rootstock (Lee, 1994). In cucumbers, the development of flowering and peel colour are major quality characteristics (Choi et al., 1992; Kang et al., 1992). These characteristics are hereditary factors that are also influenced by the rootstock (Heo,1991; Kang et al., 1992). Nevertheless, it has been demonstrated that the rootstock has a negative impact on fruit quality and other organoelectric and morphological parameters, except for size (Lee, 1994).

8. Conclusions and Future Perspectives

Throughout this chapter, we have described the role of grafted plants in horticulture, based on all the research conducted on this subject to date. Primarily, we have examined the advantages of using these plants for current agriculture, these being: resistance to evermore frequent soil diseases; tolerance of low temperatures characteristic of many latitudes of the world where intensive cultivation is economically important; tolerance to the growing problem of salinity from abuse of chemical fertilizers and desertification in many agricultural zones; enhanced water and inorganic-nutrient uptake, as well as increased endogenous-hormone production, resulting in greater development, vigour, earliness and yield in these crops.

All these advantages provide motivation for grafting in present-day world agriculture. Nevertheless, there are also a number of disadvantages which restrict the use of this technique:

1. For a high success rate, grafting and post-grafting care require the attention of specialists, while expensive conditioning chambers are needed in order to maintain saturated humidity, temperatures of 35 to 40°C and a photoperiod of 16 h of light every day.

2. Grafted plants require different practices in the field as well as a specific reduction of fertilizer application.

3. After the compatibility between varieties is determined—never a simple task—the appropriate growing season must be taken into account so as to avoid an unbalanced growth between the rootstock and the scion.

4. Frequently, after grafting, the scion may produce adventitious roots, either externally or internally. This can be avoided either by careful handling during germination or by different grafting methods to eliminate root development of the shoot in the internal cavity of the hypocotyl, where the roots are often not visible from the outside.

5. The rootstock must be selected with the prime objective of avoiding excessive uptake of water and nutrients, which would promote runaway vegetative growth and serious physiological disorders.

6. Sometimes fruit quality may deteriorate, causing inner rot, insipid flavour, low sugar content, etc. To avoid these problems, appropriate soil moisture controls, controlled foliar application of Ca and a reduction of the N supply would be required, together with other cultivation practices appropriate to the variety at hand.

7. Transplanting a grafted plant to the field or greenhouse requires somewhat different care from that of an ungrafted plant, in that brusque movements must be avoided so as not to cause breakage around the graft, and the graft scar should remain aboveground to avoid rot at that point and so that the scion does not sprout roots.

Although grafting at times appears to be a mysterious surgical procedure accessible only to highly specialized professionals, only basic instructions (although of vital importance) are needed for most people to learn successful grafting (Baldwin, 1996). It should be emphasized that grafting today is performed manually in most countries. In addition, the costs of the automated chambers to control the conditions of light, temperature and humidity, plus the price of the seeds selected for the graft, entail a substantial economic expense. The simple fact that the cut made in the hypocotyl is today made manually by specialized people using specially-designed knives implies salaries that steadily rise. At present, semi-automatic and fully automatic grafting machines are being developed (Kobayashi, 1991; Itagi, 1992; Ito, 1992; Kurata, 1994). Due to the costs involved in grafting, the use of advanced technology is becoming more attractive, as automation could raise production by 150% (Kurata, 1994).

Finally, with notable advances in molecular biology made during the previous decade, transgenic plants capable of resisting a great number of adverse factors represent competition against grafted plants.

However, the many complications in finding and introducing genes for resistance to various environmental stress factors have until now conferred an advantage of grafting over genetic engineering in plants.

8. References

Ahn, S.J., Y.J. Im, G.C. Chung, B. H. Cho, and S.R. Suh. 1999. Physiological responses of grafted-cucumber leaves and rootstock roots affected by low temperature. *Scientia Horticulturae* 81: 397-408.

Ashita, E. 1927. Grafting of watermelons (in Japanese). Korea (Chosun) *Agr. Nwsl.* 1:9-15.

Ashworth, J. 1985. Verticillium resistant rootstock research, pp. 54-56. Annual Report of the Californian Pistachio Industry. Fresno, CA.

Baldwin, B. 1996. Grafting offers endless possibilities. Bladwin B (Ed). University of Saskatchewan, Department of Horticulture. Provided as a service of the Division of Extension and Community Relations and the Department of Horticulture Science.

Behboudian, N.M., R.R. Walker and E. Torokfalvy. 1986. Effects of water stress and salinity on photosynthesis of Pistachio. *Scientia Horticulturae.* 29: 251-261.

Bernstein, N., M. Ioffe and M. Zilberstaine. 2001. Salt-stress effects on avocado rootstock growth. I. Establishing criteria for determination of shoot growth sensitivity to the stress. *Plant and Soil* 233: 1-11.

Bernstein, L.; C.F. Ehlig and R.A. Clark. 1969. Effect of grape rootstocks on chloride accumulation on leaves. *J. Amer. Soc. Hort. Sci.* 94: 584-590.

Bernstein, L., J.W. Brown and H.E. Hayward. 1956. The influence of rootstock on growth and salt accumulation in stone fruit trees and almonds. *Proc. Amer. Soc. Hort. Sci.* 68: 86-95.

Biles, C.L., R.D. Martyn and H.D. Wilson. 1989. Isozymes and general proteins from various watermelon cultivars and tissue types. *HortScience.* 24: 810-812.

Bolger, T.P., D.R. Upchurch and B.L. McMichael. 1992. Temperature effects of cotton root hydraulic conductance. *Environ. Exp. Bot* 32: 49-54.

Borochov-Neori, H. and A. Borochov. 1991. Response of melon plants to salt: 1. Growth, morphology and root membrane properties. *J. Plant Physiol.* 139: 100-105.

Bradow, J.M. 1990a. Chilling sensitivity of photosynthetic oil-seedlings. Cucurbitaceae. *J. Exp. Bot.* 41: 1595-1600.

Bradow, J.M. 1990b. Chilling sensitivity of photosynthetic oil-seedlings. Cotton and Sunflower. *J. Exp. Bot.* 41: 1585-1593.

Brown, P.H., Q. Zhang and L. Ferguson. 1994. Influence of rootstock on nutrient acquisition by pistachio. *J. Plant Nutr.* 17: 1137-1148.

Bulder, H.A.M.; Ph.R. van Hasselt, P.J.C. Kuiper and E.J. Speek. 1990. Growth temperature and lipid composition of cucumber genotypes deffering in adaptation to low energy conditions. *J. Plant Physiol.* 138: 655-660.

Bulder, H.A.M.; Ph.R. van Hasselt, P.J.C. Kuiper, E.J. Speek and A.P.M. den Nijs. 1991. The effect of low root temperature on growth and lipid composition of low temperature-tolerant rootstock genotypes for cucumber. *J. Plant Physiol.* 138: 661-666.

Cakmak, I. 1988 Morphologische und phisiologische Vernänderungen bei Zinkmangelpflanzen. Ph. D. Thesis. Universität Hohenheim. Stuttgart. Germany.

Camacho, F. and E. Fernández. 1997. Influencia de patrones utilizados en el cultivo de sandía bajo plástico sobre la producción, precocidad y calidad del fruto en Almería. *Caja Rural de Almería* (Ed.). pp 107-122.

Carlsson, G. 1963. Studies on factors influencing yield and quality of cucumbers. 2. Development and hardiness of the roots. *Acta Agric. Scand.*, 149-156.

Copper, W.C. 1961. Toxicity and accumulation of salts in citrus trees on various rootstocks in Texas. *Proc. Fla. State Hort. Soc.* 74: 95-104.

Copper, W.C. and A.V. Shull. 1953. Salt tolerance and accumulation of sodium and chloride ions in grapefruit on various rootstocks grown in a naturally-saline soil. *Proc. 7th Rio Grande Val. Hortic. Inst.* 7: 107-117.

Chaplin, M.H. and M.N. Weswood. 1980. Nutritional status of 'Barlett' Pear on cydonia and Pyrus species rootstock. *J. Am. Soc. Hort.* Sci. 105: 60-63.

Cheeseman, J.M. 1988. Mechanisms of salinity tolerance in plants. *Plant Physiol.* 87: 547-550.

Choi, J.S., K.R. Kang, K.H. Kang and S.S. Lee. 1992. Selection of cultivars and improvement of cultivation techniques for promoting export of cucumbers (in Korean with English summary). *Res. Rpt., Min. Sci. & Technol.*, Seoul, Republic of Korea. p. 74.

De Stigter, H.C.M. 1956. Studies on nature of the incompatibility in cucurbitaceous graft. *Meded. L.H. Wag.* 56: 1-51.

Delane R., H. Greenway, R. Munns and J. Gibbs. 1982. Ion concentration and carbohydrate status of the elongating leaf tissue of *Hordeum vulgare* growing at high external NaCl. I. Relationship between solute concentration and growth. *J. Exp. Bot.* 33: 557-573.

Den Nijs, A.P.M. 1980. Adaptation of the glasshouse cucumber to lower temperatures in winter by breeding. *Acta Horticulturae.* 118: 65-72.

Den Nijs, A.P.M. 1984a. Rootstock-scion interactions in the cucumber: Implications for cultivation and breeding. *Acta Horticulturae.* 156: 53-60.

Den Nijs, A.P.M. 1984b. Ervaringen met een nieuwe onderstam voor komkommer: Sicyos angulatus. G + F. April, 38-41.

Elfving, D.C. and E.D. McKibbon. 1991. Effects of rootstock on productivity and pruning requirements of 'Starkspur Supreme Delicious' apple trees in the NC-140 cooperative planting. *Fruit Var. J.* 45: 242-246.

Forner, J.B. and A. Alcaide. 1993. La mejora genética de patrones de agrios tolerantes a tristeza en España: 20 años de historia (I). *Levante Agrícola* 325: 261-267.

Forner, J.B. and A. Alcaide. 1994. La mejora genética de patrones de agrios tolerantes a tristeza en España: 20 años de historia (II) *Levante Agrícola* 329: 273-279.

Friedlander, M., D. Atsmon and E. Galun. 1977. The effect of grafting on sex expression in cucumber. *Plant Cell Physiol.* 18: 1343-1350.

Funck, R. 1929. Untersuchungen über heteroplastische Transplantationen bei Solanaceen und Cactaceen. Beitr. Biol. Pflanz. 17: 404-468.

Hartmann, H.T. and D.E. Kester. 1991. *Propagación de plantas. Cía.* Edit Continental, México.

Hartmann, H.T. and D.E. Kester. 1975. *Plant Propagation: Priciples and Practices.* 3rd ed. Prentice-Hall, Englewood Cliffs, N.J.

Hasegawa, P.M., R.A. Bressan and A.K. Handa. 1986. Cellular mechanisms of salinity tolerance. *HortScience.* 21: 1317-1324.

Hayase, H. 1966. Experimental modification of sex expression in the cucumber plant. I. Alternative grafting between sub-androecious and sub-gynoecious lines. *Jpn. J. Breeding.* 16: 213-219.

Heo, Y.C. 1991. Effects of rootstocks on exudation and mineral elements contents in defferent parts of Oriental melon and cucumber (in Korean with English summary). MS thesis, Kyung Hee Univ., Seoul, Korea. p. 53.

Herner, R. 1990. The effects of chilling temperatures during seed germination and early seedlings, growth. In: *Chilling Injury of Horticultural Crops.* C.Y. Wang (Ed.). CRC Press. Boca Raton, Fla.

Horvath, I., L. Vigh, A. Belea and T. Farkas. 1980. Hardiness dependent accumulation of phospholipids of wheat cultivars. *Physiol. Plant.* 49: 117-120.

Horvath, I., L. Vigh, Ph.R. van Hasselt, J. Woltjes and P.J.C. Kuiper. 1983. Lipid composition in leaves of cucumber genotypes as affected by defferent temperature regimes and grafting. *Physiol. Plant.* 57: 532-536.

Horvath, I., L. Vigh, Ph.R. van Hasselt, J. Woltjes and T. Farkas; P.J.C. Kuiper. 1987. Combined electron-spin-resonance, X-ray-diffraction studies on phospholipid vesicles obtained from cold-hardened wheats. *Planta.* 170: 20-25.

Itagi, T. 1992. Status of transplant production systems in Japan and new grafting technics (in Korean). *Symp. Protected Hort. Hort. Expt. Sta.,* Rural Development Admin., Suwon, Republic of Korea, pp. 32-67.

Itai, C. and H. Nirnbaum. 1991. Synthesis of plant growth regulators by roots. In: *Plant Roots.* Waisel Y., Eshel A. and Kafkati U. (Eds.). *The Hidden Half.* Marcel Dekker, New York, pp. 163-178.

Ito, T. 1992. Present state of transplant production practices in Japanese horticultural industry, p. 65-82. In: K. Kurata and T. Kozai (eds.). *Transplant production system.* Kluwer Academic Publishers, Yokohama, Japan.

Jackman, R.L., R.Y. Yada, A. Marangoni, K.L. Parkin and D.W. Stanley. 1988. Chilling injury. A review of quality aspect. *J. Food Sci.* 11: 253-277.

James, D.J. and I.J. Thurbon. 1981. Shoot and root initiation in vitro in the apple rootstock M9 and the promotive effects of phloroglucinol. *J. Hort. Sci.* 56: 15-20.

Jones, W.W., J.P. Martin and W.P. Bitters. 1957. Influence of exchangeable sodium and potassium in the soil on the growth and composition of young lemon trees on different rootstocks. *Proc. Amer. Soc. Hort. Sci.* 69: 189-196.

Kang, K.S., S.S. Choe and S.S. Lee. 1992. Studies on rootstocks for stable production of cucumber (in Korean with tables and figures in English). *Kor. Soc. Hort. Sci.* 19: 122-123 (Abstr.).

Kato, T. and H. Lou. 1989. Effect of rootstock on the yield, mineral nutrition and hormone level in xylem sap in eggplant. *J. Jpn. Soc. Gort. Sci.* 58: 345-352.

Kobayashi, K. 1991 Development of a grafting robot for the fruit-vegetables. *Plant Cell Technol.* 3: 477-482.

Kollmann, R. and C. Glockmann. 1990. Sieve elements of graft unions. En: Behnke, H.-D.; R. D. Sjolund (eds): *Sieve Elements: Comparative Structure, Induction and Development.* pp. 219-237. Springer, Berlin.

Kramer, D. 1984. Cytologycal aspects of salt tolerance in higher plants. In: *Salinity Tolerance in Plants. Strategies for Crop Improvement.* R.C. Staple; G.H. Toeniessen. (eds). Wiley & Sons, New York.

Kuiper, P.J.C. 1984. Functioning of plant cell membranes under saline conditions: Membrane lipid composition and ATPases. In: *Salinity Tolerance in Plants-Strategies for Crop Improvement.* Staples R.C. and G.A. Teinissen (Eds) John Willey and Sons, pp. 77-91.

Kurata, K. 1994. *Cultivation of Grafted Vegetables.* II. Development of grafting robots in Japan. HortScience 29: 240-244.

Kurata, K. 1992. Transplant production robots in Japan, p. 313-329. In: K. Kurata and T. Kozai (eds). *Transplant Production System.* Kluwer Academic Publishers, Yokohama, Japan.

Lazof, D.B. and N. Bernstein. 1998. The NaCl-induced inhibition of shoot growth: The case for disturbed nutrition with special consideration of calcium nutrition. *Bot. Res.* 29: 115-190.

Lee, J.M. 1989. On the cultivation of grafted plants of cucurbitaceous vegetables (in Koren with English summary). *J. Kor. Soc. Hort. Sci.* 30: 169-179.

Lee, J.M. 1994. Cultivation of grafted vegetables I. Current status, grafting methods and benefits. *HortScience.* 29: 235-239.

Leoni, S., R. Grudina, M. Cadinu, B. Madeddu and M.G. Carletti. 1990. The influence of four rootstocks on some melon hybrids and a cultivar in greenhouse. *Acta Horticulturae.* 287: 127-134.

Levitt, J. 1980. *Responses of plants to environmental stresses,* Vol. II. Academic Press, New York.

Lyons, J.M. and R.W. Breidenbach. 1979. Strategies for altering chilling sensitivity as a limiting factor in crop production. In: *Stress Physiology in Crop Plants,* pp. 179-196. Wiley and Sons, New York.

Markhart III, A.H., M.M. Peet, N. Sionit and P.J. Kramer. 1980. Low temperature acclimation of root fatty acid composition, leaf water potential gas exchange and growth of soybean seedlings. *Plant Cell Environment.* 3: 435-441.

Masuda, M. and K. Gomi. 1982. Diurnal changes of the exudation rate and the mineral concentration in xylem sap after decapitation of grafted and non-grafted cucumber (in Japanese with English summary). *J. Jpn. Soc. Hort. Sci.* 51: 293-298.

Masuda, M. and K. Gomi. 1984. Mineral absorption and oxygen consumption in grafted and non-grafted cucumber (in Japanese with English summary). *J. Jpn. Soc. Hort. Sci.* 52: 414-419.

McCully, M.E. 1983. Structural aspects of graft development. En: Moore, R. (ed.): *Vegetative Compatibility Responses in Plants.* Baylor University Press, Waco, Texas. pp. 71-88.

McWilliam, J.R., P.J. Kramer and R.L. Musser. 1982. Temperature-induced water stress in chilling-sensitive plants. *Aust. J. Plant Physiol.* 9: 343-352.

Miguel, A. 1997. In: *'Injerto en hortalizas'. Generalitat Valenciana* (Ed.). Consellería de Agricultura, Pesca y Alimentación.

Mistrik, I., M. Holobrada and M. Ciamporoda. 1992. The root in unfavourable conditions. In: *Physiology of the Plant Root System.* J. Kolek and V. Kozinka (Eds). Kluwer Academic Publishers. Dordrecht, The Netherlands

Monzer, J. and R. Kollmann. 1986. Vascular connections in the heterograft *Lophophora williamsii* Coult. on *Trichocereus spachianux* Ricc. *J. Plant Physiol.* 123: 359-372.

Moore, R. 1984. The role of direct cellular contact in the formation of compatible autografts in Sedum telephoides. *Ann. Bot.* 54: 127-133.

Moore, R. and D.B. Walker. 1981a. Studies of vegetative compability-incompatibility in higher plants. I. A structural study of a compatible autograft in *Sedum telephoides* (Crassulaceae). *Am. J. Bot.* 68: 820-830.

Moore, R. and D.B. Walker. 1981b. Studies of vegetative compatibility-incompatibility in higher plants. II. A structural study of an incompatible heterograft between *Sedum telephoides* (Crassulaceae) and *Solanum pennellii* (Solanaceae). *Am. J. Bot.* 68: 831-842.

Moore, R. and D.B. Walker. 1981c. Studies of vegetative compatibility-incompatibility in higher plants. III. The involvement of acid phosphatase in the lethal cellular senescence associated with an incompatible heterograft. *Protoplasma* 109: 317-334.

Munns, R. 1993. Physiological processes limiting plant growth in saline soils: some dogmas and hypotheses. *Plant Cell Environ.* 16: 15-24.

NC140 Cooperators. 1987. Growth and production of 'Starkspur Supreme Delicious' on 9 rootstocks in the NC-140 cooperative planting. *Fruit Var. J.* 41: 31-39.

NC140 Cooperators. 1991. Performance of "Starkspur Supreme Delicious" apple on 9 rootstock over 10 years in the NC-140 cooperative planting. *Fruit Var. J.* 45: 192-199.

O'Toole, J.C. and W.L. Bland. 1987. Genotypic variation in crop plant root systems. *Adv. Agron.* 41: 91-145.

Oda, M. 1995. New grafting methods for fruit-bearing vegetables in Japan. *Jarq* 29: 187-194.

Padgett, M. and J.C. Morrison. 1990. Changes in grape berry exudates during fruti development and their effect on mycelial growth of *Botrytis cinerea*. *J. Amer. Soc. Hort. Sci.* 115: 256-257.

Pasternak, D. 1987. Salt tolerance and crop production : A comprehensive approach. *Ann. Rev. Phytopathol.* 25: 271-291.

Patil, V.K. and J.R. Bhamboto. 1980. Salinity studies en citrus: 1. Effect of various levels of salinity on the macronutrient status of seedling rootstocks. *J. Ind. Soc. Soil Sci.* 28: 72-79.

Picchioni, G.A., S. Miyamoto and J.B. Storey. 1990. Salt effects on growth and ion uptake of pistachio rootstock seedlings. *J. Amer. Soc. Hort. Sci.* 115: 647-653.

Pulgar, G., R.M. Rivero, D.A. Moreno, L.R. López-Lefebre, G. Víllora, M. Baghour and L. Romero. 1998. Micronutrientes en hojas de sandía injertadas. In: VII Simposio nacional-III Ibérico sobre Nutrición Mineral de las Plantas. Gárate A. (Ed.), Universidad Autónoma de Madrid, Madrid, pp. 255-260.

Rachow-Brandt, G. 1987. Vergleichende transportphysiologische Untersuchungen an Pfropflingen unterschiedlicher Kompatibilität. Ph. D. Thesis, Kiel.

Raison, J.K. and J.M. Lyons. 1986. Chilling injury: A plea for uniform terminology. *Plant Cell Environ.* 9: 685-686.

Rattink, H. 1983. Introduction to soil fumigation and specific problems with a special reference to the effects of low dosages methylbromide on some fungi nematodes. *Acta Hortic.* 152: 163-169.

Reyes, E. and P.H. Jennings. 1994. Response of cucumber (*Cucumis sativus* L.) and squash (*Cucurbita pepo* L. var. *melopepo*) roots chilling stress during early stages of seedling development. *J. Amer. Hort. Sci.* 119: 964-970.

Roberts, R.H. 1949. Theoretical aspects of graftage. *Bot. Rev.* 15: 323-463.

Romero, L., A. Belakbir, L. Ragala and J.M. Ruiz. 1997. Rsponse of plant yield ad leaf pigments to saline conditions: effectiveness of different rootstocks in melon plants (*Cucumis melo* L.) Soil Sci. *Plant Nutrit.* 43: 855-862.

Ruiz, J.M., A. Belakbir, I. López-Cantarero and L. Romero. 1997. Leaf-macronutrient content ad yield in grafted melon plants. A model to evaluate the influence of rootstock genotype. *Scientia Horticulturae* 71: 227-234.

Ruiz J.M., A. Belakbir and L. Romero. 1996. Foliar level of phosphorus as its bioindicators in *Cucumis melo* grafted plants. A possible effect of rootstock. *J. Plant Physiol.* 149: 400-404.

Ryu, J.S., K.S. Choi and S.S. Lee. 1973. Effects of grafting stocks on growth, quality and yields of watermelon (in Korean with English summary). *J. Kor. Soc. Hort. Sci.* 13: 45-49.

Salveit Jr., M.E. and L.L. Morris. 1990. Overview on chilling injury of horticultural crops. In: *Chilling injury in Horticultural Crops*. Ch.Y. Wang (Ed.). CRC Press. Boca Raton, Fla.

Sánchez, M. and J. Aguirreola. 1993. *Relaciones hidricas*. En: Fisiologia -y bioquimica vegetal, Azcon-Bieto, J.; M. Talon., (eds). Interamericana-McGrow-Hill. pp.49-90.

Schneider, J.H.M., J.J. s'Jacob and P.A. van de Pol. 1995. *Rosa multiflora* 'Ludiek' a rootstock with resistant features to the root lesion nematode *Pratylenchus vulnus*. *Scientia Horticulturae* 63: 37-45.

Silberschmidt, K. 1935/36. Die Abhängigkeit des Pfropferfolges von der systematischen Verwandtschaft der Partner. *Zeitswchr. f. Bot.* 29: 65-137.

Smith, P.E. 1962. A case of sodium toxicity in citrus. *Proc. Fla. State Hort. Soc.* 75: 120-124.

Smolenska, G. and P.J.C. Kuiper. 1977. Effect of low temperature on lipid and fatty acid composition of roots and leaves of winter rape plants. *Physiol. Plant.* 41: 29-35.

Sykes, S.R. 1992. The inheritance of salt exclusion in woody perennial fruit species. *Plant Soil* 146: 123-129.

Sykes, S.R. 1985. A glasshouse screening procedure for identifying citrus hybrids which restrict cholride accumulation in shoot tissues. *Aust. J. Agric. Res.* 36: 779-789.

Tachibana, S. 1988. The influence of root temperature on nitrate assimilation by cucumber cultivars ad figleaf gourd. *J. Jpn. Soc. Hort. Sci.* 57: 440-447.

Tachibana, S. 1982. Comparison of effects of root temperature on the growth and mineral nutrition of cucumber cultivars and figleaf gourd. *J. Japan. Soc. Hort. Sci.* 51: 299-308.

Tachibana, S. 1986. Effect of root temperature on lipid and its fatty acid composition in cucumber and figleaf gourd roots. *J. Japan. Soc. Hort. Sci.* 55: 187-193.

Tachibana, S. 1987. Effect of root temperature on the concentration and fatty acid composition of phospholipids in cucumber and figleaf gourd roots. *J. Japan. Soc. Hort. Sci.* 56: 180-186.

Tagliavani, M., D. Scudellari, B. Marangoni, A. Bastianel, F. Franzin and M. Zamborlini. 1992. Leaf mineral composition of apple tree: sampling date ad effects of cultivar and rootstock. *J. Plant Nutrit.* 15: 605-619.

Tiedemann, R. 1986. Struktur und Funktion der Phloemverbindungen zwischen Reis und Unterlage beim Pfropfling *Cucumis sativus* auf *Cucurbita ficifolia*. Ph. D. Thesis, Kiel.

Tiedemann, R. 1989. Graft union development and symplastic phloem contact in the heterograft *Cucumis sativus* on *Curcubita ficifolia*. *J. Plant Physiol.* 134: 427-440.

Tudela, D. and F.R. Tadeo. 1993. Respuestas y adaptaciones de las plantas al estrés. En: Fisiologia y bioquimica vegetal, pp. 537-553. Azcon-Bieto, J.; M. Talon. ISBN: 84-486-0033-9, (eds.). Interamericana-McGrow-Hill.

Vigh, L., I. Horvarth, Ph.R. van Hasselt and P.J.C. Kuiper. 1985. Effect of frost hardenimg on lipid and fatty acid composition of chloroplast thylakoid membranes in two wheat varieties of contrasting hardiness. *Plant Physiol.* 79: 756-759.

Walker, R.R. 1986. Sodium exclusion and potassium-sodium selectivity in salt-treated Trifoliate orange (*Poncirus trifoliata*) and Cleopatra mandarin (*Citrus reticulata*) plants. *Aust. J. Plant Physiol.* 13: 293-303.

Wang, Ch.Y. 1990. *Chilling Injury of Horticultural Crops*. CRC Press. Boca Raton, Fla.

Weimberg, R., H.R. Lerner and A. Poljakoff-Mayber. 1984. Changes in growth and water-soluble concentration in *Sorghum bicolor* stressed with sodium and potassium salts. *Physiol. Plant.* 62: 472-480.

Yamakawa, B. 1983. Grafting, p. 141-153. In: Nishi (ed.). *Vegetable Handbook* (in Japanese). Yokendo Book Co., Tokyo.

Yazawa, S., T. Uemachi, T. Higashide and Watanabe H. 1996. CMV resistance developed in vigorous-growing lateral shoots from virus infected plants of *Capsicum annum* L. *Scientia Horticulturae* 65: 295-304.

Yeoman, M.M., D.C. Kilpatrick, M.B. Miedzybrodzka and A.R. Gould. 1978. Cellular interactions during graft formation in plants, a recognition phenomenon? *Symp. Soc. Exp. Biol.* 32: 139-160.

Yeoman, M.M. and R. Brown. 1976. Implication of the formation of the graft union for organisation in the intact plant. *Ann. Bot.* 40: 1265-1276.

Zekri, M. 1991. Effects of NaCl on growth and physiology of sour orange and Cleopatra mandarin seedlings. *Sci. Hortic.* 47: 305-315.

Zekri, M. and L.R. Parsons. 1990. Calcium influences growth and leaf mineral concentration of citrus under saline conditions. *HortScience* 25: 784-786.

Zekri, M. and L.R. Parsons. 1992. Salinity tolerance of citrus rootstocks: Effects of salt on root and leaf mineral concentrations. *Plant and Soil.* 147: 171-181.

Zekri, M. and Parsons L.R. 1989. Growth and root hydraulic conductivity of several citrus rootstock under salt and polyethylene glycol stresses. *Physiol. Plant* 77: 99-106.

Zijlstra, S., S.P.C. Groot and J. Jansen. 1994. Genotypic variation of rootstocks for growth and production in cucumber; possibilities for improving the root system by plant breeding. *Scientia Horticulturae.* 56: 185-196.

Zobel, R.W. 1975. The genetics of root development. In: *The Development Nad Function of Roots.* Torrey J. and Clarkson D. (Eds). Academic Press, London, pp. 261-275.

9

Domestication of Jujube Fruit Trees (*Zizyphus mauritiana* Lam.)

A.M. Bâ[1], **P. Danthu**[2], **R. Duponnois**[3] **and P. Soloviev**[4]

[1]*Laboratoire de Biologie et Physiologie Végétales, UFR Sciences Exactes et Naturelles, BP. 592, Université des Antilles et de la Guyane, 91759 Pointe-à-Pitre, Guadeloupe, France;*
[2]*CIRAD-Forêt, Programme "Arbes et Plantation", BP. 853, Antananarivo, Madagascar;*
[3]*IRD, UR IBIS "Interactions Biologiques dans les Sols des Systèmes Anthropisés Tropicaux", 01 BP. 182. Ouagadougou. Burkina Faso;*
[4]*Centre de Formation Professionnelle Horticole, BP. 3284, Dakar, Sénégal.*

1. Introduction

Zizyphus mauritiana Lam. is native of Africa (Chevalier, 1947). Amongst the numerous species of jujube tree, it is the one with the most extensive distribution, occupying the entire Sahelian zone, from Senegal to Somalia, whilst also present from Eritrea to Mozambique, as well as in Angola, Madagascar, Arabian peninsula, Indian subcontinent, South America and frequently naturalized from southern Florida and California through Mexico, Central America, and the West Indies to South America (Perrier de la Bâthie, 1924; Chevalier, 1947; Munier, 1973; Howard, 1989).

Z. *mauritinana*, often called indian jujube or *ber* is a multipurpose fruit tree that provides high quality products: fruits, medicines, fodder and fuel. Interest and attention are being focused on fruits of jujubes that assume food security, health, and provide a source of income for the people of the rural areas in Sahelian and Sudanian zones in West Africa (Bonkoungou et al., 1998, Nair, 1998). Jujube trees also perform well as live hedge species in agroforestry systems (Djimdé, 1997). However, jujube fruit trees are slow growing and little is known about their cultivation (Depommier, 1988). Domestication of this tree crop could be achieved through a combination of approaches and could include the selection of species by local people (Bonkoungou et al., 1998), selection of different provenance phenotypes based on product characteristics (Leakey and Simons, 1998; Leakey, 1999), vegetative propagation of selected trees (Vashishtha, 1997; Danthu et al., 2000 a and b; Danthu et al., 2001; Danthu et al., 2002 a,b,c and d), training and pruning (Bajwa et al., 1987), application of rock phosphate or others phophorus fertilizers in P-deficient soils, or mycorrhizal inoculation (Bâ and Guissou, 1996; Guissou et al., 1998 a and b; Bâ et al., 2000; Guissou et al., 2000; Bâ et al., 2001; Guissou et al., 2001).

This Chapter reviews the on-going research on domestication of jujubes through vegetative propagation, controlled mycorrhization, utilization of rock phosphates or others sources of phosphorus, nematode diseases, planting, training and pruning, and yield and organoleptic qualities of jujubes fruit trees.

2. Ecology

Jujube is shrub or small tree, 3-8 m tall, belonging to the Rhamnaceae family (Plate 1). It is known to be well adapted to the difficult conditions prevailing in the Sahelian zone. Even though the jujube tree can survive in extremely arid zones with an annual rainfall of 100-125 mm, it is most commonly found in areas receiving between 300 and 500 mm annual rain. Dry season lasting up to 8-10 months is well tolerated. *Z. mauritiana* resists well to extreme temperatures of up to 50°C, but is less resistant to low temperatures and specially frost. (Booth and Wickens, 1988). During the fruit set, a minimum relative air-humidity of more than 50% is required, if the relative humidity falls below this level and is accompanied by temperatures in excess of 35°C. In such cases, shedding frequently occurs (Vashishtha, 1997). The jujube tree develops well in a wide variety of soil conditions, even though neutral or slightly alkaline deep sandy soils are preferred. This species tolerates saline soil conditions (with E.C. < 9 d.m^{-1}) (Jain *et al*, 1980), as well as temporary waterlogging, lasting from 1-2 months (Vashishtha, 1997).

3. Use

In the Sahelian and Sudanian regions of Africa, *Z. mauritiana* is a non-domesticated species which is nonetheless utilized by rural populations. Its leaves, rich in lipids, iron, calcium, magnesium and zinc, are consumed by both humans and livestock (Sena et al., 1998; Arbonnier, 2000). Its wood is durable and pliable, its bark and roots have numerous applications in traditional pharmacopoeia, and the tree itself is used to set up live or dead hedges (Kerharo and Adam, 1974 ; Djimdé, 1997; Arbonnier, 2000). However, its main attraction lies in its fruit production, the jujube (Plate 2). These drupes, 1-1.5 cm in diameter, are eaten fresh

Plate 1. Young boys are harvesting fruits from a jujube tree in Senegal (photo from P. Danthu)

Plate 2. Gola fruits from Thailand is greater than Senegalese fruits (photo from P. Danthu)

or dried by the local populations and can be used for making juice or jams (Chevalier, 1947; Munier, 1973 ; Prasad et al., 1995). Jujubes are very rich in sugars, iron, calcium and vitamin C (Toury et al., 1961, 1967; Becker, 1983; Bergeret, 1986; Danthu et al., 2002b). Marketed locally, their sale makes up a sizeable part of the livelihood of many families (Becker, 1983).

On the African continent, *Z. mauritiana* is a species not subject to any selection or improvement. Local populations harvest crops from natural stands and use the species for agroforestry techniques, specially for establishing live hedges (Djimdé, 1997). However, in India and Pakistan, where the species is domesticated and cultivated in orchards (Munier, 1973; Morton, 1987; Chovatia et al., 1993; Vashishtha, 1997), a programme has been operating since the eighteenth century with the principal aim of improving fruit quality (Morton, 1987), resulting in the development of more than 90 varieties according to Morton (1987) and more than 300, according to Kaushik (2000). These cultivars vary mainly in their time taken to ripen and the size and shape of their fruit, but all have the same capacity to adapt to arid conditions (Chovatia et al., 1991, Booth and Wickens, 1988; Vashishtha, 1997, Singh et al., 1999). Of these cultivars, those that appear to adapt best to conditions in the Sahelian zone, are Gola, Seb and Unram (Vashishtha, 1997). It is these varieties that the authors of this article have attempted to transfer and adapt to conditions prevalent in the Sahelian zone.

4. Propagation

Propagation methods of *Z.mauritiana* depend most specifically on its use, its degree of domestication and, consequently, on the geographic zone of its use. In Africa, where the jujube is a wild species used in agroforestry, it is propagated mainly by seeds harvested from natural stands (Danthu et al, 1993). Seed collection is sometimes preceded by a phenotypic selection of mother trees. In India, though, the jujube tree is treated as an horticultural species and propagation is ensured by vegetative methods (Morton, 1987; Chovatia et al., 1993; Vashishtha, 1997). The authors of this article have attempted to transfer and adapt these techniques to an African context.

4.1. Seed Propagation

The stone (endocarp) from *Z. mauritiana* contains one or two seeds, making it possible to obtain up to 140 plantlets from 100 fruits (Danthu et al., 1993). The seeds remain viable for several years as long as they are preserved within the stone, within a cool, dry place. In order to obtain uniform and synchronous germination, the seeds should be sown stripped, or else the endocarp should be cracked before sowing or treated with sulfuric acid (Danthu et al., 1993; Roussel, 1995; Guissou et al., 1998). The seeds do not require any particular pre-treatment and raising them does not present any difficulties so long as they are well protected from the sun during the first two weeks. However, jujube seedlings are colonized by natural propagules of arbuscular mycorrhizal fungi in nursery conditions as indicated above (see paragraph 5 in this chapter). Within three to four months of growth, nursery seedlings can be successfully transplanted (Roussel, 1995).

4.2. Vegetative Propagation

4.2.1. Cutting and Microcutting

Z. mauritiana is difficult to propagate by cutting (Vashishtha, 1997; Booth and Wickens, 1988), an observation confirmed by the authors of this article in Senegal, where cutting tests produced poor results. A method of microcutting (*Z.mauritiana*) has been perfected in India (Mathur et al., 1995; Gupta and Srivastava, 1996), but which appears not to have been extended.

4.2.2. Horticultural Grafting

In India, different methods of vegetative propagation by horticultural grafting have been developed as this seems to be the simplest means of multiplying varieties (Vashishtha, 1997, 2001; Morton, 1987; Booth and Wickens, 1998). The standard method essentially involves shield budding or T-budding (Mawani and Singh, 1992; Morton, 1987; Vashishtha, 2001), wherein the scion is limited to a tissue fragment bearing a single bud and, according to the Indian authors, the most widely practised method seems to be patch budding, in which a rectangle of bark is severed from the rootstock and replaced by an identical fragment bearing a bud from a selected plant (for a more detailed explanation of the technique see Hartmann et al., 1997). If the grafts are performed in June, on one-year old rootstock, it is possible to achieve a 70% success rate (Mawani and Singh, 1992).

However, when applied in Senegal, these propagation methods produced success rates of less than 30% (Table 1), which would not have been adapted to large scale diffusion (Danthu et al., 2000, 2002b, 2002c). It is for this reason that taking into account the Senegalese peasants knowledge in the use of cleft grafting and side-veneer grafting to propagate mango and citrus varieties, we attempted to develop jujube tree grafting using these techniques. The results obtained from the Gola cultivar produced modest success rates of between 20% and 62% for side-veneer grafting, whereas for cleft grafting, the maximum success rate achieved was 40% (Table 1) (Danthu et al., 2002b). It seems that success rates for different grafts are related to the grafting period. Proximity to the rainy season appears to be one factor which encourages the establishment of grafts. However, it should be noted that Z.*mauritiana* is, in general, a bushy species, presenting a strong basitony tendency. Most self-rooting subjects regularly produce suckers around the collar, and the same applies to grafted plants, where vigorous shoots belonging to the rootstock develop to the detriment of the scion. These suckers, which are identifiable by foliar dimorphism between the small-leafed wild rootstock and the large-leafed scions from improved varieties, hinder the development of the tree and put the durability of plantations at risk, and as such they need to be regularly eliminated.

All these observations make it difficult to transfer propagation of jujube by horticultural grafting without large-scale optimization. In fact, these findings are in keeping with the remarks made by Munier (1973) and Mathur et al., (1995) on the limited success and the difficulty associated with horticultural grafting in Z. *mauritiana*.

Table 1. Success rate of the different horticultural grafting methods for the Gola variety. The rootstocks are young Z. *mauritiana* plants. Each test was carried out on 15 to 48 grafts.

Type of graft	Success rate (%)
Side-veneer graft	20 - 62
Cleft graft	12 - 40
Chip budding	27 - 28

4.2.3. Micrografting

In the light of these findings, CAZRI in Jodhpur has developed a method of graft miniaturization using three-month-old rootstock (Vashishtha, 1997), which is very effective. Here in Senegal, we have adopted the same line of investigation into graft miniaturization on juvenile rootstock, examining the possibilities offered by in vitro cultivation, and specifically micrografting, as a propagation system for jujube. This choice is upheld by the well-known benefits of micropropagation, and in particular, the possibility of cloning rapidly, with a high multiplication rate in those species normally recalcitrant to standard forms of vegetative propagation, while simultaneously remaining independent of seasonal cycles (Timmis et al., 1987; Hartmann et al., 1997). The method of micropropagation used is based on older research into the rejuvenation of varied ligneous species through repeated micrografting (Pliego-Alfaro and Murashige, 1987; Huang et al., 1992 a, 1992 b) and has already been adapted to several Sahelian species, including the local wild jujube trees (Danthu et al., 2001, 2002 a). It can be summed up as follows:

- One month-old rootstock from seed raised in vitro in 24 mm diameter tubes containing a polypropylene fibre support (3 cm

Milcap® plugs) immersed in 12 ml of half strength Murashige and Skoog, (1962) culture medium (MS/2), with added sucrose (20g.l) and the pH of the medium adjusted to 5.7 before autoclaving (110°C, 20 mn).

- The material to be grafted is removed from ortets established from the first plants introduced into Senegal, raised under glasshouse and regularly cut back. After removal, the fragments of hardened off stems are sterilised by brief immersion in ethanol 70° and soaking for 30 minutes in a solution of mercury chloride ($HgCl_2$, 0.1%). They are then thoroughly rinsed in sterile water and cut up into single node microcuttings, which are transferred to culture on the MS/2 with 7 g.l^{-1} of agar. The scions are supplied by the shoots derived from the development of these explants during this preculture phase.

- The micrografting consists of a miniaturization of cleft grafting. The Milcap® plug bearing the rootstock is extracted from the tube, the plantlet is beheaded at the hypocotyl and a diametral cleft is made. The base of the scion (apical or axillary node 5 to 10 mm), clipped into a simple wedge, is inserted into the cleft. The scion is then tied with a strip of sterile Parafilm® before being replaced in the tube and then grown in darkness for a week and subsequently transferred to light conditions (Plate 3).

- After a month of culture, the scion has reached the height 5 to 7cm (Plate 4).

Depending on the preferred objective, two subsequent approaches are possible: the micrograft is acclimatized and then planted out in the orchard, or else it is cut back into microscions, which in turn, are grafted onto juvenile rootstock to ensure the development of a continual provision of clones. The success rate with micrografting is consistently greater than 90%. Acclimatization under glass and then in the shade house results in minimum mortality (5–10%) and transfer to the field is accomplished without any difficulty (Table 2). Grafted plants have considerable growth and fruit from the first year (Plate 3). Finally, it should be noted that none of the micrografted plants developed suckers from the rootstock, which is probably due to the fact that grafting was carried out under the cotyledon, very close to the collar, eliminating any latent buds (Danthu et al., 2001). Throughout the course of the year 2000, in this way, around 700 micrografts of Gola were produced, as well as some 100 of the Seb variety, on an experimental basis. These first results demonstrate the feasibility of producing improved jujube plants by micrografting in the Sahelien zone. The methodology that has been perfected enables vast series of vegetative replicas (clones) of jujube to

Plate 3. Micrografting as a propagation system for Gola variety of jujube : scion is inserted into the cleft of rootstock and tied with a strip of sterile Parafilm (photo: P. Danthu)

be produced. Although, it applies principally to the Gola variety and has been validated for a second cultivar (Seb), it should be possible to extend it to include other Indian varieties whether or not they are adapted to dry zones.

Plate 4. Scion after one month of culture (P. Danthu)

5. Controlled Mycorrhization

Arbuscular mycorrhizal fungi (AMF) form the most widespread plant-microorganisms symbiosis on earth (Smith & Read, 1997). Many plants of forestry, agriculture and horticulture importance benefit in terms of

Table 2. Survival of the Z. *mauritiana* var. Gola plants grafted on Z. *mauritiana* at different stages of their production.

Stages	Survival rate (%)
Micrografting (assessed 1 month after grafting)	>95
Acclimatization (assessed one month after taken out of tube)	90-95
Transfer to the field (assessed 3 months after planting)	98
First fruiting (one-years-old plant)	100

growth, mineral nutrition and protection against root pathogens from this symbiosis association. The potential benefits of mycorrhizal inoculation can be assessed by the techniques of controlled mycorrhization, which encompasses procedures for culture isolation, isolate for selection, inoculum production of AM fungi, with the aim of producing mycorrhizal plants and evaluating their performance in the glasshouse, nursery and field conditions (Garbaye, 1991). However, the three most important factors determining the magnitude benefits from improved management of arbuscular mycorrhizas (AM) are: (1) mycorrhizal dependency (MD) of the host plants; (2) nutrient status of the soil; and (3) inoculum potential of the AMF present in the soil (Brundrett et al., 1996).

Considerations of MD of fruit trees are important because their mycorrhizal status is often unknown and may have a particular importance during their first phase of growth (Guissou et al., 1998 a; Bâ et al., 2000). For this, we used one isolate of AMF *Glomus aggregatum* Schenck and Smith emend. Schenck (IR 27) obtained from Burkina Faso (Bâ et al., 1996). Mycorrhizal inoculation of the soil in plastic bags was achieved by placing 20 g portions of a crude inoculum of AMF consisting of sand, spores, fragments of hyphae and infected roots below the seeds during transplanting. The inoculum density was calibrated by the most probable number method as 1800 infective propagules per 20 g of *G. aggregatum*. Thereafter, we evaluated the MD of thirteen woody fruit tree seedlings inoculated with five AMF, and finally, we selected jujubes as they are the most MD fruit trees when inoculated with the AMF *G. aggregatum* (Table 3, Plate 5). We also obtained a stimulation of growth in microcutted jujubes inoculated with the AMF *G. aggregatum* (Plate 6). MD of jujubes can also vary greatly from one provenance to another (Table 4). The MD of jujubes from Burkina Faso, Mali and Senegal are similar, reaching no more than 78%, while jujubes from Thaïland showed the highest MD, reaching a maximum of 95%. These MD values remain high in unsterilized soils, suggesting that *G. aggregatum* competes efficiently with indigenous fungi in the soil to improve growth of jujubes seedlings in nursery soils.

Table 3. Mycorrhizal dependency ranking of jujubes among twelve fruit tree species (Bâ et al., 2000)

Fruit tree species	Mycorrhizal dependency (MD) (%)	Occurrence
Zizyphus mauritiana	MD > 75	Very highly dependent
Tamarindus indica	50<MD<75	Highly dependent
Cordyla pinnata	25<MD<50	Moderately dependent
Dialium guineensis	«	«
Parkia biglobosa	«	«
Adansonia digitata	0<MD<25	Lowely dependent
Anarcadium occidentale	«	«
Aphania senegalensis	«	«
Sclerocaria birrea	«	«
Afzelia africana	MD = 0	None
Balanites aegyptiaca	«	«
Landolphia heudelotti	«	«
Saba senegalensis	«	«

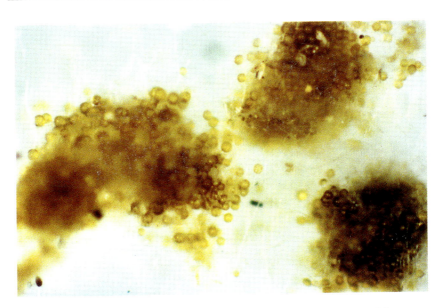

Plate 5. Aggregated spores of *Glomus aggregatum* IR 27 (photo A. Bâ)

Control and inoculated jujube seedlings from Senegal (provenance KST) were transplanted on a field located at Colobane Thiombane in the region of Thiès (Senegal) in April 1999. A randomized complete design with two factors comprized control seedlings and seedlings inoculated at sowing stage in the nursery. Plantation of jujube seedlings was made in

Plate 6. Effect of inoculation by *Glomus aggregatum* (Ga) on Gola variety of jujube 3 months after sowing in nursery forest (photo A. Bâ)

Table 4. Effect of inoculation with *Glomus aggregatum* on the growth and mycorrhizal colonization of four provenances of jujube seedlings in sterilized or unsterilized soil (n= 40) (Bâ et al., 2001).

Provenances	Soil	Mycorrhizal colonization (%)	Mycorrhizal dependency (%)
Sery (Burkina Faso)	Sterilized	nd	78 a
Bamako (Mali)	Unsterilized	nd	60 a
KST (Senegal)	Unsterilized	50	77 a
Gola (Thailand)	Unsterilized	55	95 b

Nd, not determined ; The same letters are not significantly different at P<5%

an area of 0.5 ha. This area was divided in rows of 12 m × 4.5 m separeted by a minimum distance of 4.5 m between plants and between rows. Nine replicated plots (9 × 12 seedlings) were planted by treatment. All treatment combinations were randomly assigned to the planting location between the rows. Seedlings inoculated with *G. aggregatum* showed significant differences in main shoot height and collar diameter after twenty months planting in the field (Table 5).

Table 5. Effect of inoculation with *Glomus aggregatum* on growth parameters of jujube seedlings from Senegal (provenance KST) after twenty months' planting in a field (n = 60) (Bâ et al., 2001).

Status of jujubes	Main shoot height (m)	Collar diameter (m)
Inoculated	1.55 a	2.8 a
Uninoculated	1.04 b	1.5 b

The same letters are not significantly different at P < 5%

6. Utilization of Different Sources of Phosphorus

In tropical areas, phosphorus deficiency in soils is one of the limiting factors for the establishment of tree plantations and agriculture crops (Sanchez, 1995). Organic sources and phosphorus fertilizers can be applied to increase the phosphorus content of soils depleted of this element. However, the proportion of P in organic matter recycled back to the soil is low and soluble P fertilizers are too expensive for agriculture in developing countries (Nye and Kirk, 1987). When indigenous rock phosphate (RP) is readily available, it is obviousely very important to know under which conditions it may be more profitable as an alternative source of phosphorus either to replace or complement these conventional sources. The fertilizer effectiveness of RP depends on factors relating to RP itself (mineralogy, chemical reactivity and rate of application), soil factors (pH acid, P and Ca status), and the mycorrhizal status. Plants inoculated with AMF utilize more soluble P from RP than non-AM plants. The simplest explanation is that mycorrhizas develop an extramatrical mycelium, which increases the root P-absorbing sites, as suggested by Bolan (1991).

The investigations with RP originated from the semi-arid zone of West Africa have mainly centered on their agronomic potential (Easterwood et al., 1989). In contrast, little is known about the efficiency of the direct application of RP to promote tree growth. A possible long-term solubilization of RP could favor their utilization as low external inputs for horticulture crops in semi-arid zones. Nevertheless, many West African soils are moderately acidic (pH in water > 6.2) and they cannot quicky contribute to promote a rapid dissolution of RP. In these conditions, we have compared the efficiency of three RP from West

Africa on growth and P nutrition of jujube seedlings inoculated or not with G. *aggregatum*, by using the reference soluble fertilizer SP (Table 6). The jujube seedlings grow poorly without mycorrhizal colonization and without RP applications (Plate 7). In thus, RP applications do not

Plate 7. Effects of inoculation by *Glomus aggregatum* (Ga) and fertilization with rock phosphate from Mali MP = RP (Mali) on KSP variety of jujube 3 months after sowing in glasshouse condition (photo: A. Bâ)

Plate 8. Symptoms of damage on the root system of KSP variety of jujube caused by the root-knot nematode *Meloidogyne mayaguensis* (photo: A. Bâ).

Table 6. Effects of AM inoculation with *Glomus aggregatum* and different sources of phosphorus on growth and P nutrition of jujube seedlings after 3 months (RP, rock phosphate; SP, superphosphate; RP at 250 kg of P ha^{-1}.; SP at 25 kg of P ha^{-1}.; P<5%) (Bâ et al., 2001).

Status of jujubes	Origin P	Total biomass (g)	P in shoots plus leaves (%)	Mycorrhizal colonization (%)
Uninoculated	Without P	0.38 a	0.05 a	0.00 a
	RP (Mali)	0.57 a	0.05 a	0.00 a
	RP (Senegal)	0.29 a	0.05 a	0.00 a
	RP (Burkina Faso)	0.32 a	0.04 a	0.00 a
	SP	1.80 d	0.20 e	0.00 a
Inoculated	Without P	1.50 bc	0.14 b	78.47 bc
	RP (Mali)	1.87 d	0.16 c	95.39 d
	RP (Senegal)	1.30 bc	0.17 cd	66.00 b
	RP (Burkina Faso)	1.25 b	0.18 de	87.13 cd
	SP	2.70 e	0.19 e	63.60 b

improve non-AM jujubes growth, whereas SP applications do. AM jujubes with *G. aggregatum* and without RP applications improve plant growth and increase P contents. The RP from Mali appears to be more effective in terms of biomass production than those of the RP from Senegal and Burkina Faso. However, AM jujubes achieved better results with SP applications.

In the next step, the response of jujubes to five levels of RP (Mali) application (25, 50, 75, 100 and 125 kg P/ha.) and arbuscular mycorrhizal fungus *G. aggregatum* was evaluated in a low acid sandy soil, using the reference soluble fertilizer SP (Table 7). The non-inoculated jujubes seedlings grew better with SP than RP from Mali. AM jujubes at 125 kg P/ha. of RP or SP achieved better results in terms of biomass than AM jujubes without P. From an economical point of view, it could be advised to fertilize AM jujube seedlings with RP from Mali rather than with SP.

7. Nematode Disease

Plant parasitic nematodes are a cosmopolitan and important problem affecting the production of subtropical and tropical crops being, therefore, responsible for great financial losses to the local farmers. Among these pathogens, root-knot nematodes (*Meloidogyne* spp.) are the most abundant group and the most common species are *M. javanica*, *M. incognita* and *M. mayaguensis* in West Africa. The symptoms of damage due to these pathogens are typical rounded galls on the root systems,

Table 7. Effects of AM inoculation with *Glomus aggregatum* and applications of P sources on growth and P nutrition of jujube seedlings after 3 months (RP, rock phosphate ; SP, superphosphate ; P<5%) (Bâ *et al.*, 2001).

Status of jujubes	Origin of P	Applied P (kg de P/ha.)	Total biomass (g)	P in shoots plus leaves (%)	Mycorrhizal colonization (%)
Uninoculated	Control	0	0.27 a	0.09 ab	0.00 e
	RP (Mali)	25	0.21 a	0.10 ab	0.00 e
		50	0.30 a	0.07 a	0.00 e
		75	0.28 a	0.06 a	0.00 e
		100	0.28 a	0.12 b	0.00 e
		125	0.30 a	0.11 ab	0.00 e
	SP	25	1.86 bc	0.20 cde	0.00 e
		50	2.41 cde	0.18 c	0.00 e
		75	2.37 bcde	0.25 f	0.00 e
		100	3.49 fg	0.25 f	0.00 e
		125	2.87 ef	0.31 gh	0.00 e
Inoculated	Control	0	1.67 b	0.24 ef	74.80 ab
	RP (Mali)	25	2.12 bcd	0.23 def	64.44 b
		50	2.44 cde	0.19 cde	67.98 ab
		75	2.11 bcd	0.19 cde	57.38 bc
		100	1.82 bc	0.23 def	86.29 a
		125	3.06 efg	0.19 cde	73.64 ab
	SP	25	2.88 ef	0.22 cdef	57.15 bc
		50	2.41 cde	0.26 f	76.46 ab
		75	2.65 de	0.25 f	46.02 c
		100	2.67 de	0.31 gh	48.82 c
		125	3.59 g	0.27 fg	30.74 d

white to yellowish brown becoming dark brown in aging roots. The infested plants show foliar chlorosis, leaf fall, general decline. Their growth is generally reduced and sometimes, the plants died (Netscher, 1970; Johnson & Fassuliotis, 1984).

The control of nematodes is more difficult in a perennial crop than in annual or herbaceous crops. For example, the rotation schemes, successfully used with annual crops, are not adapted to these long-term cultures. Moreover, surviving roots of excized plants or old plants left in the field can provide nutrients for nematodes and, consequently, maintain the nematodes population in the soil. This problem is particulary important with the cultural practice of agroforestry. The choice of tree species in these agricultural systems is generally based on the following properties: (i) good growth on low fertility and arid soils;

(ii) source of organic matter for the cultivated soils; and (iii) resistance to the development of pathogenic organisms. Indigenous fruit trees could be good candidates to be associated with annual plants because they are widespread throughout the Sahelian and Sudanian areas in West Africa. Although the control practices used in many forest nurseries is generally based on. (i) the production of seedlings without root-knot nematodes (i.e. disinfection of nursery soils); and (ii) the application of nematicides and/or the culture of resistant or tolerant cultivars, the susceptibility of these tree species to infestation by root-knot nematodes has been rarely assessed.

According to our knowledge, there are no data which could determine the susceptibility of Sahelian and Sudanian fruit trees to the root-knot nematodes, we report a study focussed on the interactions between *M. mayaguensis*, a common root-knot nematode species found in West Africa (Fargette et al., 1996) and several indigenous multipurpose fruit trees. In this experiment, *Acacia holosericea* (Australian acacia) has also been used as a positive control because of its high susceptibility to the root knot nematode (Duponnois et al., 1997)

In order to obtain fast and regular germination, the seeds of fruit tree species tested in this experiment were pretreated with concentrated sulphuric acid for different times. After treatment, the seeds were washed in axenic conditions with sterile distilled water until all traces of sulphuric acid were removed and then transferred aseptically in Petri dishes filled with 1% (w:v) water agar medium. These plates were incubated at 25°C. When the radicles had grown to 1 cm, the seedlings were transferred into 0.5 dm^3 polythene bags (5 cm-diameter) filled with autoclaved soil (140°C, 40 min). The physicochemical characteristics of this sandy soil were as follows: pH H_2O 7.1; fine silt 0.6%; coarse silt 1.4%; fine sand 61.6%; coarse sand 31.2%; total carbon 0.196%; total nitrogen 0.027%. The seedlings were placed in a greenhouse during the hot season (35°C day, 30°C night, 12 h photoperiod) and daily watered without fertilizer. The pots were placed in a randomized, complete block design with eight replicates per treatment.

After one month's culture, the seedlings were inoculated with 5 ml suspensions of 0, 100, 500 and 1000 second stage juveniles (J2) of M. mayaguensis. The population of *M. mayaguensis* used for the inoculum was multiplied on tomato (*Lycopersicum esculentum* Mill.) cv Roma. After 2 months' culturing, the tomato roots were harvested, cut into short lengths and placed in a mist chamber for 1 week to enable the nematode eggs to hatch (Seinhorst, 1950). The nematode density was determined from 5 ml samples and the required inoculum poured into a hole (5 mm by 100 mm) to one side of each seedling and covered with sterilized soil. Two months after nematode inoculation, when the damage associated with the different inoculum densities was observed, the plants were uprooted and the root systems gently washed. The oven dried (1 week at 65°C) weight of the shoot was measured. Each root system was cut into 2-3 cm pieces and placed in a mist chamber for 3 weeks in order to recover J2 from the hatched eggs (Seinhorst, 1950). The nematodes that hatched each week were counted. The oven-dried weight of root systems (1 week at 65°C) were then measured. Data were analyzed with one-way analysis of variance. Mean values were compared using Student's t-test (P < 0.05). Nematode numbers were $\log_{10}$ (x + 1) transformed before statistical analysis.

The root-knot nematode *M. mayaguensis* has been multiplied with 5 tree species (*A. holosericea, A. digitata, B. aegyptiaca, P. biglobosa* and *Z. mauritiana*) (Table 8). The other host plants were resistant to this plant parasitic nematode. Among these susceptible plants, their influence on the nematode multiplication was also variable. Rates of multiplication were much greater at an inoculum of 100 J2 rather than at an inoculum of 1000 J2 for *A. holosericea, A. digitata, P. biglobosa* and *Z. mauritiana*. No differences were recorded with *B. aegyptica* (Table 8). *A. holosericea* appeared to be a much better host for *M. mayaguensis* than the other tree species as it has been previously described (Duponnois *et al.*, 1995). One month after nematode inoculation, the growth of *A. holosericea, Z. mauritiana* and *A. digitata* was significantly decreased as compared with the control by all *M. mayaguensis* inoculum densities (Table 8). In contrast, no significant nematode effect has been recorded with *P. biglobosa* and *B. aegyptiaca*. We conclude from our study that in West Africa, *M. mayaguensis* may significantly decrease the potential benefits that may result from growing fruit trees such as *Z. mauritiana* (Photo 8). Tree growth are likely to be greatly decreased and the population of *M. mayaguensis* increased, to the detriment of any adjacent susceptible crops. Moreover, in these countries, the culture substrate is usually based on a mixture with sand and organic matter collected around villages and it is frequently infested with root knot nematodes.

Several research objectives can be identified to minimize the nematode effect: (i) the fruit trees used in agroforestry should be screened for their susceptibility to plant parasitic nematodes such as root-knot nematodes; and (ii) if the selected fruit tree species are susceptibles to these nematodes, then ways should be sought to minimize nematode damage. It is well known that antagonistic microorganisms may be one possibility against nematodes. Several of those have been well studied such as nematophagous fungi, the actinomycete *Pasteuria penetrans* and mycorrhizal fungi. Mycorrhizal associations between beneficial soil fungi and plant roots are formed by

Table 8. Effect of *M. mayaguensis* inoculated at different rates on the nematode development and fruit tree growth.

Plant species	Inoculum	J2/plant	Multiplication rate	Shoot biomass (mg)	Root biomass (mg)
A. holosericea	0 (control)	0 a	-	421 a	195 a
	100	6142 b	61.5 a	325 a	185 a
	500	7456 b	14.9 b	236 b	156 b
	1000	1523 c	1.5 c	152 c	125 c
A. digitata	0 (control)	0 a		2140 a	7560 a
	100	1434 a	14.3 a	2501 a	5502 a
	500	266 ab	0.5 ab	1150 b	4130 b
	1000	86 b	0.09 b	730 c	3080 c
A. occidentale	0 (control)	0	0	3920 a	1570 a
	100	0	0	5910 a	2360 a
	500	0	0	4100 a	1780 a
	1000	0	0	5730 a	2480 a
A. senegalensis	0 (control)	0	0	490 a	191 a
	100	0	0	595 a	252 a
	500	0	0	494 a	185 a
	1000	0	0	584 a	248 a
B. aegyptiaca	0 (control)	0 a		1490 b	2711 a
	100	14.4 a	0.14 a	1752 ab	4029.2 ab
	500	17.8 a	0.04 a	2070 a	5382 b
	1000	201.2 b	0.2 a	1345 b	2555 a
D. guineensis	0 (control) 100	0		529 a	321 a
	500	0	0	485 a	359 a
	1000	0	0	642 a	406 a
		0	0	543 a	389 a
L. heudilotii	0 (control)	0		1382 a	1695 a
	100	0	0	1352 a	1589 a
	500	0	0	1026 a	1356 a
	1000	0	0	1125 a	1259 a
P. biglobosa	0 (control)	0		2111 a	4520 a
	100	138.3 a	1.38 a	2120 a	3985 a
	500	66.9 a	0.12 b	2360 a	4125 a
	1000	270.9 a	0.27 b	2081 a	3896 a
S. senegalensis	0 (control)	0		856 a	762 a
	100	0	0	796 a	754 a
	500	0	0	846 a	698 a
	1000	0	0	829 a	709 a

(Table 8 Contd.)

(Table 8 Contd.)

S. birrea	0 (control)	0		1385 a	3185 a
	100	0	0	1356 a	3059 a
	500	0	0	1289 a	3156 a
	1000	0	0	1352 a	3056 a
T. indica	0 (control)	0		1240 a	2701 a
	100	0	0	1156 a	2501 a
	500	0	0	1301 a	2489 a
	1000	0	0	1258 a	2631 a
Z. mauritiana	0 (control)	0		1980 a	950 a
	100	176.1 a	1.76 a	1110 b	659 b
	500	76.1 a	0.15 b	790 c	459 c
	1000	117.2 a	0.12 b	680 c	435 c

the majority of woody plants. These symbiotic relationships enhance nutrient uptake and may protect host plants against pathogenic microorganisms (Duponnois and Cadet, 1994; Smith and Read, 1997). It has been demonstrated that endomycorrhizal fungi often increased host tolerance to nematode infection. Hence, controlled mycorrhization of susceptible fruit tree species such as Z. *mauritiana* could be a useful technique for improving the tolerance and the growth of fruit tree plantations.

8. Planting

Planting distances depend principally on annual rainfall, which should be between 7 and 8 meters between plants and between rows, provided the rainfall exceeds 750 mm. The same applies in the case of irrigated culture. For rainfall not exceeding 500 mm, distances should be reduced to 6m either way (Vashishtha, 1997). In rainy systems, planting should be carried out at the beginning of the rainy season and in irrigated culture, it should be established preferably during the hot season.

In India, the composition of the annual maintenance fertilizer is 10 kg of manure and 0.5 kg of ammonium sulphate. Some producers make intensive use of fertilizers, providing a total input of 280 kg of NPK/ha./yr fertilizer, in two apportions (Morton, 1987). In low-tech farming conditions, it can be beneficial to irrigate (or at least provide a supplement of water) during the establishment of young plantations (first year after planting) as well as in mature plantations during the fruit setting and swelling stages.

9. Training and Pruning

Research in progress in Senegal into the optimisation of arboricultural management (pruning, in particular) and fruit production has been drawn from the work carried out in India (Lal and Prasad, 1980; Bajwa et al., 1987; Chovatia et al., 1991). The result of this research, in the short term, will be the availability of a complete technological package justifying rapid propagation of the jujube tree and providing guidance on its fruit management in the Sahelian zone.

Formation pruning is essential for the jujube tree because the species has a tendency to produce shoots and also because some cultivars manifest a plagiotropic growth. Careful attention must therefore be paid, particularly during the first 2-3 years after planting, to ensure that any shoots appearing at the base of the rootstock tree are removed on a regular basis. This is all the more so in the case of plants obtained by horticultural grafting. This problem can be avoided by using plants propagated by micrografting since shoots belonging to the rootstock do not develop. The aim of formation pruning is to obtain a trunk, by eradicating all the side branches at a height of 1-1.2 metres, then allowing 4-5 well spaced supporting branches to develop, ensuring that the points from which they emerge are not at the same level. Once formation pruning is complete, it is necessary to prune for fruiting. In fact, the flowers of *Z.mauritiana* develop on the young wood and it is therefore necessary to ensure that branches are renewed.

Various studies (Bajwa, 1987; Chovatia et al., 1991; Mukherjee, 1993) tend to indicate that annually removing 25% to 50% of the tree's woody biomass is sufficient to rejuvenate the wood and obtain increased yields of good quality fruit (size, vitamin C content). Pruning should be performed after harvest, during the time of vegetative rest (Kaushik, 2000).

10. Yield and Organoleptic Qualities

Results obtained in Senegal relate exclusively to the Gola variety. This cultivar bears fruit of a weight and volume which is 9 to 17 times greater than any fruit of local origin. The weight and volume of edible pulp is 10 to 20 times greater for the Gola variety than that measured in local fruit. The average weight (18.5 g) of fruit harvested from the Gola variety in Senegal is nonetheless slightly less than that indicated in India by Chovatia et al. (1991, 1993) (20-24 g). However, it is greater than the weight range reported by Morton (1987) (14-17 g). After drying, the weight of edible dried fruit matter obtained from Gola is 5 to 6 times greater than that available from locally grown jujube. During the dehydration process, the fruits from the Senegalese batches turn a light ochre colour, whereas the Gola fruits darken, becoming almost black. This colouring of the Gola jujube during drying is most probably due to non-enzymatic browning (Maillard reaction), involving carbonyl compounds (ascorbic acid and reducing sugars) and amino nitrogen in the presence of water, accompanied by the formation of aromatic substances. This was confirmed by an organoleptic test carried out in Senegal, in which testers found the dried fruits from Gola jujube to be highly scented and with a superior flavour to the locally grown dried fruits (Danthu et al., 2002a).

11. Conclusion

Micrografting appeared to be an interesting method to propagate some exotic varieties of jujubes in the Sahelian zone. However, micrografted jujubes should be inoculated with AMF during the acclimatization phase. Controlled mycorrhization also could be a useful technique for improving the tolerance of jujube seedlings against the nematodes. Further studies should be done on jujube trees in order to determine how the combination of inoculation and fertilization could improve the fruit yield. Apart from these technical elements, several aspects, of an economic and socio-cultural nature, still remain inconclusive. Will the Sahelian populations appreciate this new form of culture from an organoleptic point of view, to the same extent as the Senegalese people? Will they see any benefit to planting these large-fruiting species of jujube, as opposed to continuing to crop the locally-growing wild jujube trees? What will be the cost of producing micrografted plants in an optimized production system and will the cost be compatible with the population's resources? This is why, with the help of various development agencies (Forestry Commission, NGOs, private arboriculturists), different initiatives are underway to inform and promote through the distribution of plants and on site testing.

12. References

Arbonnier, M. 2000. Arbres, arbustes et lianes des zones sèches d'Afrique de l'Ouest. CIRAD, MNHN, UICN, Montpellier, France, pp. 438-442.

Bâ A. M., Y. Dalpé and T. Guissou. 1996. Les Glomales d'*Acacia holosericea* A. Cunn. ex G. Don. et d'*Acacia mangium* Willd.: Diversité et abondance relative des champignons mycorhiziens à arbuscules dans deux types de sols de la zone Nord et Sud Soudanienne du Burkina Faso. *Bois et Forêts des Tropiques* 250: 5-18.

Bâ A. M. and T. Guissou. 1996. Rock phosphate and mycorrhizas effects on growth and nutrient uptake of *Faidherbia albida* (Del.) in an alkaline sandy soil. *Agroforestry Systems* 34: 129-137.

Bâ, A. M., C. Plenchette, P. Danthu, R. Duponnois and T. Guissou. 2000. Functional compatibility of two arbuscular mycorrhizae with thirteen fruit trees in Senegal. *Agroforestry Systems* 50, 95-105.

Bâ A. M., T. Guissou, R. Duponnois, C. Plenchette, O. Sacko, D. Sidibé, S. Kondé and V. Baba. 2001. Mycorhization contrôlée et fertilisation phosphatée: applications à la domestication du jujubier. *Fruits* 56 (4) : 261-269 (Article de synthèse).

Bajwa G.S., H.S. Sandhu and J.S. Bal. 1987. Effect of different pruning intensities on growth, yield and quality of ber (*Ziziphus mauritiana* Lamk.). *Haryana Journal Horticultural Science* 16 (3-4): 209-213.

Becker, B. 1983. The contribution of wild plants to human nutrition in the Ferlo (Northern Senegal). *Agroforestry Systems* 1: 257-267.

Bergeret, A. 1986. Nourritures de cueillette en pays sahélien. *Journal d'Agriculture Traditionnelle et de Botanique Appliqueé* 33 : 91-130.

Bolan, N. S. 1991. A critical review on the role of mycorrhizal fungi in the uptake of phosphorus by plants. *Plant Soil* 134 : 189-207.

Bonkoungou, E.G., M. Djimdé, E.T. Ayuk, I. Zoungrana and Z. Tchoundjeu. 1998. Taking stock of agroforestry in the Sahel-harvesting results for the future : end of phase report 1989-1996. ICRAF, 57pp.

Booth, F.E.M. and G.E. Wickens. 1988. Non–timber uses of selected arid zone trees and shrubs in Africa. FAO Conservation Guide 19, FAO, Rome, Italie, pp 164-176.

Brundrett, M., N. Bougher, B. Dell, T. Grove and N. Malajczuk. 1996. Working with mycorrhizas in forestry and agriculture. *ACIAR Monograph series*, 374 pp.

Chevalier, A. 1947. Les jujubiers ou *Ziziphus* de l'Ancien monde et l'utilisation de leurs fruits. *Revue de Botanique Appliqueé* 27: 470-483.

Chovatia, R.S., D.S. Patel and G.V. Patel. 1993. Performance of ber (*Ziziphus mauritiana* Lamk) cultivars under arid conditions. *Annals of Arid Zone* 32: 215-217.

Chovatia, R.S., D.S. Patel, G.V. Patel and A.T. Patel. 1991. Pruning studies in ber (*Ziziphus mauritiana* Lamk) under dryland conditions. *Annals of Arid Zone* 30: 353-356.

Danthu, P., A. Gaye, J. Roussel, J. and A. Sarr. 1993. Quelques aspects de la germination des semences de *Ziziphus mauritiana* Lam.. In: *Tree Seed Problems, With Special Reference to Africa*, L.M. Somé and M. de Kam (eds.). Backhuys Publishers, Leiden, the Netherlands, pp. 192-197.

Danthu, P., A. M. Bâ, I. Diallo, G. Delhove, E.V. Coly and P. Sall. 2000a. Domestication and improvement of jujube tree (*Ziziphus mauritiana*) - State of art in Senegal. In: *Proceedings of the XXI IUFRO World Congress*, Kuala, Lumpur, Malaysia, Vol III, pp. 271-272.

Danthu, P., P. Soloviev and M.A. Touré. 2000b. La domestication du jujubier (*Ziziphus mauritiana* Lam.) au Sénégal : quelques résultats concernant sa propagation végétative. Bulletin de Liaison de la coopération régionale pour le développement des productions horticoles en Afrique : 18: 29-32.

Danthu, P., P. Soloviev, M.A. Touré and A. Gaye. 2002c. Propagation végétative d'une variété améliorée de jujubier (*Ziziphus mauritiana* Lam.) introduite au Sénégal. *Bois et Forêts des Tropiques* 272 : 93-96.

Danthu, P., P. Soloviev, A. Totté, E. Tine, N. Ayessou, A. Gaye, T.D. Niang, M. Seck and M. Fall. 2002d. Caractères physico-chimiques et organoleptiques comparés des jujubes de la variété Gola introduite au Sénégal et des fruits sauvages. *Fruits* (57: 173-182).

Danthu, P., B. Hane, M. Touré, P. Sagna, M. Sagna, S. Bâ, M.A. de Troyer and P. Soloviev. 2001. Microgreffage de quatre espèces ligneuses sahéliennes (*Acacia senegal, Faidherbia albida, Tamarindus indica* et *Ziziphus mauritiana*) en vue de leur rajeunissement. *Tropicultura* 19 : 43-47.

Depommier, D. 1988. *Zizyphus mauritiana* Lam., culture et utilisation en pays Kapsiki (Nord-Cameroun). *Bois et Forêts des Tropiques* 218 : 57-63.

Djimdé, M. 1997. *Ziziphus mauritiana* a live hedge species. In: *Proceedings of Atelier pan-african sur le Ziziphus mauritiana*, ICRAF, IPALAC, 17-20 july 1997, Bamako, 18 pp.

Duponnois, R. and P. Cadet. 1994. Interactions of *Meloidogyne javanica* and *Glomus* sp. on growth and N$_2$ fixation of *Acacia seyal*. *Afro-Asian Journal of Nematology*, 4: 228-233.

Duponnois, R., P. Cadet, K. Senghor and B. Sougoufara. 1997. Etude de la sensibilité de plusieurs acacias australiens au nématode à galles *Meloidogyne javanica*. *Annales des Sciences Forestières*, 54 : 181-190.

Easterwood, G. W., J. B. Sartain and J. J. Street. 1989. Fertilizer effectiveness of three carbonate apatites on an acid ultisol. Commun. *Soil Science and Plant Analysis* 20 : 789-800.

Fargette, M., R. Duponnois, T. Mateille and V. Block. 1996. Characterization of *Meloidogyne mayaguensis* and its relationship to other tropical root-knot nematodes. Third International Nematology Congress. Pointe-à-Pitre. July 7-12, 1996.

Gupta, N. and P. S. Srivastava. 1996. *In vitro* regeneration and isozymes patterns in *Ziziphus mauritiana*. *Journal of Plant Biochemistry & Biotechnology* 5: 87-90.

Garbaye, J. (1991). Les relations entre les plantes et les champignons. In : Strullu DG, Garbaye J, Perrin R, Plenchette C, eds. *Les mycorhizes des arbres et plantes cultivées*. Lavoisier, Paris, 1991 : 9-49.

Guissou, T. A. M. Bâ, Ouadba J. M., Guinko S. and Duponnois R. 1998a. Responses of *Parkia biglobosa* (Jacq.) Benth, *Tamarindus indica* L. and *Zizyphus mauritiana* Lam. to vesicular mycorrhizal fungi in a phosphorus-deficient sandy soil. *Biology and Fertility of Soils* 26 : 194-198.

Guissou T, Bâ AM, Guinko S, Duponnois R and Plenchette C 1998b. Influence des phosphates naturels et des mycorhizes à vésicules et à arbuscules sur la croissance et la nutrition minérale de *Zizyphus mauritiana* Lam. dans un sol à pH alcalin. *Annales des Sciences Forestières* 55: 925-931

Guissou T, Bâ AM, Guinko S, Plenchette C & Duponnois R. 2000. Mobilisation des phosphates naturels de Kodjari par des jujubiers (*Zizyphus mauritiana* Lam.) mycorhizés dans un sol acidifié avec de la tourbe. *Fruits* 55 (3) : 187-194.

Guissou T, Bâ A.M., Plenchette C, Guinko S & Duponnois R 2001. Effets des mycorhizes à arbuscules sur la tolérance à un stress hydrique de quatre arbres fruitiers *Balanites aegyptiaca* (L.) Del., *Parkia biglobosa* (Jacq.) Benth., *Tamarindus indica* L. et *Zizyphus mauritiana* Lam. *Sécheresse* 12 (2) : 121-127.

Howard, R.A. 1989. *Flora of the Lesser Antilles*. Vol. 5, Dicotyledoneae, part 2, 604 p.

Jain, B.L., R.S Goyal, and O.P. Pareek. 1980. Effect of saline irrigation water on soil and performance of ber. *Indian Journal of Horticulture* 45: 203-207.

Johnson, A.W. and G. Fassuliotis. 1984. Nematode parasites of vegetable crops. In: *Plant and Insect Nematodes*, pp. 323-372. Ed. W.R. Nickle. New York and Basel: Marcel Dekker Inc.

Kaushik, R.A. 2000. Ber (*Ziziphus mauritiana*) in Thar desert. In: *Proceedings of Trees for Arid Lands Workshop*, IPALAC, Beer Sheva, Israel, 115-119.

Kerharo, J. and J.G. Adam. 1974. *La pharmacopée sénégalaise traditionnelle*. Vigot Frères, Paris, 1011pp.

Lal, H. and A. Prasad. 1980. Pruning in Ber (*Ziziphus mauuritiana* Lamk.). III. Effect on yield and fruit quality. *Punjab Horticulture Journal* 20: 162-166.

Leakey, R. R. B. and A. J. Simons. 1998. The domestication and commercialization of indigenous trees in agroforestry for the alleviation of poverty. *Agroforestry Systems* 38 : 165-176.

Leakey, R. R. B. 1999. Potential for novel food products from agroforestry trees : a review. *Food Chemistry* 66 : 1-14.

Mathur, N., K.G. Ramawat and D. Nandwani. 1995. Rapid *in vitro* multiplication of jujube through mature stem explants. *Plant Cell Tissue Organ Cult* 43: 75-77.

Mawani, P.B., S.P. Singh. 1992. Effect of method and time of budding on budding success in ber (*Ziziphus mauritiana* Lamk) cv Gola. *Horticultural Journal* 5: 31-35.

Morton, J. 1987. Indian jujube. In: *Fruits of warm climates,* J.F. Morton (ed.), Miami, Florida, pp. 272-275.

Mukherjee, S. and A.K. Soni. 1993. Growth and yield of ber (*Ziziphus mauritiana* Lamk) cv. Seo under different pruning severities. *Annals of Arid Zone* 32(3): 165-166.

Munier, P. 1973. Le jujubier et sa culture. *Fruits* 28: 377-388.

Murashige, T. and F. Skoog. 1962. A revised medium for rapid growth and bioassays with tobacco tissue culture. *Physiology Plantasum* 15: 473-497.

Nair, P.K.R. 1998. Directions in tropical agroforestry research : past, present and future. *Agroforestry Systems* 38 : 223-245.

Netscher, C. 1970. Les nematodes parasites des cultures maraîchères du Sénégal. Cahier ORSTOM Série Biologie, 11 : 209-229.

Nye P.H. and G.J.D. Kirk. 1987. The mechanism of rock phosphate solubilization in the rhizosphere. *Plant and Soil* 100 : 127-134.

Perrier de la Bâthie, H. 1924. Sur quelques plantes non cultivées de Madagascar à fruits comestibles ou utiles et sur la possibilité de leur culture. *Revue de Botanique Appliquée & d'Agriculture Coloniale* 4: 652-663.

Plenchette C. J.A. Fortin and V. Furlan. 1983. Growth responses of several plant species to mycorrhizae in a soil of moderate P-fertility. I. Mycorrhizal dependency under field conditions. *Plant and Soil* 1983 ; 70 : 199-209.

Pliego-Alfaro, F. and T. Murashige. 1987. Possible rejuvenation of adult avocado by graftage onto juvenile rootstocks in vitro. *HortScience* 22: 1321-1324.

Prasad, R.N., G.J. Bankar and B.B. Vashishtha. 1995. Products from ber fruits. *Scientific Horticulture* 5: 47-49.

ecosystems. In : Macfayden A, Begon M, Fitter AH, eds. *Advances in Ecological Research.* Ademic Press, London, 1991 ; 21 : 171-313.

Roussel, J. 1995. *Ziziphus mauritiana* Lam.. In: Pépinières et plantations forestières en Afrique tropicale sèche, ISRA, CIRAD, Dakar, Senegal, pp. 417-421.

Sena, L.P., D.J. Vanderjagt, C. Rivera, A.T.C. Tsin, I. Muhamadu, O. Mahamadou, M. Millson, A. Pastuszyn and R.H. Glew. 1998. Analysis of nutritional components of eight famine foods in the Republic of Niger. *Plant Food for Human Nutrition* 52: 17-30.

Sanchez, P.A. 1995. Science in agroforestry. *Agroforestry Systems* 30 : 6-55.

Seinhorst, J.W. (1950). De betekenis van de toestand van de grond voor het optreden van aantasting door het stengelaaltje (*Ditylenchus dipsaci* (Kühn) Filipjev). *Tijdschrift over Plantenziekten,* 56: 289-349.

Singh, R.S., R.N. Prasad, J.P. Gupta, B.B. Vashistha, and Y.S. Ramakrishna. 1999. Thermal time requirement for fruit development and maturity of jujube (*Ziziphus mauritiana*) grown under rainfed conditions in Indian hot desert. *Annals of Arid Zone* 38: 161-166.

Smith S. and J. Read. 1997. *Mycorrhizal Symbiosis* (2nd edition). Clarendon Press, 605 pp.

Timmis, R., Abo El-Nil, M.M. and Stonecypher, R.W. 1987 Potential genetic gain through tissue culture. In: Bonga JM and Durzan DJ (eds) *Cell and Tissue Culture in Forestry. Volume 1. General Principles and Biotechnology,* Martinus Nijhoff Publishers, Dordrecht, The Netherlands, pp 198-215.

Toury, J., P. Lunven and R. Giorgi. 1961. Aliments de cueillette et de complément au Sénégal et en zone sahélienne. *Qualitas Plantarum et Materiae Vegetabiles* 8: 139-156.

Toury, J., R. Giorgi, J.C. Favier and J.F. Savina. 1967. Aliments de l'Ouest africain. Tables de composition. *Annales de la Nutrition et de l'Alimentation* 27: 73-127.

Vashishtha, B.B. 1997. *Ziziphus* for drylands - a perennial crop solving perennial problems. *Agroforestry Today* 9: 10-12.

Vashishtha, B.B. 2001. Le jujubier reste toujours le jujubier. *Sahel Agroforestry* 1: 1-3.

Index

Acid rain 10, 25

Acidification 3, 7, 31, 35, 36, 41, 42, 43, 47

Agriculture 93, 94, 98, 99, 112

Analysis of plants 85

Anthocyanin 135

Apple 77

Arbuscular mycorrhizal fungi 259, 263

Ash particles 203, 205, 206, 207

Assessment 30

Assimilate 126

Assimilate transport 126

ATP synthesis 120

B Distribution 83

B Forms in Soil 79

B Toxicity 78

Banana 146

Biological activity 23, 107

Biomass 102, 103, 107, 110

Bioregulators 56, 62, 63, 71

Boron 77, 139

By-product 3, 24

C/N ratio 209, 212

Calcium 138

Carbon 137

Carbon balance 117

Carbon metabolism 118

Carotenoid 135

Chemical properties 204, 208, 212, 226, 227

Chlorine 140

Chlorophyll 135

Citrus 153

CO_2 compensation point 121

Composting 23, 24

Consumption 35

Contamination 201, 223, 224

Conventional farming 27, 34

Copper 139

Cultivation 93, 94, 96, 110

Deficiency 78

Denitrification 7, 8, 12, 13, 40, 44

Denitrification/nitrification 6

Depletion in the non-renewable resources 42

Depletion of non-renewable resources 41, 42, 47

Depletion of ozone 10, 42, 44, 46, 47

Diluting effect 118

Dry matter accumulation 118

Ecotoxicity 31

Emissions 3, 5, 6, 7, 8, 9, 10, 13, 17, 25, 28, 29, 30, 32, 35, 36, 38, 44, 45, 47

Energy 35

Energy consumptions 30, 35

Environmental factors 57, 59, 70

Environmental hazards 201

Environmental impacts 3, 12, 30, 33, 29, 34, 36

Eutrophication 16, 24, 30, 31, 35, 36, 41, 43, 47

Farm 220, 224

Fertilization 60, 61, 64, 69, 70, 73, 75, 85, 86

Fertilizer 94, 95, 96, 97, 98, 99, 102, 103, 104, 107, 110, 112, 113

Fertigation 38, 39, 42, 44, 47

Flavor Volatiles 135

Foliar B sprays 87

Food 93, 94

Forest 201, 208, 222, 223, 224, 225, 226, 227, 228

Formation of ozone 10

Fruit maturity 126

Fruit trees 241, 243, 245, 248

Functional unit 30, 37

Fungi 207

Gas exchange 121

Genetic variability 56, 63, 70

Global effects 31

Global impacts 35

Global warming 3, 31, 35, 36

Goal definition 31

Grafting 56, 59, 259, 260, 261, 264, 275

Grapes 180

Greenhouse effect 7, 10, 30, 41, 44, 47

Groundwater 224

Growth medium 232

Guava 162

Harvest index 124

Heavy metal 19, 36, 52, 201, 223, 225

Human and terrestrial toxicity 41

Human toxicity 44, 45

Hydrogen 137

Impact assessment 30, 32, 41

Improvement analysis 32

Integrated 41

Integrated agricultural production 36

Integrated farming 35

Integrated production 34, 36, 37

Inventory analysis 32, 41

Iron 139

Jujubes 255, 258, 256, 264, 265, 267, 269, 270, 276, 277

K fertilizers 20, 23

LCA 3, 31, 32, 33, 34, 36, 3.', 47

Leaching 2, 6, 10, 11, 12, 13, 17, 19, 22, 24, 26, 39, 208, 217, 218, 219, 220, 224, 227, 228

Leaf analyses 116, 117

Leaf area 121

Leaf P concentration 123

Life Cycle Assessment (LCA) 3, 34, 30

Life cycle inventory 35

Litchi 173

Local effects 31, 34

Macro-elements 1, 137

Magnesium 139

Mango 140

Manure 2, 6, 23, 24, 25, 26, 27, 28, 29, 33, 35, 36, 38, 40, 95, 99, 111, 112

Matter 6

Micro-elements 1

Micronutrients 214, 215, 216

Mineral and organic fertilizers 7

Mineral fertilizers 2, 7, 8

Mineralization 5, 7, 12, 22, 23, 26, 39

Molybdenum 140

Mycorrhizal dependency 264, 265, 279

N and P balances 35

N cycle 5

N fertilizer 5, 6, 11, 12, 13

Nematode 256, 268, 269, 271, 272, 272, 273, 274, 278

Net photosynthesis 120, 121

Net photosynthetic rate 121

Nitrification 6, 7, 13

Nitrogen 138

Nitrogen balance 5

Nitrogen cycle 5, 6

Nitrogen fertilizers 3, 4, 5, 21, 28, 33, 34, 35

Non-climacteric 133

NPK fertilizer 96, 207

Nutrient balance 3, 23, 25, 32, 38, 40

Nutrient cycle 32

Nutrient status 117

Nutrient uptake 118, 125

Nutrients 201, 206, 207, 209, 210, 213, 214, 215, 217, 218, 221, 222, 224, 228
Nutrition 133

Organic and integrated productions 36
Organic compounds 203, 206, 223
Organic farming 27, 34, 35, 36
Organic fertilizers 2, 7, 24, 26, 27
Organic matter 7, 8, 16, 19, 21, 22, 23, 25, 29, 34, 39, 40
Oxygen 137
Ozone depletion 31

P availability 117
P balance 16
P deficiency 116, 124
P fertilizers 16, 17, 19, 26, 36
P leaching 27
P requirement 116, 118
P status 117
Papaya 177
Pesticides 58
Phenolics 135
Phosphate and potash fertilizers 33
Phosphate fertilizers 13, 14, 16, 19
Phosphorus 116, 138
Phosphorus Cycle 14
Photochemical oxidant 42, 46, 47
Photo-oxidant 31
Photosynthesis 117
Photosynthetic system 121
Pineapple 167
Plant growth 55, 56, 67, 74, 211, 217, 221, 222, 224, 228
Plastic mulches 66, 67, 72, 73, 74
Postharvest quality 126
Potash fertilizer 20, 22
Potassium balance 22
Production 41

Quality 133

Recycling 201, 225, 227, 228
Residues 99, 105, 107, 109, 111
Respiration rate 117
Ripening 136

Rock phosphate 19, 256, 267, 268, 269, 270, 277, 279
Rootstock 83
Rowcovers 68
Run-off 6, 10, 13, 16, 17, 25, 26

Salinity 59, 62, 71, 73, 74
Sink strength 126
Soil 93, 94, 95, 96, 97, 98, 99, 100, 101, 102, 103, 104, 105, 106, 107, 108, 110, 111, 112, 113
Soil additive 202, 226
Soil analysis 84
Soil application 86
Soil Moisture 81
Soil pH 82
Soil/plant system 6
Soil solution 208, 209, 210, 211, 215, 217, 218, 222, 224, 225, 226, 228
Solid fertilization 38, 42, 44
Starch accumulation 121, 125
Starch synthesis 126
Stomatal conductance 121
Strawberry 116
Strawberry leaves 122
Sucrose/starch ratio 125
Sulphur 139

Terrestrial ecotoxicity 44, 45
The uptake process 80
Titratable acidity 126
Total soluble solids 126
Toxicity (human) 31
Transpiration rate 121
Transport 83
Tropical Horticulture 133

Urea 3, 6, 26, 27, 36

Vegetative propagation 256, 259, 260
Volatilization 6, 7, 13, 26, 28, 36, 40

Waste management 23
Wood 201, 202, 203, 204, 205, 206, 207, 208, 209, 210, 211, 212, 213, 214, 215, 216, 217, 218, 219, 220, 221, 222, 223, 224, 225, 226, 227, 228

Yield 95, 96, 98, 99, 100, 101, 103, 104, 105, 106, 107, 108, 109, 110, 111, 112

Zinc 139

Zizyphus mauritiana 255, 265, 278